Design Aids for Offshore Topside Platforms under Special Loads

Design Aids for Offshore Topside Platforms under Special Loads

Srinivasan Chandrasekaran, Arvind Kumar Jain, Nasir Shafiq, and M. Mubarak A. Wahab

CRC Press
Taylor & Francis Group
Boca Raton London New York

CRC Press is an imprint of the
Taylor & Francis Group, an **informa** business

MATLAB® is a trademark of The MathWorks, Inc. and is used with permission. The MathWorks does not warrant the accuracy of the text or exercises in this book. This book's use or discussion of MATLAB® software or related products does not constitute endorsement or sponsorship by The MathWorks of a particular pedagogical approach or particular use of the MATLAB® software.

First edition published 2022
by CRC Press
6000 Broken Sound Parkway NW, Suite 300, Boca Raton, FL 33487-2742

and by CRC Press
2 Park Square, Milton Park, Abingdon, Oxon, OX14 4RN

© 2022 Srinivasan Chandrasekaran, Arvind Kumar Jain, Nasir Shafiq, M. Mubarak A. Wahab

CRC Press is an imprint of Taylor & Francis Group, LLC

Reasonable efforts have been made to publish reliable data and information, but the author and publisher cannot assume responsibility for the validity of all materials or the consequences of their use. The authors and publishers have attempted to trace the copyright holders of all material reproduced in this publication and apologize to copyright holders if permission to publish in this form has not been obtained. If any copyright material has not been acknowledged please write and let us know so we may rectify in any future reprint.

Except as permitted under U.S. Copyright Law, no part of this book may be reprinted, reproduced, transmitted, or utilized in any form by any electronic, mechanical, or other means, now known or hereafter invented, including photocopying, microfilming, and recording, or in any information storage or retrieval system, without written permission from the publishers.

For permission to photocopy or use material electronically from this work, access www.copyright.com or contact the Copyright Clearance Center, Inc. (CCC), 222 Rosewood Drive, Danvers, MA 01923, 978-750-8400. For works that are not available on CCC please contact mpkbookspermissions@tandf.co.uk

Trademark notice: Product or corporate names may be trademarks or registered trademarks and are used only for identification and explanation without intent to infringe.

Library of Congress Cataloging-in-Publication Data

Names: Chandrasekaran, Srinivasan, author. | Jain, Arvind Kumar, author. | Shafiq, Nasir, author. | Wahab, M. Mubarak A., author.
Title: Design aids for offshore topside platforms under special loads / Srinivasan Chandrasekaran, Arvind Kumar Jain, Nasir Shafiq, M. Mubarak A. Wahab.
Description: First edition. | Boca Raton : CRC Pess, [2022] | Includes bibliographical references and index.
Identifiers: LCCN 2021027711 (print) | LCCN 2021027712 (ebook) | ISBN 9781032136844 (hardback) | ISBN 9781032139159 (paperback) | ISBN 9781003231424 (ebook)
Subjects: LCSH: Offshore structures--Design and construction--Handbooks, manuals, etc.
Classification: LCC TC1665 .C4565 2022 (print) | LCC TC1665 (ebook) | DDC 627/.98--dc23
LC record available at https://lccn.loc.gov/2021027711
LC ebook record available at https://lccn.loc.gov/2021027712

ISBN: 978-1-032-13684-4 (hbk)
ISBN: 978-1-032-13915-9 (pbk)
ISBN: 978-1-003-23142-4 (ebk)

DOI: 10.1201/9781003231424

Typeset in Times
by Deanta Global Publishing Services, Chennai, India

Contents

Preface xi
About the Authors xiii

Chapter 1 Materials and Loads on Topside 1

    1.1   Offshore Topside: Loads and Analysis Methods 1
    1.2   Geometric Configuration 3
    1.3   Materials 5
            1.3.1   Wire Arc Additive Manufacturing 8
    1.4   Wind and Blast Loads 13
    1.5   Impact Load 15
    1.6   Modal Analysis 16
    1.7   Pushover Analysis 19
    1.8   FEED: Basics 24
    1.9   Essentials of FEED 25
    1.10   Basic Engineering Requirements of FEED 27
    1.11   Factors Influencing Design of Topside 28
    1.12   Advanced Level in FEED Studies 30
            1.12.1 Plant Design and Model Studies 32
    1.13   EPC Execution Planning 33
    1.14   Overall FEED Phases 34
    1.15   Axial Force-Bending Moment Interaction 35
            1.15.1 Properties of Concrete 37
    1.16   Mathematical Development of P-M Interaction 39
    1.17   Example Studies and Discussions 49
    1.18   Fire Load 51
    1.19   Classification of Fire 52
            1.19.1 Pool Fire 52
            1.19.2 Jet Fire 53
            1.19.3 Fireball 54
            1.19.4 Flash Fire 54
    1.20   Steel at Elevated Temperature 54
    1.21   Fire Load on Topside 55
            1.21.1 Time-Temperature Behavior 59
    1.22   Parametric Fire Curve 60
    Credits 64
    Exercise 65

Chapter 2 Basic Design Guidelines 67

    2.1   Design Methods and Guidelines 67
    2.2   Design Loads 69

v

		2.2.1	Dead Loads 69
		2.2.2	Live Loads 69
	2.3	Design Stages 73
		2.3.1	Static In-Place Analysis 73
		2.3.2	Load-Out Analysis 75
		2.3.3	Lifting Analysis... 76
		2.3.4	Transportation Analysis 78
		2.3.5	Analysis of Miscellaneous Items 80
	2.4	Weight Control.. 80
	2.5	Numerical Tools 80
	2.6	Design Considerations 82
		2.6.1	Design Acceptance Criteria 83
	2.7	Design Methods... 83
	2.8	Plastic Design 86
		2.8.1	Shape Factor 89
		2.8.2	Depth of Elastic Core 90
	2.9	Shape Factors Used in Offshore Topside.............................. 92
	2.10	Moment-Curvature Relationship 99
	2.11	Load Factor.. 101
	2.12	Stability of the Structural System 102
	2.13	Euler's Critical Load.. 104
	2.14	Standard Beam Element, Neglecting Axial Deformation.......106
		2.14.1	Rotational Coefficients 112
	2.15	Stability Functions under Axial Compression 115
		2.15.1	Rotation Functions 115
		2.15.2	Rotation Functions under Zero Axial Load 119
		2.15.3	Rotation Functions under Axial Tensile Load 121
		2.15.4	Translation Function under Axial Compressive Load 122
	2.16	Lateral Load Functions under Uniformly Distributed Load........ 131
	2.17	Fixed Beam under Tensile Axial Load................................ 135
	2.18	Lateral Load Functions for Concentrated Load 136
	2.19	Exercise Problems on Stability Analysis Using MATLAB®....... 140
	2.20	Critical Buckling Load 174
	Exercise 189
Chapter 3	Special Design Guidelines 191
	3.1	Thin-Walled Sections 191
		3.1.1	Torsion in Open, Thin-Walled Section.................... 195
	3.2	Buckling 197
		3.2.1	Global Buckling Modes... 197
	3.3	Lateral-Torsional Buckling 198
	3.4	Mechanisms behind LTB 202
		3.4.1	Torsional Effect 203

3.5	Measurements against LTB	205
	3.5.1 Effects of the Point of Application of Load	206
	3.5.2 Effects of Lateral Restraint on LTB	206
3.6	Behavior of Real Beam	206
	3.6.1 Factors Causing Reduction of Capacity	209
3.7	LTB Design Procedure	209
	3.7.1 Three-Factor Formula for M_{cr}	209
	3.7.2 Moment Correction Factors (C_1, C_2, and C_3)	211
3.8	Design Check for LTB	212
	3.8.1 General Method	214
	3.8.2 Buckling Curves	214
	3.8.3 Alternative Method	215
3.9	Example of LTB Using Euro Code	218
	Estimation of maximum bending moment, M_{ED}	218
	Steel section properties (referred UK steel sections)	219
	Yield strength of steel grade S275	220
	Section classification	220
	Calculation of section moment of resistance, $M_{b,Rd}$	221
	Stability Check against LTB	221
3.10	Design Check for LTB Using Indian Code (IS 800-2007)	223
3.11	Unsymmetrical Bending	226
	3.11.1 Bending Stresses under Unsymmetrical Bending	227
Exercises		233
	Example 1	233
	Example 2	234
	Example 3	235
	MATLAB Code	237
	Output	238
	Example 4	238
	MATLAB Code	240
	Output	241
	Example 5	241
3.12	Curved Beams	242
	3.12.1 Bending for Small Initial Curvature	242
	3.12.2 Deflection for Small Initial Curvature	245
	3.12.3 Curved Beams with Large Initial Curvature	246
	Sign convention	250
	3.12.4 Simplified Equations	250
Exercise problems		251
	Example 1	251
	MATLAB Code	252
	Output	253
	Example 2	253
	MATLAB Code	253
	Output	254

		Example 3 254
		MATLAB Code 255
		Output 255
		Example 4 256
		MATLAB Code 256
		Output 257
	Exercise 257
Chapter 4	Risk, Reliability, and Safety Assessment 259
	4.1	Background for Reliability Assessment 259
	4.2	Overview of Safety 260
	4.3	Lessons Learned from the Past 260
	4.4	Role of Safety in Offshore Plants 261
		4.4.1 Risk and Safety 262
		4.4.2 Measurement of Accident or Loss 262
	4.5	Quantitative Risk Analysis 263
		4.5.1 Logical Risk Analysis 263
	4.6	Risk Assessment 264
		4.6.1 Chemical Risk Assessment 265
		4.6.2 Application Issues of Risk Assessment 265
	4.7	Safety in Design and Operation 265
		4.7.1 Offshore Hazards 266
	4.8	Organizing Safety 267
		4.8.1 Hazard Groups 267
	4.9	Hazard Evaluation and Control 269
		4.9.1 Hazard Evaluation 269
		4.9.2 Hazard Control 270
	4.10	Quantitative Risk Assessment 271
		4.10.1 Initiating Events 271
		4.10.2 Cause Analysis 271
	4.11	Fault Tree Analysis 272
		4.11.1 Probability of Final Event 274
		4.11.2 Analysis Using the Fault Tree Method 274
	4.12	Event Tree Analysis 274
	4.13	Risk Characterization 276
		4.13.1 Principles of Risk Characterization 277
	4.14	Failure Mode and Effect Analysis 277
		4.14.1 Failure Mode Effect Analysis Variables 280
	4.15	Risk Acceptance Criterion 280
		4.15.1 UK Regulations 281
		4.15.2 Acceptable Risk 281
	4.16	Reliability 281
	4.17	Importance of Reliability Estimates 282
		4.17.1 Types of Uncertainties 283

Contents

4.18	Formulation of the Reliability Problem	283
	4.18.1 Time-Invariant Problem	284
	4.18.2 Time-Variant Problem	284
	4.18.3 Reliability Framework	284
4.19	Ultimate Limit State and Reliability Approach	286
4.20	Levels of Reliability	287
4.21	Reliability Methods	288
	4.21.1 First-Order Second-Moment (FOSM) Method	288
	4.21.2 Advanced FOSM Method	289
Exercise		291

Bibliography 293

Index 303

Preface

Design Aids for Offshore Topside Platforms under Special Loads explores the methods of design and the application of novel materials in offshore topside structures. A combination of various loads that act on the topside can result in extreme stresses, making analysis and design a complicated process. Several geometric forms of the topside have been deployed successfully for both oil exploration and oil production as a result of research and scientific developments. Chapter 1 deals with the basic components of the topside of an offshore platform while highlighting the use of different construction materials. Functionally Graded Materials (FGMs), a novel and recent innovation to topside, are discussed in detail together with their salient advantages and suitability. Fire load characteristics and their effects on structures are presented in detail, with examples. Details of the numerical analysis of topside under hydrocarbon fire are presented through numerical investigations. Details of Front-End Engineering Design (FEED) and its application to offshore projects are also discussed. Chapter 2 deals with the basic design guidelines of the topside of an offshore platform while highlighting offshore industry adopted practices in analysis and design.

Commonly utilized international design codes, steel material, and loading types encountered in designing the topside, together with the design stages and software programs commonly used, are examined. A detailed methodology of the stability analysis of beam-column connections is discussed and illustrated with a few examples and MATLAB® codes. The discussion of their analysis and design in this chapter will prove useful to practicing engineers. Chapter 3 deals with the special design guidelines of the topside. Lateral-torsional buckling is an inevitable consequence of large-span beams in the offshore topside and this chapter deals with its causes, occurrences, and design implications. Examples are provided for a better understanding of the conceptual design. Equally important concerns, not commonly addressed in textbooks, such as eccentric loads on beam columns, curved sections for crane hooks, etc. are considered. The derivation of design equations for estimating stresses in curved beams is derived from first principles, which will prove very useful. The chapter pays particular attention to the concept of unsymmetrical bending and analysis procedures using MATLAB® code.

Chapter 4 deals with the risk, reliability, and safety assessment of offshore industries. Various quantitative and qualitative methods of risk assessment tools and risk characterization techniques are discussed in detail. There are also detailed discussions of reliability assessment and safety practices applicable to offshore industries. The recent development of structural forms of offshore compliant structures is unique and expensive in terms of design, installation, commissioning, and operability (for example, the buoyant leg storage and regasification platform (BLSRP) and triceratops). Ocean structures are more focused on being form-dominant rather than strength-based designs. When coupled with risers and moorings, complicated and floating structures show a complex behavior under environmental loads.

The immense experience of the authors, in both academia and offshore structural consultancies, is combined to investigate the topics presented in this book. The authors express their immense gratitude to all their research scholars, graduate students, and colleagues for their support in various capacities. In particular, the work of Dr. Pachaiappan and Dr. Hari Srinivasan needs special mention. The lead author thanks the Chairman, Centre of Continuing Education, Indian Institute of Technology Madras, India, for extending administrative support in preparing the manuscript of this book. Authors also thank the support extended by the Indian Institute of Technology, Delhi, India, and the University Technology Petronas, Malaysia.

Srinivasan Chandrasekaran

Arvind Kumar Jain

Nasir Shafiq

M. Mubarak A. Wahab

MATLAB® is a registered trademark of The MathWorks, Inc. For product information, please contact:

The MathWorks, Inc.
3 Apple Hill Drive
Natick, MA 01760-2098 USA
Tel: 508 647 7000
Fax: 508-647-7001
E-mail: info@mathworks.com
Web: www.mathworks.com

About the Authors

Professor Srinivasan Chandrasekaran has maintained a balance between academic and practical experience over the past 30 years. He has interests in structural dynamics and earthquake engineering, nonlinear dynamics of offshore structures under environmental loads, and structural health monitoring and control. Professor Chandrasekaran has a combined experience of teaching, research, and industrial consultancy and has completed many research-based industrial consultancy projects resulting in the design and development of new design principles/mechanisms applied to buildings and offshore structures. He is an active researcher who has successfully supervised some 30 research theses and published 170 papers in refereed journals and conference proceedings. He has published around 17 textbooks on the topic of offshore structural engineering for leading international publishers.

Professor Arvind Kumar Jain obtained his B.E. in civil engineering from the then University of Roorkee, now the Indian Institute of Technology (IIT) Roorkee, India, and his Ph.D. in structural engineering from IIT Delhi, India. His areas of research are the earthquake-resistant analysis of structures, the wind-load analysis of structures, structural dynamics, offshore structures, tall structures, and reliability engineering. He has published a large number of papers in international journals and contributed to many international and national conferences. He co-authored (with Srinivasan Chandrasekaran) the textbook, *Ocean Structures: Construction, Materials and Operation* (Routledge, 2017), which is popular with graduate students of and consultants in offshore engineering. He has supervised several Ph.D., M.Tech., and B.Tech. theses in both civil and structural engineering. He is member of the American Society of Civil Engineers (ASCE) and the International Society of Offshore and Polar Engineers.

Professor Nasir Shafiq is a professor of structural engineering, construction, and materials in civil engineering in the University Technology Petronas, Malaysia. He has 30 years' experience as an academic, researcher, and structural design consultant. His expertise includes structural reliability, high performance and geopolymer concrete, building information modelling, and sustainable and low carbon living. He has published more than 170 papers in ISI- and Scopus-indexed journals and proceedings (Scopus ORCID: 0000-0002-9496-5430) and is the co-author (with Kah Yen Foong) of *"Green" & "Sustainable": The Future Global Construction Industry* (LAMBERT Academic Publishing, 2013). In addition to his professorial position, from July 2014 through November 2017, he was the Director of Sustainable Resources Mission-Oriented Research. He has secured many research grants from various agencies in Malaysia and overseas, totalling some US$2 million. He has supervised the research of 27 Ph.D. and 22 M.Sc. students and has served as visiting professor in ENSAM France, Najran University Saudi Arabia, and Pakistan.

Professor M. Mubarak A. Wahab has worked as an engineer for more than 20 years. He holds a Master's degree in structural engineering from the University of Melbourne, Australia, and a Ph.D. from the University Technology Petronas, Malaysia. He has had six solid years of progressively responsible supervisory, managerial, and business development on construction and design projects. An academic since 2007, his research interests are in the domains of structures and materials, onshore and offshore steel structures, and materials integrity and reliability. Keeping abreast of industry developments and becoming involved in industrial projects, Professor Wahab has maintained a balance between academic work and practical industry exposure. He has published some 50 research papers and has completed research-based industrial consultancies and research projects valued at US$0.25 million, and industry projects valued at US$20 million.

1 Materials and Loads on Topside

1.1 OFFSHORE TOPSIDE: LOADS AND ANALYSIS METHODS

Offshore platforms differ from other conventional structures generally deployed deep-sea for exploratory drilling and the production of natural gasses and oil (Chandrasekaran & Srivastava, 2018; Chandrasekaran, 2015, 2017; Chandrasekaran & Nagavinothini, 2020). The topside comprises several facilities that support platform operations. Offshore platforms are vulnerable to a combination of environmental and special loads, including impact and blast load, resulting in higher stress concentration on structural elements (API RP 2L, 1986; Chandrasekaran, 2014, 2019, 2020). Analysis and design of the topside from hydrocarbon explosion and impact load require innovative materials to take care of stresses and to satisfy functional requirements (API RP 2A, 2014; Chandrasekaran & Bhattacharyya, 2012). Pushover analysis of the topside using functionally graded material (FGM) and X52 steel is performed to compute the topside's performance under special loads (Chandrasekaran & Pachaiappan, 2020). The topside view and details of several components contributing to the load during operation are shown in Figure 1.1.

To understand the dynamic response behavior of topside structures, we will compare this to a typical multi-story complex building under lateral and gravity loads. A multi-story complex building is normally examined for its story displacement and force distribution using static pushover analysis. However, the response spectrum method underestimates the structure's response computed using modal pushover analysis (Chandrasekaran & Roy, 2006; Amador et al., 2008). Nonlinear static pushover analysis helps compute the natural frequencies, mode shapes, modal period, lateral displacement, deformed shape, hinge locations, and pushover curves for such structural systems under different bracings (Abu-Lebdeh & Voyiadjis, 1993; Kaley & Baig, 2017; Branci et al., 2016; Candappa et al., 2001; Chandrasekaran et al., 2021; Shulka et al., 2020). To inherit the advantages of higher redundancy, steel moment-resisting frames of a tall building configuration are provided with three different connection arrangements, namely i) reduced beam section (RBS), ii) reduced web section (RWS), and iii) fully fixed moment connection. Studies have shown that RWS performs better than the other two connections in terms of inter-story drift and plastic hinge formations (Carreira & Chu, 1986; Challamel & Hjijaj, 2005; Naughton et al., 2017; Chandrasekaran & Roy, 2004; Chandrasekaran et al., 2008a; Chandrasekaran et al., 2003; Chao et al., 2006; Chopra & Goel, 2002). The structural system's shear capacity is increased by introducing bracings in suitable locations, which was ascertained through nonlinear static pushover analysis (Khan & Khan, 2014; Elnashai, 2001; Esra & Gulay, 2005; Foire et al., 2016; Fragiacomo et al., 2011). This method is

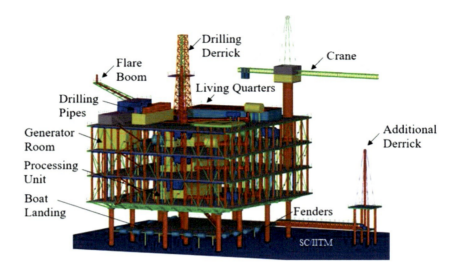

FIGURE 1.1 Typical view of a topside. (Courtesy of Chandrasekaran & Pachaiappan, 2020.)

capable of estimating the structural response at various seismic performance levels, accounting for both the first-order elastic and second-order geometric stiffness properties (Ganserli et al., 2000; Ghobarah, 2001; Giordano et al., 2008; Hasan et al., 2002; Jirasek & Bazant, 2002).

One of the uncertainties associated with the nonlinear static pushover analysis is the effect of resistance distribution and hinge mode contribution (Elnashai, 2001; Chandrasekaran et al., 2010; Chandrasekaran et al., 2008b; Fan Sau & Wang, 2002; Gilbert & Smith, 2006). Alternatively, the inelastic behavior is captured well through model pushover analysis, while the method retains its simplicity. Upon comparison with that of the results of the response spectrum analysis (RSA), the former was giving a better result in terms of the practical application to analysis and design (Chopra & Goel, 2002; Fiore et al., 2016). A multi-story building with a cross-laminated timber panel was analyzed by nonlinear static pushover analysis and seismic performance, and the influence of connection ductility on global structure ductility was also investigated (Fragiacomo et al., 2011; Hsieh et al., 1982; Hognestad et al., 1955; Jirasek & Bazant, 2002; Kaveh et al., 2010; Khan et al., 2007).

Furthermore, pushover analysis is capable of exploiting the nonlinear behavior of a steel structural frame. It accounts for the material nonlinearity, developed due to both single and combined stress states (Hasan & Grierson, 2002). Alternatively, the ant colony optimization (ACO) method effectively determines the performance of steel frames under lateral loads; it also compares well with the genetic algorithm approach (Kaveh et al., 2010). Recent experimental studies carried out on a steel column in a blast tunnel under long-duration blast loads and thermal loads highlighted that high-temperature effects increased the damage potential, causing columns to fail under buckling (Clough & Clubley, 2019; Denny & Clubley, 2019). Estimates of the modifications factor proposed for concentrically braced steel frames and buckling restrained braced frames are useful to predict the frame's response. Their dependence on ductility, over-strength, the height of the frame, and the number of bracing frames is well illustrated (Mahmoudi & Zaree, 2010).

Increased complexities that arise from the geometric layout of a topside can be handled using a modified, dynamic-based pushover (MDP) analysis. Alternative studies carried out on multi-story steel, moment-resisting frames showed their successful application and closer agreement with the dynamic-based story force distribution (DSFD) method (Mirjalili & Rofooei, 2017; Mwafy & Elnashai, 2001). A more simplified form, a floor response spectra (FRS) method, is also suitable for a multi-degree-of-freedom (MDOF) system. It uses an equivalent single-degree-of-freedom system and considers the first mode's participation to yield higher modes. This simplified method is equally efficient compared to the computationally tedious conventional model pushover analysis (Pan et al., 2017). The conservative model pushover (CMP) analysis is used to determine the seismic demand by enveloping the peak nonlinear responses resulting from the single-stage and multi-stage pushover analysis. It is shown that the CMP procedure accounts for higher mode effects but amplifies the seismic displacements of unsymmetric, plane tall structures (Poursha et al., 2011; Soleimani et al., 2013). Capacity factor, geometric irregularity, and varying stiffness should be considered when applied to special moment-resisting frames (Branci et al., 2016; Tarta & Pintea, 2012). Offshore topside falls under this category.

The structural behavior of topside frames becomes more complex due to eccentric bracings. Typical bracings provided on topside to avoid lateral-torsional buckling of the long-span deck girders need a special mention. Support from a few studies is worth mentioning in this context. The shear force demand is computed by the shell and frame element model and compared to the lumped plasticity model, which underestimates the shear force demand (Simonini et al., 2012). Flag-shaped bracings experience lower residual response drift and bilinear braces exhibit lesser floor acceleration. The brace hysteresis backbone curve controls the frame's response and the energy dissipation has an insignificant effect on the response (McInerney & Wilson, 2012). Therefore, pushover analysis is an effective and useful tool to evaluate the topside's response under lateral loads. Detailed investigations are carried out on the topside's performance using X52 steel and FGM under a combination of wind and special loads. These results will be discussed in further sections of this chapter.

1.2 GEOMETRIC CONFIGURATION

A typical topside comprises a multi-tier deck system used for various purposes based on its functionality. Figures 1.2 and 1.3 show the topside plan and elevation, illustrating the arrangements of structural members. The geometric of long-span beams and columns is inevitable in order to facilitate the function of the topside. Topside details are given in Table 1.1, while Table 1.2 shows the cross-sectional dimensions of structural members obtained from the preliminary design (Chandrasekaran & Pachaiappan, 2020).

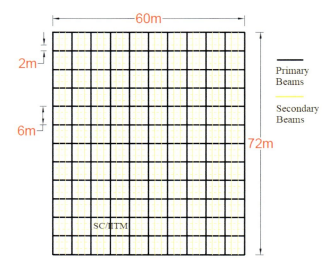

FIGURE 1.2 Topside plan.

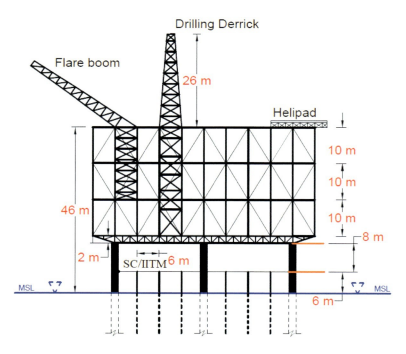

FIGURE 1.3 Topside elevation.

TABLE 1.1
Topside Details

Parameters	Values
Total number of stories	3
Height between floors.	10 m
Topside height.	46 m
Drilling derrick height.	26 m
Story height (boat landing).	6 m
Materials.	X52 steel and FGM
Topside geometry.	Symmetrical
Platform dimension.	60 m × 72 m
Columns center-to-center distance.	6 m
Beams center-to center distance.	2 m
No. of bays along the width (X-axis).	10
No. of bays along the length (Y-axis).	12
Total dimensions of the deck plate.	60 m×72 m
Deck plate dimension (taken for impact load analysis).	6 m×6 m
Deck plate thickness.	17 mm
Stiffener dimension (T-bar).	100 mm×6 mm/36 mm×6 mm
Airgap.	6 m

Source: Chandrasekaran & Pachaiappan, 2020.

1.3 MATERIALS

The design of offshore topside under hydrocarbon explosion or high-velocity impact loads is a major challenge. Material selection plays a crucial role under such conditions, as the strain rate increases rapidly over a finite time, resulting in permanent deformation or distortion of the material. Both material and geometric nonlinearities will be called out to provide adequate resistance. Conventional material, like X52 steel, despite its advantages, does not cater to special requirements as demanded. Alternatively, FGMs have a better application perspective. FGM is common to a wider range of applications in various fields, such as biomaterials, optics, and energy conversion. But the application of FGM in the offshore industry is much less and is scarce in the literature.

Composites are a parallel competitor, but delamination poses a threat to their use under the combination of thermal and mechanical loads of an extreme nature (Asian & Daricik, 2016; Kawasaki & Watanabe, 1987). It is inferred that delamination has a significant influence on both the flexural and compressive strengths of materials (Kawasaki & Watanabe, 1987). Under high temperature and pressure, thermal stresses occur at the metals and ceramics bonding interface due to the difference in their coefficients of thermal expansion. It leads to a crack formation and results in delamination at the interface. Researchers (Albino et al., 2018) have proposed a graded interlayer for metal and ceramic bonding and highlighted

TABLE 1.2
Section Details of Structural Members

Members	Parameters	Cross-sectional Dimensions
Columns (primary) (rectangular tubular section).	Depth	0.82 m
	Width	0.46 m
	Flange thickness	0.042 m
	Web thickness	0.042 m
Beams (primary) (rectangular tubular section).	Depth	0.62 m
	Width	0.42 m
	Flange thickness	0.03 m
	Web thickness	0.03 m
Bracings (channel section).	Depth	0.42 m
	Flange width	0.16 m
	Flange thickness	0.025 m
	Web thickness	0.025 m
Columns supporting boat landing (tubular section).	Diameter	0.56 m
	Wall thickness	0.045 m
Deep beams (rectangular tubular sections).	Depth	1.0 m
	Width	0.62 m
	Flange thickness	0.05 m
	Web thickness	0.05 m
Columns on buoyant legs (tubular section).	Diameter	2.1 m
	Wall thickness	0.18 m
Topside deck.	Thickness	0.018 m
Stiffeners (T-section).	Flange	37 mm × 7 mm
	Web	100 mm × 6 mm

their superiority in certain properties like heat resistance and toughness to arrest crack propagation. However, FGMs are designed to overcome the damage posed by delamination as they possess no distinct material interfaces. FGMs are increasingly used in aerospace, defense, and medical applications; recent attempts have also been made to explore their use in offshore engineering (Koizumi, 1997). They are also used to create thermal barriers, anti-oxidation coatings, and cemented carbide cutting tools. Thermoelectric materials are fabricated using FGM by grading their carrier concentration (Kan et al., 2018). Recent studies have highlighted the improved resistance possessed by FGM to second-order vibrations, buckling, and bending (Chandrasekaran et al., 2019, 2020). FGMs have a continuous variation or a step-wise grading of materials, which is application-specific. Their composition and microstructure are changed along the cross-section to generate a property gradient with the combined materials. Depending on the type of fabrication, FGM material can be continuously varying or stepped type.

Functional grading utilizes the salient advantages of the composition of materials. Research has been carried out on FGM composites, a functionally graded combination of an alloy and a ceramic (Kan et al., 2018; Madec et al., 2018; Yeo et al., 1998). Several additive manufacturing techniques successfully fabricate metallic FGMs using nickel-based super alloys (Yeo et al., 1998), titanium (Gao & Wang, 2000), and stainless steel (Rajan et al., 2010; DNV, 2018). Most of the studies are based on ceramic and metal-based FGMs when compared to intermetallic-based material. Conventional methods of fabricating an FGM can be achieved by centrifugal casting (Yuan et al., 2012; Shishkovsky et al., 2012), spark plasma sintering (Jin et al., 2005), laser deposition (Ubeyli et al., 2014), and powder metallurgy (Chen et al., 2017; Bermingham et al., 2015). There are several additive manufacturing techniques in which FGMs can be manufactured, such as electron beam melting, selective laser melting, direct laser melting, and wire arc additive manufacturing. Advances in material science and improved additive techniques have resulted in the realization of printing complex alloy parts. The current focus is on metal additive manufacturing techniques, as most of the components are metallic for engineering applications.

A few sets of samples are fabricated by functionally grading different materials to suit offshore topside applications: sample 1 is a combination of duplex stainless steel and carbon–manganese steel; sample 2 comprises nickel, carbon–manganese steel, and duplex stainless steel; and sample 3 is graded using titanium, carbon–manganese steel, nickel, and duplex stainless steel. Among three different samples, FGM sample 1 is presented in more detail regarding mechanical and structural properties. Comparisons of the mechanical properties of FGM samples and X52 steel (Debroy et al., 2018) are shown in Table 1.3, while their properties at elevated temperatures are summarized in Table 1.4. Figure 1.4 shows the stress–strain plots of X52 steel and the FGM samples (Chandrasekaran et al., 2019, 2020). The various structural members made of FGM proposed for the topside design are illustrated in Figure 1.5.

TABLE 1.3
Comparison of the Mechanical Properties

Material/Parameters	X52 Marine Riser Steel	FGM
Youngs modulus (GPa).	210	209.66 ± 4.48
Yield strength (MPa).	358	390.66 ± 12.23
Ultimate strength (MPa).	453	587.66 ± 12.76
Strength ratio.	1.265	1.50 ± 0.02
Ductility ratio.	32.207	45.47 ± 0.82
Tensile toughness (J/m3).	104.92	120.50 ± 2.84
Poisson's ratio.	0.3	0.30 ± 0.07
% elongation.	21	22.31 ± 0.11

Source: Chandrasekaran et al., 2019, 2020.

TABLE 1.4
Comparison of Mechanical Properties During High-Temperature Test

Material/Parameters	X52 Marine Riser Steel	FGM	FGM at 200° C
Yield strength (MPa).	358	390.66±12.23	339.93
Ultimate strength (MPa).	453	587.66 ± 12.76	468.19
% elongation.	21	22.31 ± 0.11	18.9

Source: Chandrasekaran et al., 2019, 2020.

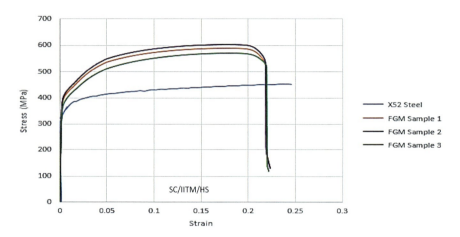

FIGURE 1.4 Stress–strain plots of X52 and FGM samples.

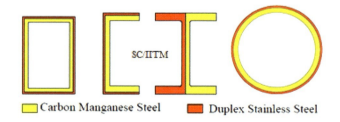

FIGURE 1.5 Structural components proposed for a topside design using FGM.

1.3.1 Wire Arc Additive Manufacturing

Wire arc additive manufacturing (WAAM) enables the step-wise addition of material layers to manufacture a three-dimensional component aided by a digital model. During the layer deposition process, component wires can be changed to form a single FGM. For example, ER 70S-6 carbon–manganese steel and ER 2209 duplex

Materials and Loads on Topside

stainless steel wires of 1.2 mm diameter are used in this WAAM process to obtain the functionally graded combination of sample 1. WAAM was introduced to overcome the limitations of deposition rate, equipment and powder cost, and part size in the powder bed process. The heat source is an electric arc, which is cheaper, and the expensive metallic powder is replaced with wire feedstock. WAAM is economical and capable of achieving a higher deposition rate, leading to rapid production of parts with larger size and minimum material wastage (Williams et al., 2016; Martina et al., 2012). WAAM also finds application in the fabrication and repair of metallic components to their full density, reducing the processing time. The arc-based welding approach is used in WAAM, in which the wire is melted due to the arc formed and deposited in layers. The metal transfer precision and guidance of the motion system determine the accuracy of the part fabricated. In building layers, wires can be changed to obtain a different material locally, resulting in a single FGM.

A stainless steel plate of 330 mm × 30 mm × 45 mm is used as a substrate to deposit the materials. Fronius™ cold metal transfer (CMT) TransPuls Synergic 4000 is used as the power source for the deposition process. A 3-axis computer numeric control (CNC) machine is integrated with the welding torch, as shown in Figure 1.6. The welding torch is programmed using G-codes in the desired X, Y, and Z coordinates to obtain the weaving pattern of deposition. To obtain multiple layers, a bidirectional weaving strategy was employed. The bidirectional strategy enables columnar grain disruption during the solidification process (Debroy et al., 2018).

FIGURE 1.6 Setup used to manufacture FGM on lab scale. (Courtesy of Srinivasan Chandrasekaran, 2020.)

The filler wire diameter is 1.2 mm, and the contact-tip-to-work distance (CTWD) is 15 mm. The stainless steel substrate surface is neatly prepared and cleaned with acetone before carrying out the metal deposition process. Before the FGM fabrication, the deposition parameters of the individual materials are determined in terms of the current, voltage, wire feed rate, and weld speed. The weaving width used for the deposition process is taken as half of the optimized transverse bead width obtained in a straight weld. Using the deposition parameters, three layers of duplex stainless steel are deposited on the stainless steel substrate, as shown in Figure 1.7. Using the optimized parameters, five layers of carbon–manganese steel are deposited over the duplex stainless steel forming an FGM of sample 1, as shown in Figure 1.8. Figure 1.9 shows the boundary between the two materials deposited on the substrate.

Tension tests are carried out on three sets of sample 1 to determine their mechanical properties. ASTM E8 dimensions are maintained in the test specimens. The samples are cut out from the build after operations like cutting, milling, and wire electro discharge machining. A gauge width of 17.47 mm is maintained in the sample, as shown in Figure 1.10, which consists of 3 mm of duplex stainless steel and 14.47 mm of carbon–manganese steel. It is important to note the direction of loading, which is kept parallel to the boundary. This is done to assess the separation between the layers, if any, even under the failure of the specimen.

A universal testing machine, the ZwickRoell Z100, integrated with a video extensometer, is used to determine the longitudinal and transverse strain and Poisson's ratio. Four white dots are painted on the test specimen's surface on a background of black paint, which serves as an optical marker. Two horizontal white dots represented by a blue line are used to calculate the longitudinal tensile strain. Figure 1.11 shows the stress–strain curve obtained for FGM sample 1 in the transverse direction.

FIGURE 1.7 Initial deposit of stainless steel layer of FGM, sample 1. (Courtesy of Srinivasan Chandrasekaran, 2020).

Materials and Loads on Topside

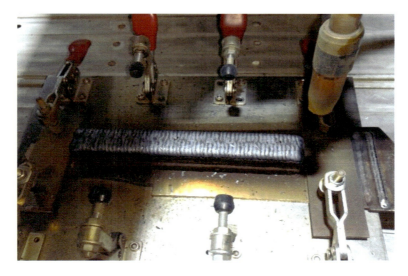

FIGURE 1.8 Successive deposit of carbon–manganese steel over stainless steel layer of FGM sample 1. (Courtesy of Srinivasan Chandrasekaran, 2020).

FIGURE 1.9 Functionally graded combination of duplex stainless steel and carbon–manganese steel sample 1. (Courtesy of Srinivasan Chandrasekaran, 2020).

X-ray computed tomography was carried out using a General Electric™ Phoenix V|tomeIx s on the sample in three different configurations, namely top, front, and side orientations. It is seen that the sample is free from porosity and micro-cracks in all three orientations. The X-ray computed tomography confirms that the manufacturing process is suitable for building a completely solid material that can be utilized for engineering applications.

The mechanical properties of the FGM samples are obtained from the stress–strain curves. The tensile toughness or deformation energy of the FGM sample 2 is measured as 123.61 Jm^{-3}, while that of X52 steel is 104.92. This indicates a clear advantage of the FGM material over the conventional X52 steel for offshore topside applications. Fractography is carried out on the sample after the tensile test with scanning electron microscope Quanta 200F using a secondary electron imaging mode. The FGM is also tested under elevated temperature to establish its suitability for high-temperature applications, as per American Society for Testing and

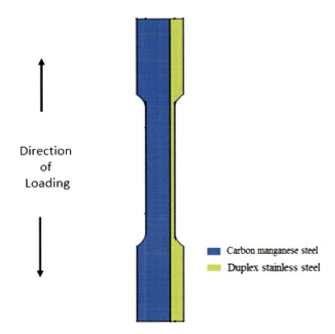

FIGURE 1.10 ASTM test sample examined under UTM. (Courtesy of Srinivasan Chandrasekaran, 2020)

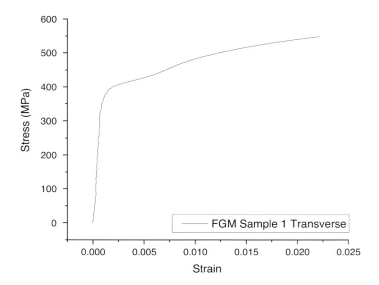

FIGURE 1.11 Stress–strain curve for FGM sample 1 in transverse direction. (Courtesy of Srinivasan Chandrasekaran, 2020).

Materials (ASTM) E21. The high-temperature tensile test is carried out at 200° C. Apart from the samples fabricated for the conventional tensile test, three more samples were fabricated for the high-temperature tensile test. The test specimens were soaked for 30 minutes at 200° C before the test was carried out. The fractured surface consists of duplex stainless steel and carbon–manganese steel parts along the tensile samples' cross-section. Fractography analysis is carried out in these two regions. The dimples are prominent and larger on the carbon–manganese side than the duplex stainless steel, as shown in Figure 1.12 (a) and (b).

A few smooth cleavages are also observed on the duplex stainless steel side. This observation indicates that carbon–manganese steel has contributed to improving the FGM's ductility compared to duplex stainless steel. Both carbon–manganese steel and duplex stainless steel show ductile failure. Under the load application, a stress partitioning occurs between the two materials in the functionally graded combination.

1.4 WIND AND BLAST LOADS

The topside of an offshore platform is prone to the combined action of wind and special loads. Wind loads can be computed on the topside at every story level using the gust factor method (GFM). A typical mean wind speed of 55 m/s for the Gulf of Mexico can be used for the preliminary studies. The variation of wind force on the topside is shown in Figure 1.13. Wind force on the XZ plane is relatively lesser than that of the YZ plane due to the topside's smaller exposure area on the XZ plane. Deformation of beams and columns due to hydrocarbon explosion is computed by using Abaqus/Explicit. Payloads were taken by referring to IS 875 (1987) Parts 1–2 and 5.

Further, topside beam-column connections are analyzed under the blast load, expected to be caused by a typical hydrocarbon explosion on the topside. The blast load is idealized as a linearized triangular impulse, as seen in Figure 1.14. The blast pressure is applied on the beams and column surfaces, and suitable boundary

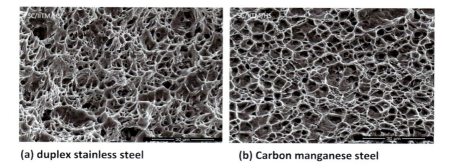

(a) duplex stainless steel (b) Carbon manganese steel

FIGURE 1.12 Fractography images of FGM sample 1. (Courtesy of Srinivasan Chandrasekaran, 2020.)

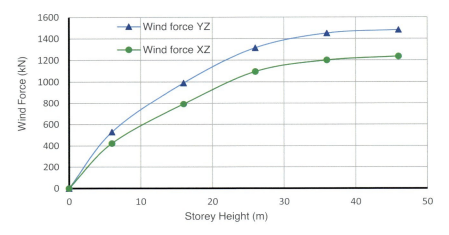

FIGURE 1.13 Wind load variation on topside. (Chandrasekaran & Pachaiappan, 2020.)

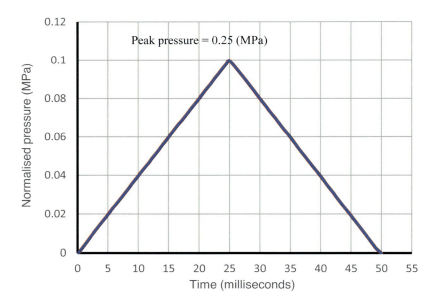

FIGURE 1.14 Idealized blast load on topside. (Chandrasekaran & Pachaiappan, 2020.)

conditions are applied to the model. The displacement contour of the beams and columns under hydrocarbon explosion using X52 steel and FGM are shown in Figure 1.15. It is seen that the structural members made of FGM sample 1 showed displacements lesser than that of X52 steel members. This can be attributed to the enhanced mechanical properties of the FGM sample 1. Figure 1.16 shows the deformed shape of the topside under the combined action of wind and blast loads for both FGM and X52 steel.

Materials and Loads on Topside

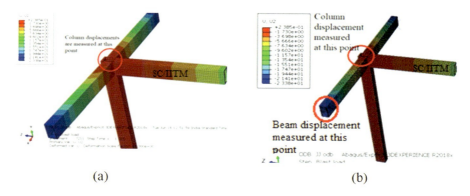

FIGURE 1.15 Displacement of beams and columns (a) X52 steel; (b) FGM sample 1. (Chandrasekaran & Pachaiappan, 2020.)

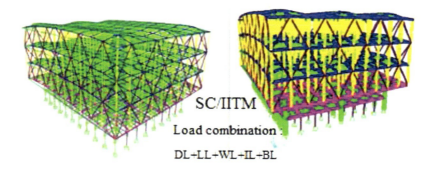

FIGURE 1.16 Topside deformed shape using X52 steel and FGM in a prone region.

1.5 IMPACT LOAD

Impact load on the topside decks generally arises from the fall of machinery, crane hook, equipment, drill pipes, tools, etc. The fall of objects from considerable heights may lead to permanent deformation or distortion of the topside deck (Shivakumar et al., 1985; Liew et al., 2009; Zucchelli et al., 2010). Figure 1.17 shows the numeric model of the topside deck and stiffeners considered for impact analysis. In this example study, impact load is considered to arise from the fall-off of a solid square block from a height of 20 m; it is a typical case of dropping off objects from an offshore crane during the erection/commissioning process of the topside. Figure 1.18 shows a typical offshore crane and a dropped object on the offshore deck.

The topside deck plate with stiffeners is analyzed under the impact load due to a typical equipment fall. Four-node shell elements were used to model the deck plate as the 3D deformable, and eight-node elements were used to model the stiffeners. Stiffeners are placed at the bottom of the deck plate using tie constraints, and sides of the square plate are assigned with the fixed boundary condition, as shown in

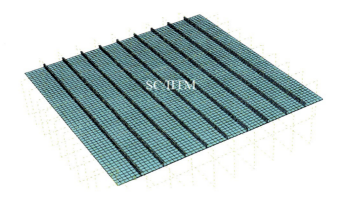

FIGURE 1.17 Numeric model of the deck plate.

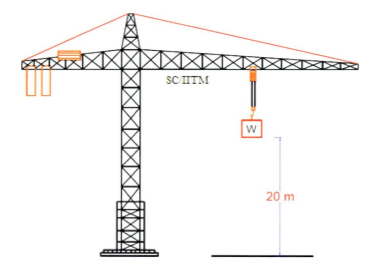

FIGURE 1.18 Typical offshore crane and drop object.

Figure 1.19. A comparison of the deformed shape of the topside deck under impact loads shows FGM as a better alternative to X52 steel (Figure 1.20).

1.6 MODAL ANALYSIS

Modal analysis helps to determine the dynamic characteristics of a structural system. Natural frequencies, mode shapes, and modal participation mass ratios of the topside are determined using SAP 2000 nonlinear analysis. Two examples are examined using FGM and X52 steel. In case 1, the topside is modeled using FGM (used in prone regions) and X52 steel. In case 2, X52 steel is used to model the entire topside. A preliminary design of the topside must withstand a combination of gravity loads,

Materials and Loads on Topside 17

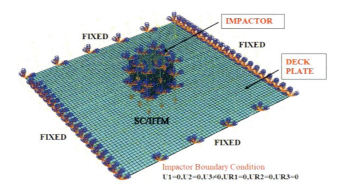

FIGURE 1.19 Boundary conditions of the deck plate and impactor model. (Chandrasekaran & Pachaiappan, 2020.)

FIGURE 1.20 Comparison of central deflection of deck under impact load. (Chandrasekaran & Pachaiappan, 2020.)

impact loads, and blast loads. Table 1.5 shows the dynamic characteristics of the topside for cases 1 and 2. It is seen that the natural frequency of topside for case 1 is marginally higher than that of case 2, indicating an increase in modal stiffness of the topside due to FGM. Furthermore, the frequency difference is significantly less. Both cases' modal participation mass ratio is higher (90 percent) in the first mode, indicating a good structural design. The collapse pattern of the topside for both the cases is shown in Figures 1.21 and 1.22. It can be seen that in both cases the first two modes are subjected to lateral vibration; the third mode shows the lateral and torsional vibrations, indicating an acceptable design.

TABLE 1.5
Dynamic Characteristics of the Topside

	Case 1			Case 2		
Mode	Period (s)	Frequency (rad/s)	Modal Participating Mass Ratio	Period (s)	Frequency (rad/s)	Modal Participating Mass Ratio
1	0.360	17.406	0.810	0.384	16.330	0.81947
2	0.328	19.117	1.95E-06	0.348	18.005	1.48E-06
3	0.311	20.145	1.88E-10	0.328	19.106	1.99E-11
4	0.177	35.480	0.06688	0.188	33.274	0.0572
5	0.167	37.525	2.94E-07	0.182	34.504	1.82E-07
6	0.147	42.595	1.09E-09	0.162	38.551	1.60E-09
7	0.137	45.752	0.01023	0.156	40.253	0.01413
8	0.123	50.851	2.46E-06	0.137	45.708	2.34E-07
9	0.123	50.936	2.09E-07	0.136	45.935	1.74E-07
10	0.113	55.385	1.35E-10	0.123	50.965	2.52E-10
11	0.101	61.696	1.34E-10	0.115	54.382	2.51E-10
12	0.093	67.117	0.00373	0.105	59.348	0.00507

Materials and Loads on Topside

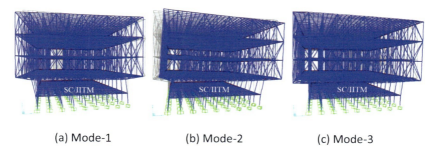

(a) Mode-1 (b) Mode-2 (c) Mode-3

FIGURE 1.21 Collapse pattern of topside (case 1: FGM+X52 steel).

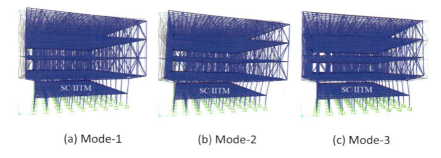

(a) Mode-1 (b) Mode-2 (c) Mode-3

FIGURE 1.22 Collapse pattern of topside (case 2: X52 steel).

1.7 PUSHOVER ANALYSIS

Pushover analysis is an effective method of analyzing the structure subjected to a lateral force that increases monotonically with an invariant height distribution until a target displacement is reached. Pushover analysis comprises a series of consecutive elastic analyses, superimposed to get an approximate force-displacement curve of the entire structure. A bilinear or trilinear load-deformation curve of all the lateral force-resisting elements is created for a 2D or 3D model (Khan & Khan, 2014; FEMA 440, 2005; FEMA 450, 2004). The structure is subjected to gravity loads and predefined lateral load combinations. The lateral forces are distributed along with the structure's height and increased until the yielding of structural members. The structural system is modified to account for reduced stiffness caused due to the yielding, and lateral forces are increased until successive members yield. The process is continued until the structure becomes unstable. The static pushover curve is obtained by plotting the displacement against base shear. An example problem deals with the nonlinear static pushover analysis of an offshore platform's topside using FGM and X52 steel under special loads. The displacement-controlled pushover analysis (FEMA-356) is used to compute the topside capacity, the position of plastic hinge formation, and failure modes. The preliminary analysis and design of an offshore platform's topside are carried out using SAP 2000 nonlinear analysis.

The topside is designed to withstand gravity loads, wind loads, blast loads (hydrocarbon explosion), and impact loads (ISO EN, 2007). The topside response is now investigated under monotonic displacement controlled lateral loading by nonlinear static pushover analysis. Figure 1.23 shows an idealized pushover curve for different states, namely immediate occupancy (IO), life safety (LS), collapse prevention (CP), and collapse (C) as per ACT 40 and FEMA 356. A comparison of the capacity curve (pushover curve) for cases 1 and 2 is shown in Figure 1.24. It can be seen that case 1 shows a higher capacity in comparison to case 2 (8 percent more). The resistance of the structure against pushover loading depends on the material strength and the member's cross-sectional dimensions. Improved performance of the topside is achieved by the enhanced mechanical properties of the FGM. Bottom-story columns experienced larger displacement due to the absence of bracing and stiffness variation, resulting in the formation of plastic hinges. Tables 1.6 and 1.7 show the hinge formations on the topside for cases 1 and 2, respectively. Figures 1.23 and 1.24 show

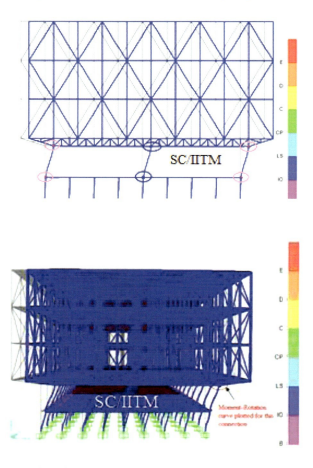

FIGURE 1.23 Location of plastic hinges for case 1.

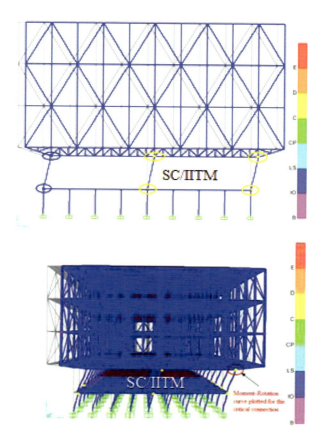

FIGURE 1.24 Location of plastic hinges for case 2.

the corresponding hinge locations and collapse pattern of the topside. The use of FGM in a vulnerable region or critical location of the topside reduces the number of plastic hinge formations compared to X52 steel, as seen in these figures. However, FGM is not used for the entire topside after taking into account the complexities of the fabrication process and economic considerations.

A comparison of the moment–rotation curve of critical beam-column connection distinctly shows that FGM possesses higher resistance to plastic rotation (Figure 1.25). Corresponding values are given in Table 1.8. It is seen that FGM offers higher resistance to plastic rotation of about 21.4 percent compared to X52 steel. Results comparison for cases 1 and 2 is given in Table 1.9.

The different ranges of performance levels beyond yield limit, such as immediate occupancy (IO), life safety (LS), collapse prevention (CP), and collapse (C), are defined as per Applied Technology Council, Federal Emergency Management Agency's (ATC), FEMA 356. In case 1, the maximum number of hinges beyond the yield limit falls under IO. It states that the topside experiences marginal damage but

TABLE 1.6
Hinge Details for Case 1

Step	Displacement (m)	Base Shear (kN)	A to B	B to IO	IO to LS	LS to CP	CP to C	C to D	Beyond E
0	0	0	6812	0	0	0	0	0	0
1	0.200	4962	6812	0	0	0	0	0	0
2	0.400	9925	6812	0	0	0	0	0	0
3	0.465	11543	6812	0	0	0	0	0	0
4	0.657	15920	6800	12	0	0	0	0	0
5	0.708	16584	6683	129	0	0	0	0	0
6	0.909	17878	6110	634	68	0	0	0	0
7	1.231	19287	5958	208	646	0	0	0	0
8	1.432	20045	5748	266	798	0	0	0	0
9	1.735	21001	5738	106	968	0	0	0	0
10	1.945	21657	5536	226	1050	0	0	0	0
11	2.000	21822	5486	264	1062	0	0	0	0

TABLE 1.7
Hinge Details for Case 2

Step	Displacement (m)	Base Shear (kN)	A to B	B to IO	IO to LS	LS to CP	CP to C	C to D	Beyond E
0	0	0	6812	0	0	0	0	0	0
1	0.200	3411	6812	0	0	0	0	0	0
2	0.400	6822	6812	0	0	0	0	0	0
3	0.600	10233	6800	12	0	0	0	0	0
4	0.677	11544	6332	452	28	0	0	0	0
5	0.895	15101	6110	634	68	1	0	0	0
6	1.059	16584	5862	258	692	1	0	0	0
7	1.259	17658	5738	106	968	3	0	0	0
8	1.427	18540	5514	244	1054	7	0	0	0
9	1.674	19286	5486	102	1224	9	1	0	0
10	1.880	19975	5414	98	1300	9	1	0	0
11	2.000	20100	5360	152	1300	11	2	0	0

is able to retain its original strength. Beam-column connection experiences minor yielding at very few locations. There is no visible buckling and no complete distortion of structural members. Followed to IO, case 1 comprises LS hinges. Therefore, the topside experiences moderate damage, and all stories of the topside retain some residual strength. Hinges are formed in beam-column connection; however, the shear connection remains sound. In case 2, the maximum number of hinges falls under LS;

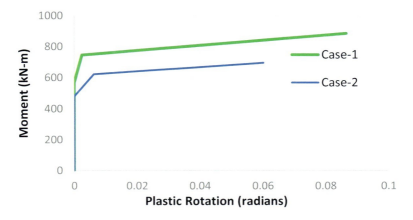

FIGURE 1.25 Moment–rotation curve of the connection for cases 1 & 2.

TABLE 1.8
Moment–Rotation Values for Cases 1 and 2

	Case 1		Case 2	
Steps	Moment (kN m)	Rotation (radians)	Moment (kN m)	Rotation (radians)
1	3.3328	0	2.256198	0
2	241.2092	0	157.8793	0
3	486.3068	0	318.1379	0
4	566.3312	0	478.52	0
5	746.042	0.00246	478.52	0
6	750.5084	0.00517	621.883	0.00619
7	768.446	0.016053	632.8165	0.014214
8	804.1636	0.037724	650.1947	0.026967
9	827.1488	0.05167	665.357	0.038094
10	858.0372	0.070411	679.3993	0.048399
11	879.1944	0.083248	691.8169	0.057512

TABLE 1.9
Results Comparison for Cases 1 and 2

Cases	Topside Capacity under Lateral Loading Conditions (kN)	Moment Resistance at Connection (kN m)	Range of Plastic Hinge Formations
Case 1 (FGM + X52 Steel).	21822	879	IO, LS
Case 2 (X52 Steel).	20100	691	IO, LS, CP, C
% difference	8	21.4	NA

however, the range of hinges that reached CP indicates that the topside is damaged severely and retains little residual stiffness and strength. The topside suffers permanent deformation, and beams and columns are distorted heavily. Followed to CP, case 2 comprises few critical joints, which come under collapse (C), and affects the topside's total integrity.

The example study presented here discusses the superiority of FGM in an offshore topside subjected to special loads and displacement-controlled pushover loading conditions. Topside response under blast load (hydrocarbon explosion) and impact load are computed using Abaqus/Explicit and are designed to withstand the special loads using FGM and X52 steel. The design of the topside under special loads satisfies the code recommendations. Nonlinear static pushover analysis is performed to compute the topside capacity using SAP 2000. The capacity of the topside is determined for two different cases. In case 1, the topside is modeled using FGM (used in prone regions only) and X52 steel, while case 2 comprises X52 steel used for the entire topside. The results obtained from the numerical studies showed that using FGM in a prone region decreases the number of plastic hinge formations and hinge rotations and increases the topside capacity under pushover loading conditions. However, FGM is not used for the entire topside after taking into account the complexities of the fabrication process and economic considerations. Case 1 is recommended for the topside design, which offers an 8.5 percent improved topside capacity against lateral loading conditions. The range of hinge formations is within immediate occupancy (IO) and life safety (LS), indicating that the topside experiences light damage and retains its original strength. In case 2, the range of hinges reached collapse level indicating that the topside was damaged severely and retained little residual stiffness and strength. Similarly, the critical beam-column connection's moment–rotation curve shows that FGM possesses higher resistance to plastic rotation of 21.4 percent than X52 Steel. From the present study, it is observed that FGM's enhanced mechanical properties play a crucial role in maintaining the topside's integrity and avoids collapse (C) hinge formations.

1.8   FEED: BASICS

Front-end engineering design (FEED) is basic engineering, an outcome of a conceptual design or feasibility study. FEED focuses on the technical requirements and an approximate CAPEX (capital investment cost) of the proposed project. FEED can be divided into separate packages covering different portions of the proposed offshore project. The FEED package is used as the basis for bidding the execution phase contracts (engineering, procurement, and construction, or EPC; engineering, procurement, construction, and installation, or EPCI, etc.). It is also used as the design basis for the offshore project (Chandrasekaran, 2014). A good FEED report will reflect all the technical requirements of the project, and financial constraints, if any. A detailed FEED will assist the project group in avoiding significant changes during the execution phase. FEED contracts take about a year to complete for a larger-sized offshore project. During the FEED phase, there is close communication between the project owners, the operators, and the engineering contractors to arrive at the specific requirements of the proposed project. FEED also refers to basic engineering conducted after completion of the conceptual design or a feasibility study. Before starting the EPC, various studies are carried out to address the technical issues anticipated for the project. An estimate of rough investment cost is also presented as a part of the FEED report. During the preparation of the FEED report, all the intentions of the client raised during frequent meetings are documented and addressed. Detailed reporting, which includes the feedback/reviews of the client and the contractors, will end up with minimum deviation during the execution stage. Any details that are overlooked in the FEED stage will lead to a significant change during the EPC phase, which will have a serious financial consequence on the project. So it is essential to maintain close communication with the client and it is common practice for the client to station at the contractor's office during the FEED report.

FEED includes robust planning and an effective design towards the project's life cycle. It is important to note that there is a higher probability of intervening with the proposed design at the time of FEED preparation. It will not affect the project cost much, as any changes proposed during the FEED stage are relatively low-end costs. It typically applies to projects with high capital investment and a long life cycle. Offshore projects are classic examples where FEED is an important stage. Though it often adds a small amount of time and cost to the early portion of a project, these costs are insignificant compared to the alternative of the costs and effort required to make changes later in the project.

1.9 ESSENTIALS OF FEED

There is a need to define the basic scope, parameters, and economic impact(s) in the early stages of a new or a conceptual project. FEED is a study used to analyze the various technical options for new developments and to define the facilities required. With pre-FEED (preliminary front-end engineering design) work, one typically provides the client with a complete overview of the design of the project's overall development. This is very important for offshore plants as they are unique in a number of ways, namely i) the conceptualization of the structural geometry (FORM), as discussed below, ii) the use or choice of innovative materials for the plant, iii) understanding the complexities related to the environmental loads and conditions of the sea state to model them for analysis and design mathematically, iv) detailed topside design to include the layout of machinery that are customized for specific functional requirements of the project, and v) uncertainties during the construction stage. A FEED report conceptualizes the proposal of the offshore plant, which is neither available previously nor properly undertaken at the conceptual design stage.

At pre-FEED, engineers emphasize developing a basis for the process design and specify the required process parameters on which the main FEED work depends. FEED, therefore, does not focus on the timeline of the project but on the details of the engineering design and the process/mechanisms. When the design basis is complete, one can typically have the following information defined in the report, namely i) material specifications, ii) plant capacity requirements, iii) product specifications, iv) critical parameters that will influence the plant in operation, v) specifications of the available utilities, vi) performance requirements of every unit in the plant, vii) a pre- energy efficiency study, viii) a pre-greenhouse gas (GHG) demonstration study, in support of ALARP (as low as reasonably practicable), ix) process regulatory requirements, and x) all other operating goals and constraints desired by the plant owners/operators/engineers.

Front-end engineering design (FEED) is also known as pre-project planning (PPP), front-end loading (FEL), feasibility analysis, conceptual planning, programming/schematic design, and early project planning. It is expressed in a draft format and describes the various stages in the proposed project of the offshore plants, such as the industry-phase upstream. One of the basic objectives of the FEED is to collect strategic information on the perceived risks through a detailed risk analysis (either a quantitative or a qualitative risk analysis, or QRA). It will allow strategic planning by the responsible persons in the initial stage itself to reduce the risk to the ALARP level. FEED deals with an early stage of a project and enables moderate design changes without invoking high costs. It is also an important prerequisite of all major projects in order to limit the deviations in the design and cost investment. It is, therefore, mandatory for capital-intensive projects with a long life cycle. Although the approach is relatively more expensive and takes more time, a good FEED report will avoid additional costs in proportion to any significant changes in the design at a later stage of the project.

For good practice, FEED can be divided into six phases. First three phases, namely FEED 1, FEED 2, and FEED 3. FEED 1 addresses issues related to material balance, energy balance, and project charter. FEED 2 focuses on the preliminary design of equipment, the layout of the process plant, the preliminary schedule of works, and the preliminary cost estimate. FEED 3 includes detailing equipment specifications to make them purchase-ready, a detailed project execution plan, a definite estimate of cost, a project execution plan, a layout of electro-mechanical equipment, and a timeline for commissioning the plant (Chandrasekaran, 2014). The FEED study is a vital starting point for any offshore development. It lets one assess both the technical and financial viability of the project while addressing technical uncertainties. Key deliverables for different FEED levels are identified for different phases of execution of the project. FEED is a clean guideline from the initial concept to implementation through detailed design. Detailed analysis in FEED also includes the following, namely i) mechanical and structural engineering, ii) equipment design, iii) electrical engineering (low voltage, LV; medium voltage, MV; and high voltage, HV), iv) load flow analysis and protection studies, v) submarine cable engineering, vi) marine engineering, vii) subsea engineering, viii) details related to the field development, ix) installation analysis, and x) cost–benefit analysis (commercial analysis).

FEED provides a detailed insight into the main equipment's mechanical data sheets, including the process specifications issued during the basic engineering design (BED) stage. It also incorporates the specific code requirements, as applicable to the respective project. Data also include the thermal rating of heat exchangers, which are used in the process line. Development of the piping and utility diagrams for the process plants and their verification for the successful execution is undertaken through hazard and operability studies (HAZOP) (Srinivasan, 2016), an important component of the FEED report. Preparation of the tender packages for the purchase order of the main equipment is also included in the FEED study. A FEED report enables the engineers to identify the hazard areas in the plant and implement the necessary safety measures to control the envisaged risk. The cost factor in mitigating probable hazards and the degree of control achievable through design and risk characterization are a few of the vital outputs of the FEED study. A detailed engineering study, which follows the FEED report, covers the development of the layout of pipes and works related to instrumentation, electrical, and other infrastructure facilities. Detailed drawings of the piping and instrumentation diagram (P&ID), which are good for construction, are released only after a thorough review of the FEED report.

1.10 BASIC ENGINEERING REQUIREMENTS OF FEED

In an offshore project, engineering studies will include all or part of the following. Basic engineering design (BED), covering conceptual process studies (material balances, process flow sheets, etc.); a preliminary plan of the plot; preliminary piping and instrument diagrams; definition and sizing of the main equipment; specification of effluents; and definition of control and safety devices. The design and construction of an offshore plant involve many interdependent actions carried out at different stages (Chandrasekaran, 2014). These are largely classified as i) those related to the criteria of selecting the FORM for the plant, ii) those related to the analysis, design, sizing of members, construction, and installation methods, and iii) those related to cost implications of the project, like bidding, etc. The operational criteria of the proposed plant are examined thoroughly. They involve the number of wells to be drilled, the type of drilling equipment required for drilling, and the material to be used. The floor area required on each deck and the number of decks required for the topside of the plant should be determined based on the specific requirements of various operations to be carried out on the deck. The geometric form of the platform chosen for the said design/operation and sea state should meet the operational requirements. Subsequently, it is necessary to determine the environment to which this equipment will be subjected. All environmental loads that are imposed on the platform must be estimated. Forces arise from waves, tidal conditions, storm wave heights, storm and wind velocities, the current, and, in many instances, earthquakes. The major stages in the design and development of an offshore plant, such as, for example, a template structure, are as follows 1) the preliminary phase, 2) the design phase, 3) the bidding phase, and 4) the construction phase.

In the preliminary phase, the necessity of the plant is established for a chosen sea state and geographical location. Environmental forces acting on the members are computed. Feasibility studies are carried out to calculate the cost estimates of a variety of options. For the chosen scheme, financing arrangements and monetary allotments are discussed. In the design phase, preliminary geotechnical investigations are carried out to finalize the site selection based on soil characteristics. This phase also includes selecting the derrick and transportation barges required for floating (wet-towing) prefabricated modules. A timescale is worked out based on crew transportation arrangements and the availability of specific types of cranes capable of operating in the chosen weather window. Detailed designs and drawings for the foundation and the structure are prepared. A detailed document is prepared containing the following: i) specifications of members and materials, ii) contracts, iii) bid reply sheets, iv) rental estimates and contracts for derricks and transportation barges, and v) rental estimates and contracts for tug boats and work boats. In the bidding phase, the preparation of technical bid and financial bid sheets, the short-listing of bidders, the evaluation of bids (both technically and then financially), and contracts for the framed terms and conditions are carried out. Finally, in the construction phase, onshore fabrication works, including ordering and receiving materials, fabrication of joints, connections, and members, coating components for corrosion protection, and other corrosion protection measures are primarily addressed. Further, erection details, including placement of underwater components, the installation of the pile foundation, and the setting of equipment above water, are included

1.11 FACTORS INFLUENCING DESIGN OF TOPSIDE

In the design of major projects like offshore plants, engineers know how to design the required facilities but are not well versed in problem definition. Most successful designs are produced for a pre-defined problem. The crux lies, therefore, in problem definition. It is important to understand the correlation between design alternatives and cost factors at the design stage. Based on the decision to choose a specific FORM for an offshore plant, the cost will control the project development. As the design process is taken forward, costs increase sharply for the design's given alternatives (Chandrasekaran, 2019). In this loop of discussions, problem development is considered as the core of FEED. Six categories characterize FEED as the initial start of the concept design alternatives: team management, market assessment for the cost–benefit analysis, user feedback on the design alternatives, product exploration, development plans, and target specifications. Therefore, FEED is modified as various members participating in the loop of discussions evaluate it. The whole process is then reviewed at an early stage to select the design, which is competitive so as to yield an early return on the investment. In the design of offshore plants, a systematic approach in the development is required at the early stage itself (Chandrasekaran, 2019). Unlike in other projects, offshore projects do not have the liberty of diversifying the design objectives, as the cost involvement is significantly high. Therefore, it is vital that the design development or generation of ideas occurs in a predefined engineering manner. For example, let us consider developing a new structural FORM for offshore plants in the ultra-deep sea for oil and gas exploration. It is very important to note that the FORM development cannot be a wild choice of the engineer but is formulated from existing platform designs.

More time is spent on FEED, which is generally justified by an improved return on the investment (RoI). It is important to note that most aspects of a project are decided early on in the design process. Figure 1.26 shows the FEED layout as a solution-oriented process. A typical FEED report consists of the following

1. Introduction.
2. Purpose and goals of the project.
3. Offshore plant scope.

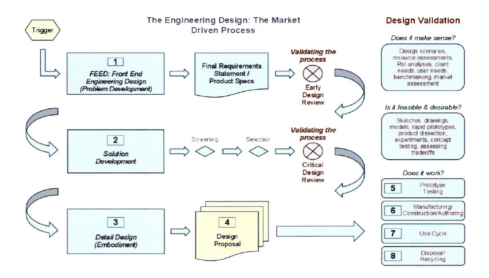

FIGURE 1.26 FEED-solution model of the design process. (Chandrasekaran, 2014.)

4. FEED study for the scope of services offered by the plant.
5. Equipment definition, selection, pricing, and bidding process.
6. Engineering, procurement, and construction (EPC) planning and execution.
7. Capital cost estimates.
8. Operating cost estimates.
9. Health, safety, and environmental issues.
10. Conclusions.

One of the major challenges in offshore plant projects is for the technology providers to execute the chosen kind of deployment of the design alternatives. It is vital to understand that the original equipment manufacturers and the EPC contractors do not have the first-hand information or experience to bid for the plant at the scale required. It necessitates a costly FEED study to define the costs and performance with enough certainty. It will be helpful in order to establish commercial guarantees and attract equity and debt financing. A typical initial FEED study report will ensure the following

1. Confirmation of the expected performance, which is backed by commercial guarantees.
2. Sufficient details of the technical and execution planning to support high-accuracy EPC cost estimates.
3. Affirmation that the EPC price is consistent with the indicative bid used in the selection process for execution of the project.
4. A design (chosen from various alternatives) that complies with international standards for integrity, operability, availability, safety, and environmental profiles.

5. An estimate of the annual operation and maintenance cost of the offshore plant.
6. Confirmation of the specifications of all the major equipment and their availability in the market.
7. Confirmation that the plant will yield the desired product for the required period. An early start of production will be the intensive focus of the design, for obvious reasons.
8. Confirmation of constructability, indicating the methods of transporting, commissioning, and erection of the prefabricated modules of the plant.

A FEED report will also contain additional information, as listed below.

1. FEED study results, including the cost and performance.
2. Challenges handled in the initial FEED study and their probable consequences.
3. Confirmation of construction and operation of the large-scale plant based on the design alternatives and construction methodology discussed in the initial FEED report.
4. Confirmation of health, safety, and environmental impacts associated with the commissioning of the offshore plant.
5. Problems unaddressed in the initial FEED report, their reviews, and correction measures/recommendations.
6. Capital cost estimate accuracy of the chosen design alternatives as a complete solution.

The FEED study is a critical phase in the development of large projects like offshore plants. It is important as these projects are unique in terms of scale, technology, or configuration, for which reference cost and performance information are not available. The main objective of FEED is to establish and define the technical scope of the work and the project execution with sufficient details to determine the project cost and the commercial terms of the EPC proposal. It amounts to the completion of about 15–20 percent of the plant engineering and planning effort, which reduces the project risk before moving into the detailed engineering and construction of the offshore plant. The FEED project team consists of a multi-disciplinary task force that includes team members with various specializations in the engineering field and functions. These are project management, process engineering, mechanical engineering, civil and structural engineering, piping engineering, electrical engineering, instrumentation and control engineering, Health and Safety Executive (HSE) management, scheduling and planning, procurements, contracts, construction engineering, commissioning, and start-up, and cost estimating. There should be seamless coordination between the team members of the FEED group through routine communication and multiple design reviews.

1.12 ADVANCED LEVEL IN FEED STUDIES

The FEED study follows a typical development sequence. The sequence includes the eight phases listed below, with key deliverables associated with each phase listed in parentheses. A Pre-FEED analysis is necessary to select a particular FORM for the desired functions of the offshore plant. The chosen FORM is carefully analyzed for the probable sea states of the project site to determine its survivability under the worst combination of environmental loads. In continuation of the selected basic design, the FEED team continues to investigate in the following phases:

- Phase 1. Process Engineering (design basis, process flow diagrams (PFDs), heat and material balances (HMBs)).
- Phase 2. Equipment Definition (equipment data sheets, specifications).
- Phase 3. Equipment Pricing (bid lists, bid packages, bid tabs).
- Phase 4. Plant Layout and Model (plot plan).
- Phase 5. Material Quantification (material takeoffs (MTOs), physical drawings).
- Phase 6. EPC Execution Planning (EPC schedule, contract plan).
- Phase 7. Cost Estimating (capital, operating costs).
- Phase 8. Commercial Proposal (cost, schedule, performance guarantees).

Process Engineering starts with the preparation and establishment of the formal project design basis. The economic evaluation factors that influence the project are also considered at this stage. The potential of these factors is to allow the FEED team to evaluate various design options and determine the best economic decision for the project. Details of the equipment required for drilling, processing, transportation, and storage are prepared with the relevant data sheets and specifications. Site-specific design criteria are chosen for the analysis. They are ambient temperature range, critical and operational sea states, variations in the sea surface elevation, atmospheric pressure, rainfall, design wind speed, etc. Once the equipment listing is complete, a possible layout of the main process is drawn up. The process flow diagram (PFD) forms the basis for developing both material selection diagrams (MSDs) and piping and instrumentation diagrams (P&IDs). MSDs form the basis for the mechanical equipment, piping, and instrumentation specifications. The final activity in the process engineering phase is the preparation of flow stream summaries, including utility system and environmental summaries and pipeline flow summaries. A design review is carried out on the chosen layout of the process diagram and layout plan of the equipment in terms of production convenience, effective utilization of the topside space, weight minimization during erection and operations, hazard evaluation, risk characterization, and loss control/prevention. An experienced team carries out hazard identification (HAZID) for each area of the plant. The whole plant is divided into various segments for which HAZID is carried out. Based on the findings of the HAZID study, P&ID is reviewed. During this review, a few critical factors of the plant are finalized: valve types, equipment connections, line sizes, piping connectivity, and various other operating scenarios.

A hazard and operability study (HAZOP) is a structured and systematic technique for process hazard analysis and risk management. Many international regulations insist on a HAZOP as a mandate. The scope of HAZOP includes identifying potential hazards in the system and operability problems that are likely to lead to undesirable events. This method assumes that the deviations from the intended design or operating conditions cause risk events. The study identifies such deviations using guide words, which is a systematic identification of deviations. The process engineering studies conducted earlier now initiate discipline-specific engineering in mechanical, structural, piping, control systems, and electric areas. Mechanical engineers use the process data sheet to develop the equipment designs further. After sizing, the equipment is created and piping engineers generate the plot plan of piping layout in the topside, which the electrical engineers will coordinate to design electrical systems.

Major equipment and systems pricing is a key element in developing the project cost estimate, excluding the offshore plant's structural cost. The pricing effort starts with identifying high-value equipment and systems that comprise most capital costs for the project. The FEED team prepares a detailed commercial term document with the technical attachments to produce a request for quotation (RFQ) bid packages for each identified purchase order. RFQ packages are issued to each bidder from the approved bidder list, and bid submission deadlines are set depending on the complexity of the equipment and systems. During this period, the FEED team supports the bidding process by providing clarifications to the vendors' questions. They also hold bid-conditioning meetings to discuss technical and commercial terms. This ensures a common understanding of project objectives. Once received, the FEED team evaluates the bids on both commercial and technical bases. The evaluations will result in the preparation of formal bid tabulations. These tabulations provide the basis for the selection of the recommended vendor and the associated cost. The recommended bidders and the associated cost of the priced equipment lists are then included in the project cost estimate. After agreement on the preferred vendor, the FEED engineering team incorporates the vendor-specific equipment information into the design.

1.12.1 Plant Design and Model Studies

The plan for the chosen structural FORM of the platform is prepared, including the horizontal and vertical zoning of the equipment layout. The FEED team, as described earlier, has already prepared details of the equipment and their specifications. The equipment sizing and the requirements become available along with the P&IDs and the proposed plot plan will be modified with the combined effort of the engineering group. The approved plot plan now sets the basis for the development of a 3D model of the plant. This is essential for offshore construction projects as the scale involved is very large. The 3D plots enable the FEED team to ensure the integrity of final interfaces and to minimize interferences and errors. Computer-enabled 3D modeling allows the engineering team to work in a common design platform and reduces construction interface problems. By linking the database of material specifications, structural steel catalog data, electrical connectivity systems, mechanical systems, and their operational features/specifications electronically, the 3D model becomes more intelligent. The 3D medium shows the construction sequencing, operator accessibility, and safety features present in the plant's overall structure. With the help of such a linked database, it is convenient to order material procurement according to its requirements so that over-buy can be avoided. Apart from making the project management and construction engineering more transparent, 3D models provide a good opportunity for the operators and training personnel to become familiar with the plant configuration.

1.13 EPC EXECUTION PLANNING

Project execution should be as efficient and cost-effective as possible. Some of the major considerations that must be addressed in the formulation of engineering, procurement construction, and commissioning (EPC) execution planning include the following: i) contract terms, ii) labor availability, iii) site logistics, iv) technical complexity, v) major equipment and material sources, and vi) extent of filed fabrication. It is important to note that most of the project's execution planning depends upon the FEED study. Hence, execution planning cannot begin until the FEED study of the early project has been accomplished. The overall project schedule should be prepared, which will include detailed engineering and procurement efforts regarding materials for fabrication and equipment with high-end specifications. Procurement should generally focus on a long lead-time for the delivery of items as critical points. Scheduling of the construction/fabrication of the platform takes many factors into consideration, such as the extent of field fabrication activities, subcontracting plans, craft workforce availability, the hiring time of cranes required for installation, weather window estimates for wet towing, minimizing hook-up time, etc. Schedule development also considers work sequences, concentration of craft personnel, inspection, and testing procedures. The FEED team should hold periodic reviews regarding construction planning to design the platform facility with due consideration of the available construction methods. There should be a strong interface established between the construction design and the capability/expertise that matches the method of execution chosen for the platform type.

Construction planning should focus on determining the split of work to be performed by the directly hired workforce and the work to be completed by any subcontractor, which will lead to the preparation of the construction schedule. Construction work allocation is then prepared for the development of the overall project engineering, procurement, construction, and commissioning (EPCC) schedule. Vital stages that influence the EPCC schedule are the following:

i) The procurement plan, as it determines when the vendor will be available to support the required engineering effort (equipment delivery planning, including the inspection hold points, shipping, and logistics, needs to be consistent with the construction sequence to meet the overall scheduled target).
ii) The availability of detailed engineering drawings will determine the engineering work sequence.
iii) Site mobilization timings.
iv) Raft levels.
v) Work sequences.

1.14 OVERALL FEED PHASES

FEED is generally undertaken in six phases.

Phase 1. Definition of scope and strategy for engineering execution.
Phase 2. Commencement of FEED design.
Phase 3. Initial FEED development and hazard identification.
Phase 4. Initial design audit.
Phase 5. FEED design approval.
Phase 6. FEED report compilation.

In the first phase, the scope definition should be apposite for preparing the commencement of the FEED design. The main objective under this phase is to evaluate the FEED deliverables that are sufficiently mature for the detailed design process. Table 1.10 shows the activities under the first phase of FEED.

In the second phase, the commencement of the FEED design is carried out. In this phase, the engineering design process is formally a part of the project execution plan. Table 1.11 shows the activities that are generally addressed in the second phase.

The third phase is focused on the initial design development and on detailed hazard identification. The initial design deliverables are subjected to inter-disciplinary

TABLE 1.10
Activities under the First Phase of FEED

No.	Activities/Deliverables	Description
1	Engineering execution plan and engineering organization chart.	Create engineering execution plan with key supporting documents/information.
2	Preparation/validation of conceptual study report/basic engineering design.	Confirm maturity of the conceptual design report.
3	Project baseline standards.	Establish the list of codes and standards to be used for the design specifications.
4	Engineering Audit schedule.	Create a periodic schedule for technical audits of the reports, design reviews, HAZID and HAZOP reports, etc.
5	Technical risk register.	Create a risk register that will record all possible risks associated with the engineering design.
6	Integrated engineering systems.	Document and record the engineering execution plan.
7	Qualitative risk analysis (QRA) and fire risk assessment (FRA) strategies.	Establish status for quantitative risk assessment, fire risk assessment, and emergency escape & rescue assessment.
8	Discipline job design specification.	List key design parameters and operational constraints that are applicable to appropriate discipline. Assign roles and responsibilities to the team that should address these issues. Also, list the codes and standards to be followed.

Materials and Loads on Topside   35

TABLE 1.11
Activities under the Second Phase of FEED

No.	Activities/Deliverables	Description
1	Safety-critical systems/elements and performance standards.	Confirm that safety-critical systems/elements are set as per the appropriate performance standards, namely functionality, reliability, and survivability.
2	Document distribution matrix.	Create a formal document distribution matrix to ensure all disciplines and departments receive all necessary design information on time.
3	Structural integrity interface.	Ensure structural integrity interfaces are identified and visible to project. See, for example, AMEC/client focal points, technical authorities, independent verification bodies, etc.
4	Material selection report.	Compile report defining the requirements and specifications of the necessary material for structural members, pipelines, topside details, etc.
5	Hazard management plan.	Create hazard management plan identifying all key hazards, technical safety activities, and the approach to managing major accident hazards (MAHs) in the design.
6	Equipment layout plan of the topside.	Develop/confirm baseline layout of equipment of the plant

review to identify the key hazards. Table 1.12 shows the list of activities under the third phase.

The fourth phase is focused on confirming the initial design development. Table 1.13 shows the list of activities under the fourth phase.

The fifth phase is to freeze the process design and develop detailed design and procurement activities. Table 1.14 shows the list of activities under the fifth phase.

The sixth phase is to ensure that the final design deliverables are completed. Client and statutory obligations are satisfactorily resolved. Table 1.15 shows the list of activities under the sixth phase.

1.15 AXIAL FORCE-BENDING MOMENT INTERACTION

The limit state design procedure for reinforced concrete elements has undergone a major revision in recent times, emphasizing a performance-based engineering approach. Revised design methodologies prescribed by several international codes include desirable features such as ultimate strength and working stress procedures to ensure a ductile response. Moreover, the increasing necessity of the structural assessment of buildings that do not qualify under the recent update of seismic codes demands a thorough understanding of axial force-bending moment (P-M) yield interaction of elements. Such studies on reinforced concrete (RC) circular columns are relatively limited in the literature; on the other hand, circular columns are better

TABLE 1.12
Activities under the Third Phase of FEED

No.	Activities/Deliverables	Description
1	HAZID.	Perform hazard identification studies, as identified on engineering audit schedule/technical safety events schedule.
2	(ENVID).	Make a formal assessment of environmental aspects associated with a modification or a change, and propose prevention or mitigating measures.
3	Process flow diagrams (PFDs) for inter-disciplinary review.	Issue PFDs on formal inter-disciplinary check (IDC).
4	Piping and instrumentation diagram (P&ID) for inter-disciplinary review.	Issue PFDs on formal inter-disciplinary check (IDC).
5	Plot plans.	Issue equipment plot plans/layouts on formal inter-disciplinary check (IDC).
6	High integrity pressure protection systems (HIPPS) review.	Initial review of HIPPS for the plant.
7	Dropped objects study.	Identify and assess the potential risks to drilling, production, process equipment, and staffed areas from dropped or swinging objects. Where the consequence of a dropped object is considered significant, propose prevention or mitigating measures.
8	Mechanical handling strategy.	Develop mechanical handling strategy including general approach/constraints, platform crane limits, installation philosophy, operation, and maintenance, etc.
9	Cause–effects studies.	Initiate cause & effects survey for formal inter-disciplinary review.
10	Hazardous area layout.	Issue hazardous area layout for inter-disciplinary review.
11	Fire risk assessment.	Determine fire & explosion scenarios and identify requirements for prevention, control, and mitigation measures.
12	Blast design philosophy.	Develop a philosophy that assesses potential explosion hazards and identify appropriate blast loadings to be applied to the engineering design process, e.g., vessels, pipe supports, structural steel, etc.
13	Equipment list and data sheets.	Create a master project equipment list identifying all major items of tagged equipment, e.g., vessels, pumps, motors, generators, hydro-cyclones, chemical injection packages, electrical switchgear, control panels, etc.
14	Emergency escape & rescue assessment.	Identify requirements for emergency escape & rescue, e.g., lifeboats, escape routes, plant layout, etc.
15	Passive and active fire protection.	Develop philosophy that identifies requirements for passive fire protection (PFP) and active fire protection (AFP).
16	Preliminary weight report.	Establish data on the total preliminary weight report for review of the design of the platform and constructability.

TABLE 1.13
Activities under the Fourth Phase of FEED

No.	Activities/Deliverables	Description
1	Hazardous area review.	Undertake formal engineering discipline review of hazardous area layout to confirm the requirements for the correct selection and location of each piece of equipment.
2	Piping and instrumentation diagram (P&ID) review.	Undertake formal engineering discipline review of P&IDs to confirm design basis.
3	Plot plans and 3D model reviews.	Ensure that the layout design complies with the required technical, safety, operability, and maintainability standards, make sure the layout is cost effective, and certify the constructability of the work.
4	Cause–effects review.	Draw up cause & effects diagrams to confirm design basis and that they are in alignment with hazard management plan.
5	Technical audit.	Obtain confirmation by a sampling of project documentation that the engineering of the project is to the required standard and free from errors.

aesthetically, and possess certain structural advantages. Detailed mathematical modeling of P-M yield interaction of RC circular columns based on Euro Code is important. Five sub-domains defining the boundary of P-M yield interaction are classified based on the strain limit condition in concrete and reinforcing steel. A complete set of analytical expressions should be proposed and illustrated through relevant examples. Results obtained for the failure interaction curve of RC circular sections show that by adopting Euro Code strain limits the boundary curve is divided into two main parts, namely i) tension failure with weak reinforcement resulting in yielding of steel, and ii) compression failure with strong reinforcement resulting in crushing of concrete. The curves are given in analytical form for every feasible coupling of bending moment and axial force. With the help of the mathematical model and proposed expressions for P-M yield interaction, the structural design of the top side of the reinforced concrete topside deck with circular columns can be performed with better understanding and improved accuracy.

1.15.1 PROPERTIES OF CONCRETE

Concrete is a heterogeneous, cohesive-frictional material exhibiting a complex non-linear inelastic behavior under multi-axial stress states (Mo, 1992). Wide use of concrete as a primary structural material in several complex structures demands a detailed understanding of the material response under a combination of different loads (Abu-Lebdeh & Voyiadjis 1993; Candappa et al. 2001; Park & Kim, 2003). Sufficient ductility ensured in the design procedure is an important prerequisite for the suitability of reinforced concrete structures to resist seismic loads (IS 13920-1993); seismic design philosophy demands energy dissipation/absorption by post elastic

TABLE 1.14
Activities under the Fifth Phase of FEED

No.	Activities/Deliverables	Description
1	Final design review.	Verify and validate that the design (structural, process, and layout plans) are prepared following the relevant standards and feedback/reviews given by the inter-disciplinary team.
2	HAZOP.	Undertake hazard and operability studies—systematic multi-discipline studies based on the application of guidewords to identify causes of potential hazards and operability constraints in the plant.
3	Best available techniques (BAT) report.	Analyze the environmental issues that have been considered a part of the process to select the final FORM and plant layout plan. It should be ensured that any prevailing/effective alternatives are not overlooked.
4	Constructability review.	Conduct formal constructability reviews to ensure the safe and efficient installation of plant and equipment.
5	Piping and instrumentation diagram (P&ID) approval for design.	Issue approved-for design (AFD) to the process and instrumentation diagrams.
6	Causes–effects approval for design.	Issue approved-for design regarding causes and effects.
7	Safety integrity level (SIL) assessments.	Conduct assessment to determine target SIL. For SIL 3, there is a likelihood that HIPPS is required.
8	High integrity pressure protection systems (HIPPS) study.	Conduct a study to formally identify HIPPS and demonstrate that it is robust.
9	Quantitative risk assessment.	Conduct multi-disciplinary study of hazards, the likelihood of their occurrences, and potential consequences in terms of societal risk and individual risk. Estimate the possible fatality accident rate for the perceived risks.

TABLE 1.15
Activities under the Sixth Phase of FEED

No	Activities/deliverables	Description
1	FEED report.	Collate all inputs from various disciplines and finalize the study report.

deformation for collapse prevention during major earthquakes (Chandrasekaran et al., 2003; Chandrasekaran et al., 2008a). Ductility also ensures effective redistribution of moments at critical sections as the collapse load is approached (Bangash 1989; Papadrakakis et al., 2007). Ductility, a measure of energy dissipation by inelastic deformation during major earthquakes, depends mainly on the moment-curvature relationship at critical sections where plastic hinges are expected/imposed to be formed at collapse. RC structures have the facility of changing, within certain limits, the ultimate moment as the designer pleases, without changing the overall dimensions of the cross-section. As a result, it is sometimes suggested that reinforcement steel areas should be adjusted to distribute the ultimate moment in the same members as the elastic bending moment diagram for the factored (ultimate) load. It is a critical aspect of (intended) performance-based design of the structure, leading to several advantages, namely i) the necessary elastic analysis will be more laborious, ii) the resulting design will address the required performance criteria set by the designer, and iii) plastic hinges are made to form on the selected structural components of the desired choice (for example, on the beam and not on the column), thus ensuring the required performance of buildings under seismic loads. In other words, the structures should be able to resist earthquakes in a quantifiable manner and to present levels of desired possible damage (Ganzerli et al., 2000; Ghobarah, 2001). Conducted studies (Paulay & Priestley, 1992) indicate that the behavior of statically indeterminate RC structures depends on the cross-section area of reinforcing steel-to-concrete ratio. For smaller values of this ratio, reinforcement yields plastically before the concrete is crushed in compression. At the same time, for larger values, it may initiate the crushing of concrete before the yielding of reinforcing steel. However, this ratio becomes critical when tensile steel reaches the yield limit simultaneously as the extreme compressive fiber of concrete reaches its crushing strain. Increasing concern regarding the structural safety of existing buildings not complying with current seismic codes demands performance assessment to evaluate their seismic risk, which is a major task ahead for structural designers.

Thus, the objective of ensuring safe buildings intensifies the above-stated concerns, for which pushover analysis can be seen as a rapid and reasonably accurate method (ATC 40, 1996; Naughton et al., 2017). Pushover analysis accounts for the inelastic behavior of the building models and provides a reasonable estimate of deformation capacity while identifying critical sections likely to reach a limit state during earthquakes (Sinan & Asli, 2007; Chopra & Goel, 2002; Chao et al., 2006). Researchers have used pushover analysis successfully for seismic evaluation and have compared other detailed analysis procedures (for example, Esra & Gulay, 2005; Chandarsekaran & Anubhab, 2004, 2006; Chandrasekaran et al., 2008b; Srinivasan et al., 2008). Researchers emphasize that basic inputs strongly influence i) the accuracy of results obtained from stress–strain relationship of constitutive materials, ii) P-M yield interaction, and iii) the moment–rotation capacity of members (see, for example, Srinivasan Chandrasekaran et al., 2010; Sumarec et al., 2003; Szalai & Papp, 2010; Zhang et al., 1993). A qualitative insight of these inputs, namely M-θ and P-M interaction, in particular, for rectangular cross-sections with different tensile and compressive steel accounting for nonlinear properties of constitutive materials is relatively absent in the literature (SAP, 2000). This chapter presents a mathematical development of the nonlinear behavior of reinforced concrete circular columns and derives P–M yield interaction while describing their five sub-domains.

1.16 MATHEMATICAL DEVELOPMENT OF P-M INTERACTION

Concrete under multi-axial compressive stress exhibits significant nonlinearity, which can be successfully represented by nonlinear constitutive models

(Hognestad et al., 1955; Chen & Chen, 1975; Chen, 1994a, b). Many researchers have reported different failure criteria in stress space by various independent control parameters (see, for example, Hsieh et al., 1982; Menetrey & William 1995; Sankarasubramaniam & Rajasekaran 1996; Nunziante et al., 2007; Nunziante & Ocone, 1988). The nonlinear elastic response of concrete is characterized by a parabolic stress–strain relationship in the current study shown in Figure 1.27. Elastic limit strain and strain at cracking are limited to 0.2 percent and 0.35 percent, respectively (DM.9 Gennaio, 1996; DM.14, 2005). Tensile stresses in concrete are ignored in the study. Design ultimate stress in concrete in compression is given by

$$\sigma_{c0} = \frac{(0.83)(0.85)R_{ck}}{\gamma_c} \tag{1.1}$$

The stress–strain relationship for concrete under compression stresses is given by

$$\sigma_c(\varepsilon_c) = a\varepsilon_c^2 + b\varepsilon_c + c 0 < \varepsilon_c < \varepsilon_{c0} \tag{1.2}$$

$$\sigma_c(\varepsilon_c) = \sigma_{c0} \varepsilon_{c0} < \varepsilon_c < \varepsilon_{cu} \tag{1.3}$$

where compression stresses and strains are assumed to be positive in the analysis. Constants a, b, and c in Equation (1.2) are determined by imposing the following conditions:

$$\sigma_c(\varepsilon_c = 0) = 0 \tag{1.4}$$

$$\sigma_c(\varepsilon_c = \varepsilon_{c0}) = \sigma_{c0} \tag{1.5}$$

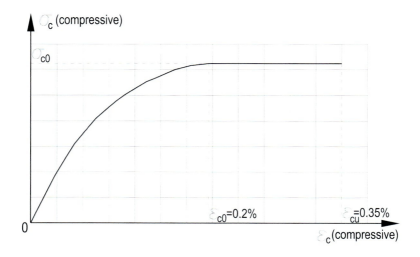

FIGURE 1.27 Stress–strain relationship for concrete.

$$\left[\frac{d\sigma_c}{d\varepsilon_c}\right]_{\varepsilon_c=\varepsilon_{c0}} = 0 \Rightarrow c = 0 \qquad (1.6)$$

$$a\varepsilon_{c0}^2 + b\varepsilon_{c0} = \sigma_{c0} \qquad (1.7)$$

$$2a\varepsilon_{c0} + b = 0 \qquad (1.8)$$

By solving the above equations, we get

$$a = -\frac{\sigma_{c0}}{\varepsilon_{c0}^2}, \quad b = \frac{2\sigma_{c0}}{\varepsilon_{c0}}, \quad c = 0 \qquad (1.9)$$

By substituting in Equations (1.2 and 1.3), we get

$$\sigma_c(\varepsilon_c) = -\frac{\sigma_{c0}}{\varepsilon_{c0}^2}\varepsilon_c^2 + \frac{2\sigma_{c0}}{\varepsilon_{c0}}\varepsilon_c, \quad 0 < \varepsilon_c < \varepsilon_{c0} \qquad (1.10)$$

Steel is an isotropic and homogeneous material exhibiting a stress–strain relationship, as shown in Figure 1.28. While the ultimate limit strain in tension and compression are taken as 1 percent and 0.35 percent, respectively (DM. 9 gennaio, 1996),

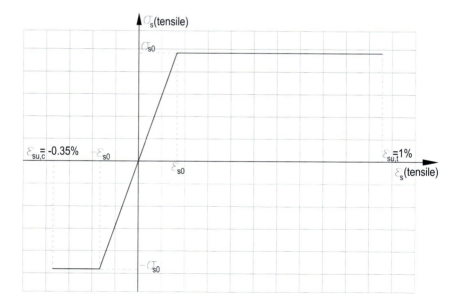

FIGURE 1.28 Stress–strain relationship for reinforcing steel.

elastic strain in steel in tension and compression is considered the same. The ultimate design stress in steel is given by

$$\sigma_{s0} = \frac{\sigma_y}{\gamma_s} \qquad (1.11)$$

The stress–strain relationship for steel is given by

$$\sigma_s(\varepsilon_s) = E_s \varepsilon_s \quad -\varepsilon_{s0} < \varepsilon_s < \varepsilon_{s0} \qquad (1.12)$$

$$\sigma_s(\varepsilon_s) = \sigma_{s0} \rightleftarrows \rightleftarrows \rightleftarrows \varepsilon_{s0} < \varepsilon_s < \varepsilon_{su,t} \left(\varepsilon_{su,t} = \varepsilon_{su} \right) \qquad (1.13)$$

$$\sigma_s(\varepsilon_s) = -\sigma_{s0} \quad -\varepsilon_{su,c} < \varepsilon_s < -\varepsilon_{s0} \qquad (1.14)$$

The reinforced concrete column of the circular cross-section shown in Figure 1.29 is examined for axial force-bending moment yield interaction. The fundamental Bernoulli hypothesis of linear strain over the cross-section, both for elastic and elastic-plastic responses of the beam under bending moment combined with axial force, is assumed. Interaction behavior becomes critical when one of the conditions applies, namely i) strain in reinforcing steel reaches ultimate limit, as well as ii) strain in concrete in extreme compression fiber reaches the ultimate limit. Figure 1.30 shows the axial force-bending moment limit domain consisting of five sub-domains, as described below. Only the upper boundary curves (corresponding to positive bending moment M) will be examined as there exists a polar symmetry of the domains for the center. Figure 1.31 shows the strain and stress profile in steel and concrete for sub-domains (1), (2), and (3), where collapse is caused by yielding of steel, whereas for sub-domains (4) to (5), collapse is caused by crushing of concrete.

Sub-domains (1)–(3): collapse caused by yielding of steel.

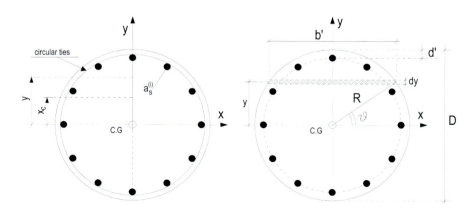

FIGURE 1.29 Typical cross-section of circular column of topside.

Materials and Loads on Topside

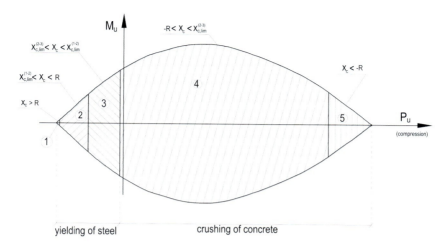

FIGURE 1.30 P-M interaction domains for RC circular column.

In sub-domain (1) (see Figures 1.30 and 1.31), the position of the neutral axis measured from the origin placed at the center of gravity (CG) of the circular section varies in the range [R, + ∞]. The strain in rebar reaches the ultimate limit, and the corresponding stress reaches ultimate design stress for this range of the neutral axis. Ultimate axial force and bending moment are given by

$$P_u = A_s \sum_{i=1}^{n} \sigma_s^{(i)} \quad \text{(tensile)} \tag{1.15}$$

$$M_u = A_s \sum_{i=1}^{n} \sigma_s^{(i)} \cdot d^{(i)} \quad \text{(tensile)} \tag{1.16}$$

$$d^{(i)} = (R - d')\sin(\alpha \cdot i + \alpha_0) - x_c \quad \forall i \in \{0, 1, 2, \ldots (n-1)\} \tag{1.17}$$

where d(i) is the distance of the rebar from the neutral axis. It can be seen from the above equations that only rebars contribute to yield (P–M) interaction as there is no axial force in the concrete section.

In sub-domain (2), reinforcing steel yields while strain in concrete remains within elastic limits. Depth of neutral axis lies in the range $x_{c,lim}^{(1-2)} < x_c < R$ and is given by

$$x_{c,lim}^{(1-2)} \frac{\varepsilon_{su} R - \varepsilon_{c0} \left[(R - d')\sin(\alpha \cdot \psi + \alpha_0) \right]}{(\varepsilon_{su} + \varepsilon_{c0})} \tag{1.18}$$

$$\alpha = \frac{2\pi}{n} \quad \text{for } 0 < \alpha_0 < \alpha \tag{1.19}$$

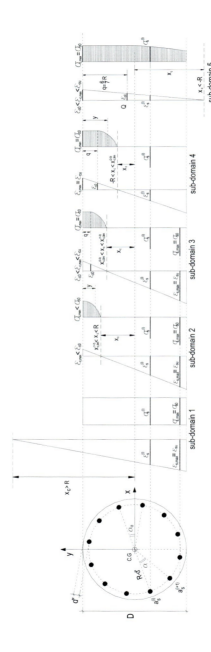

FIGURE 1.31 Strain diagrams for different sub-domains.

Materials and Loads on Topside

where n is the number of rebars in the column, ψ is the index number of the bar located as distance d from the neutral axis, in which the tensile strain is the maximum and is given by

$$\psi = \frac{n}{2\pi}\left[\frac{3}{2}\pi - \alpha_0\right] \tag{1.20}$$

where Equation (1.20) gives a truncated integer number. Regarding Figure 1.28, we have

$$x = R\cos\theta \tag{1.21}$$

$$y = R\sin\theta \tag{1.22}$$

$$b'dy = dA_c = 2R^2\cos^2\theta\,d\theta \tag{1.23}$$

Ultimate axial force and bending moment due to concrete in sub-domain (2) is given as

$$P_u = \int_{\mathrm{Arcsin}\left(\frac{x_c}{R}\right)}^{\frac{\pi}{2}} 2R^2\cos^2\theta \cdot \left(a\varepsilon_c^2(\theta) + b\varepsilon_c(\theta)\right)d\theta \tag{1.24}$$

$$M_u = \int_{\mathrm{Arcsin}\left(\frac{x_c}{R}\right)}^{\frac{\pi}{2}} 2R^3\cos^2\theta \cdot \sin\theta \cdot \left(a\varepsilon_c^2(\theta) + b\varepsilon_c(\theta)\right)d\theta \tag{1.25}$$

where, ε_c is the strain in concrete in generic fiber at distance y from the CG, and is given by

$$\varepsilon_c = \chi(y - x_c) = \chi(R\sin\theta - x_c) \tag{1.26}$$

The solution for the above equations in closed form is given by

$$P_u = \frac{1}{12}R\chi \left[\sqrt{1 - \frac{x_c^2}{R^2}}\left\{4b\left(2R^2 + x_c^2\right) - ax_c\left(13R^2 + 2x_c^2\right)\chi\right\} + 3R\left\{-4bx_c + a\left(R^2 + 4x_c^2\right)\chi\,\mathrm{Arc}\cos\left(\frac{x_c}{R}\right)\right\} \right] \tag{1.27}$$

$$M_u = \frac{1}{60}R\chi \left[\sqrt{1 - \frac{x_c^2}{R^2}}\left[5bx_c\left(-5R^2 + 2x_c^2\right) + 2a\left(8R^4 + 9R^2 x_c^2 - 2x_c^4\right)\chi\right] + 15R^3\left\{b - 2ax_c\chi\,\mathrm{Arc}\cos\left(\frac{x_c}{R}\right)\right\} \right] \tag{1.28}$$

where the curvature in sub-domains (1, 2, and 3) is given by

$$\chi = \frac{\varepsilon_{su}}{x_c - (R - d')\sin(\alpha \cdot \psi + \alpha_0)} \quad (1.29)$$

Strain in the ith bar in sub-domains (1, 2, and 3) is given by

$$\varepsilon_s^{(i)} = \chi d^{(i)} = \left[\frac{\varepsilon_{su}}{x_c - (R - d')\sin(\alpha \cdot \psi + \alpha_0)}\right] d^{(i)} \quad \forall i \in \{0,1,2,\ldots(n-1)\} \quad (1.30)$$

Distance of the ith bar from the neutral axis is given by

$$d^{(i)} = (R - d')\sin(\alpha \cdot i + \alpha_0) - x_c \quad \forall i \in \{0,1,2,\ldots(n-1)\} \quad (1.31)$$

The total ultimate axial force and bending moments can be obtained as the sum of results obtained from the above equations, respectively.

In sub-domain (3), reinforcing steel yield while strain in concrete reaches the ultimate limit. Depth of neutral axis in this sub-domain lies in the range $x_{c,lim}^{(2-3)} < x_c < x_{c,lim}^{(1-2)}$ and is given by

$$x_{c,lim}^{(2-3)} \frac{\varepsilon_{su} R - \varepsilon_{cu}\left[(R-d')\sin(\alpha \cdot \psi + \alpha_0)\right]}{(\varepsilon_{su} + \varepsilon_{cu})} \quad (1.32)$$

Ultimate axial force and bending moment due to concrete is given by

$$P_u = \int_{\theta_a}^{\theta_b} 2R^2 \cos^2\theta \cdot \left(a\varepsilon_c^2 + b\varepsilon_c\right) d\theta + \int_{\theta_b}^{\frac{\pi}{2}} 2R^2 \cos^2\theta \cdot \sigma_{c0}\, d\theta \quad (1.33)$$

$$M_u = \int_{\theta_a}^{\theta_b} 2R^3 \cos^2\theta \cdot \sin\theta \cdot \left(a\varepsilon_c^2 + b\varepsilon_c\right) d\theta + \int_{\theta_b}^{\frac{\pi}{2}} 2R^3 \cos^2\theta \cdot \sin\theta \cdot \sigma_{c0}\, d\theta \quad (1.34)$$

$$\theta_a = \text{Arc}\sin\left(\frac{x_c}{R}\right) \quad (1.35)$$

$$\theta_b = \text{Arc}\sin\left(\frac{R-q}{R}\right) \quad (1.36)$$

The total ultimate axial force and bending moments can be obtained as the sum of results obtained from Equations (1.33–1.36) and (1.27–1.28), respectively. The maximum strain reached in extreme compression fiber in concrete in sub-domains (1, 2, and 3) is given by

Materials and Loads on Topside

$$\varepsilon \frac{\varepsilon_{su}(R-x_c)}{x_c-(R-d')\sin(\alpha\cdot\psi+\alpha_0)}_{c,\max} \tag{1.37}$$

Depth of plastic kernel of concrete, q in sub-domains (1, 2, and 3) is given by

$$q_{(1,2)} = 0 \tag{1.38}$$

$$q_{(3)} = R - x_c - \frac{\varepsilon_{c0}}{\varepsilon_{su}}\left[x_c - (R-d')\sin(\alpha\cdot\psi+\alpha_0)\right] \tag{1.39}$$

Ultimate axial force and bending moment due to concrete in sub-domain (3) is given in closed form as

$$\begin{aligned} P_{(u,concrete)} = &\ 1/24 R^2 [4R\chi(b-2ax_c\chi)(3\cos(\theta_a)+\cos(3\theta_a)-4\cos^3(\theta_b))] \\ &+ \sin(\theta_a-\theta_b)\frac{1}{8}R^2\chi\Big[8x_c(b-ax_c\chi)\cos(\theta_a+\theta_b) \\ &\quad + aR^2\chi\big[\cos(3\theta_a+\theta_b)+\cos(\theta_a+3\theta_b)\big]\Big] \\ &+ \frac{R^2}{4}\Big[2\sigma_{c0}(\pi-2\theta_b-\sin(2\theta_b))+4bx_c(\theta_a-\theta_b)\chi \\ &\quad + a(R^2+4x_c^2)(\theta_b-\theta_a)\chi^2\Big] \end{aligned} \tag{1.40}$$

$$\begin{aligned} M_{(u,concrete)} =&\ \frac{R^3\chi}{4}\Big[-2bx_c + a(R^2+2x_c^2)\chi\Big]\cos(2\theta_a) \\ &+ \frac{R^3\chi}{24}\Big[-4bx_c + a(R^2+4x_c^2)\chi\Big]\cos(3\theta_a) \\ &+ \frac{R^3}{4}\Big[2\sigma_{c0}+2bx_c\chi - a(R^2+2x_c^2)\chi^2\Big]\cos(\theta_b) \\ &+ \frac{R^3}{24}\Big[4\sigma_{c0}+4bx_c\chi - a(R^2+4x_c^2)\chi^2\Big]\cos(3\theta_b) + \frac{R^5\chi^2}{40}a\cos(5\theta_b) \\ &- \frac{R^4\chi}{80}\Big[2aR\chi\cos(5\theta_a) - 5(b-2ax_c\chi)(-4\theta_a+4\theta_b+\sin(4\theta_a))\Big] \\ &+ \frac{R^4\chi}{16}\Big[-b+2ax_c\chi\sin(4\theta_b)\Big] \end{aligned} \tag{1.41}$$

In sub-domains (4) and (5), collapse is caused by the crushing of concrete. In sub-domain (4), collapse occurs when maximum strain in concrete reaches crushing strain ($\varepsilon cu_{c,\max}$), while strain in reinforcement varies in the range $[\varepsilon_{s0}, \varepsilon_{su}]$ Stress in tensile steel is σ_{s0} and position of neutral axis varies in the range $\{-R < x_c < x_{c,\lim}^{(2-3)}\}$.

Strain in the ith bar in sub-domain (4) is given by

$$\varepsilon_s^{(i)} = \left[\frac{\varepsilon_{cu}}{R - x_c}\right] d^{(i)} \quad \forall i \in \{1, 2, 3, \ldots n\} \quad (1.42)$$

Curvature and depth of plastic kernel of concrete for sub-domain (4) are given by

$$\chi = \frac{\varepsilon_{cu}}{(R - x_c)} \quad (1.43)$$

$$q_{(4)} = \left[\frac{\varepsilon_{cu} - \varepsilon_{c0}}{\varepsilon_{cu}}\right](R - x_c) \quad (1.44)$$

Ultimate axial force and bending moment due to concrete in sub-domain (4) are given by the same expressions valid for sub-domain (3).

In sub-domain (5), tensile steel gains compressive stress progressively. Strain in the ith bar in sub-domain (5) is given by

$$\varepsilon_s^{(i)} = \chi d^{(i)} \quad \forall i \in \{1, 2, 3, \ldots n\} \quad (1.45)$$

Depth of plastic kernel of concrete is given by

$$q_{(3)} = \left[\frac{\varepsilon_{cu} - \varepsilon_{c0}}{\varepsilon_{cu}}\right] 2R \quad (1.46)$$

Ultimate axial force and bending moment due to concrete is given by

$$P_u = \int_{-\frac{\pi}{2}}^{\text{Arcsin}\left(\frac{1}{7}\right)} 2R^2 \cos^2\theta \cdot \left(a\varepsilon_c^2 + b\varepsilon_c\right) d\theta + \int_{\text{Arcsin}\left(\frac{1}{7}\right)}^{\frac{\pi}{2}} 2R^2 \cos^2\theta \cdot \sigma_{c0} \, d\theta \quad (1.47)$$

$$M_u = \int_{-\frac{\pi}{2}}^{\text{Arcsin}\left(\frac{1}{7}\right)} 2R^3 \cos^2\theta \cdot \sin\theta \cdot \left(a\varepsilon_c^2 + b\varepsilon_c\right) d\theta$$

$$+ \int_{\text{Arcsin}\left(\frac{1}{7}\right)}^{\frac{\pi}{2}} 2R^3 \cos^2\theta \cdot \sin\theta \cdot \sigma_{c0} d\theta \quad (1.48)$$

The total ultimate axial force and bending moments can be obtained as the sum of results obtained from the above equations and Equations (1.27–1.28), respectively.

Materials and Loads on Topside

These expressions, in closed form, can be readily obtained from Equations (1.40–1.41) by substituting the following values

$$\theta_b = \text{Arc}\sin\left(\frac{1}{7}\right) \quad (1.49)$$

$$\theta_a = -\frac{\pi}{2} \quad (1.50)$$

1.17 EXAMPLE STUDIES AND DISCUSSIONS

Using the above expressions, axial force-bending moment yield interaction is now studied for RC beams of different diameters and reinforced in tension and compression zones. The cross-section dimensions and other relevant data can be seen from the captions to the figures. All the six sub-domains are traced and plotted, as seen in Figures 1.32–1.35. The sample plots are shown for relevant practical cases, namely i) for different diameters, ii) for a varying number of rebars, iii) for different percentages of tension steel, and iv) for different shapes of the cross-section.

The results obtained for the RC failure interaction curve of circular cross-section under axial force-bending moment yield interaction showed that, by adopting Euro Code strain limits, the boundary curve is first divided into two parts based on the type of failure, namely i) tension failure with weak reinforcement resulting in yielding of steel, and ii) compression failure with strong reinforcement resulting in crushing of concrete. The expression for different sub-domains is also given in the analytical form for every feasible coupling of bending and axial force. The boundary curve for the steel failure, in which by definition the tensile steel is in its ultimate yielding condition, can be further subdivided into three sub-domains (1, 2a, 2b). When subjected to increasing compressive axial force, these parts correspond to compression of concrete reaching the ultimate limit. Subsequently, for the concrete

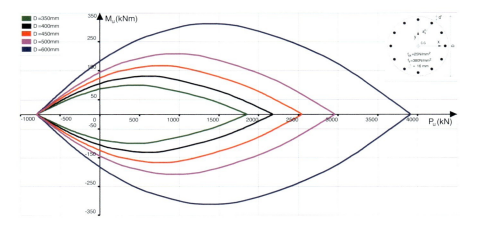

FIGURE 1.32 P-M interaction domains for circular columns with different diameter.

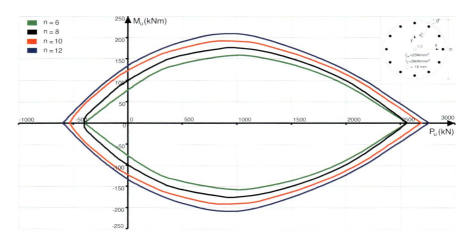

FIGURE 1.33 P-M interaction for circular columns with different number of rebars.

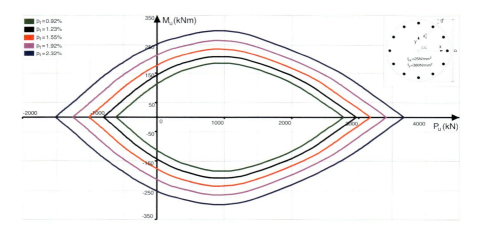

FIGURE 1.34 P-M interaction for circular columns with different tension reinforcement.

failure part, for which by definition concrete is in crushing, the curve can be subdivided into four sub-domains (3, 4, 5, 6), for which, by increasing the compressive axial force, strain in steel varies between the tensile failure limit and the tensile elastic limit, until the elastic limit in compression for concrete all over the cross-section. The sharp bend seen in the boundary of sub-domains (3)–(4) corresponds to the fact that the stress–strain relationship for steel is bilinear. The procedure proposed for the bending moment-curvature relationships, in the presence of constant axial force, is very simple and furnishes expressions in closed form for elastic-plastic regions of RC sections after verification with the numerical results. Though the applied principle may not be new, the presented expressions in the closed form will be very useful for practical nonlinear static analyses like a pushover. The analysis of these responses

Materials and Loads on Topside

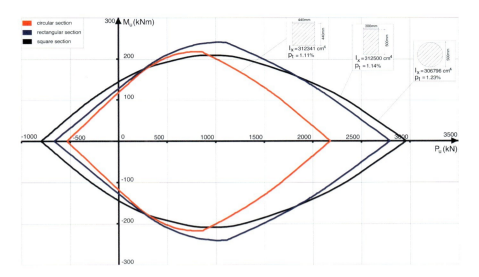

FIGURE 1.35 P-M interaction domains for columns of different cross-sections.

shows that for every kind of reinforcement, due to the small bending strength increment between the elastic and the failure limits, the moment-curvature response of rectangular RC sections is bilinear. After the first linear response, a further linear plastic branch is present, with a small slope. The whole response is very close to an elastic-plastic response, characterized by a sort of small "hardening effect."

It is important to note that the sub-domains classified for P-M interaction are based on strain limit conditions imposed by the codes. For these prescribed strain limits, all sections' points are not in failure condition; this implies that the stress and strain increments of the points lying along the P-M boundary depend on both the elastic and plastic increments. Hence, the normality rule that is valid for the true plastic domain does not hold completely true. Detailed discussions on the validity and applicability of flow rule can be seen at Chandrasekaran et al., 2009. Subsequent verifications made by the authors show that the plastic flow rule is completely satisfied in the sub-domains where failure is caused by steel yield and not verified for sub-domains where failure is caused by crushing of concrete.

1.18 FIRE LOAD

Steel is one of the favorite construction materials for industrial structures and offshore platforms. Design procedures under conventional loads—like dead load, live load, and wind load—and environmental loads—like wave, current, and earthquake loads—are well discussed in the literature. However, special attention is needed to understand the design procedures under special loads like blast, fire, and impact. Industrial structures and offshore structures are susceptible to accidental explosions. Although such incidents are rare, they may lead to severe consequences, thereby causing very high risk. They result in both financial and personal loss and a detrimental impact on the public and the environment. Alongside several accidents that have occurred in infrastructure industries, the two best-known accidents—Piper Alpha (June 1988) and Deepwater Horizon (April 2010)—emphasize the severe impacts of accidents on offshore structures caused by fire and explosion (Morin et al., 2017; Oltedal, 2012).

Fire is the rapid, exothermal oxidation of an ignition fuel. Fuel can be in solid, liquid, or gaseous states. The onset of fire releases energy in the form of exothermal reaction, while with time the released energy reaches peak intensity. Fire can also result from an explosion, which is a rapid expansion of gases caused by pressure or shock waves. These waves propagate very fast, and their rapidity results in adiabatic expansion. The explosion resulting from the fire can be either a mechanical or chemical process. It is evident that about 70 percent of the accidents that have occurred in offshore facilities are due to hydrocarbon explosion and fire, whose consequences are very serious (Donegan, 1991; Jin & Jang, 2015). A major concern of designers is to make offshore facilities fire resistant (Paik & Czujko, 2013; Paik et al., 2013). If the platform deals with liquefied natural gas (LNG), then the potential risk due to fire and explosion is more severe as the physical and chemical conditions of LNG are different from those of liquid hydrocarbon (Quiel & Garlock, 2010). The most common oversights that can lead to fire accidents are as follows (Manco et al., 2013):

- All heat transfer systems should be thoroughly inspected for no leaks or smoke.
- In case of smoke, one should not disconnect the smoldering insulation as this will allow an excess of air inside and result in auto-ignition.
- Do not let fluid drop on any heat sources, as this will ignite the fluid.
- If fluid leaks and gets trapped within a system, it can get oxidized, which results in heat (exothermal).

By avoiding the above, one can avert fire accidents. However, fire-resistant design is still imperative for structures that are susceptible to fire accidents (Soares & Teixeira, 2000; Soares et al., 1998; Chandrasekaran & Srivastava, 2018).

1.19 CLASSIFICATION OF FIRE

Fire is triggered when leakage (or spill) of any flammable mixture occurs in the presence of a potential ignition source. Fire can be classified as pool fire, jet fire, fireball, and flash fire (Srinivasan, 2016). The sub-classification of fire includes flares, fire on the sea surface, and running liquid fire. This sub-classification can be grouped to the main classification as i) flares can be treated as a jet fire in modeling, and ii) fire on the water surface and running liquid fire can be treated as a pool fire.

1.19.1 POOL FIRE

A pool fire is a turbulent diffusion fire that burns above the pool that vaporizes hydrocarbon, which has less momentum. The release of liquid fuel forms a pool on the surface. It vaporizes and causes pool fire by ignition. The probability of pool fire

Materials and Loads on Topside

in an offshore platform is very high due to the continuous handling of hydrocarbons. Liquid fuel, released accidentally during the overfilling of storage tanks, may also cause pool fire. It may also occur due to the rupture of pipelines and cracks in the storage tanks caused by the corrosion of metal. The pool diameter is equal to that of the diameter of the bund, which is constructed to contain the spread of pool fire and is given by the following relationship

$$D_p = \sqrt{\frac{4A}{\pi}} \quad (1.51)$$

where A is the area of the bund in m² and D_p is the diameter of the pool. The following relationship gives pool fire length

$$L = 42 D_p \left[\frac{\text{Burning Rate}}{\rho_{air}\sqrt{9.81 D_p}} \right]^{0.61} \quad (1.52)$$

1.19.2 Jet Fire

A jet fire is classified by the turbulent diffusion of flame resulting from the combustion of fuel, which is continuously released. It has a significant momentum to propagate in a downwind direction. It can affect offshore installations very seriously, even if they are located far away from the potential source of the fire. Jet fire releases gases while propagating forward, which may be either in the horizontal or vertical direction. Between the two, a horizontal jet fire is more catastrophic as it is capable of causing extensive damage on the downwind side. It may result in the following consequences: structural failure, storage vessel failure, and/or pipework failure. The heat flux released during a jet fire is about 200–400 kW/m², dependent on the type of fuel released. Pressurized gas pipelines are one of the potential sources of a jet fire. In the case of a leak, the initial gas release rate is given by

$$Q_o = C_D A p_o \sqrt{\frac{MV}{RT_o}\left(\frac{2}{r+1}\right)^{(r+1)(r-1)}} \quad \text{if} \quad p_o > p_a \left(\frac{2}{r+1}\right)^{\left(\frac{r-1}{r}\right)} \quad (1.53)$$

where, C_D is the discharge coefficient, A is the area in m², p_o is the operating pressure of the gas, M is the molecular weight of gas in g/mol, V is the rate of specific heat, R is the universal gas constant (= 8314 J/kg mol k), T_o is the operational temperature in Kelvin, and p_a is the absolute pressure. The Chamberlain equation gives the flame length of a jet fire

$$m = 11.14 (Q_o)^{0.447} \quad (1.54)$$

where, Q_o is the initial release rate in kg/s. Jet fire length and the corresponding time frame are estimated based on the following relationship

$$Q_t = Q_o e^{\left(\frac{Q_o}{MG}\right)t} \quad (1.55)$$

where,

$$M_G = \frac{PM}{0.08314} \pi r^2 L \qquad (1.56)$$

where M_G is the mass of the gas in kg, P is the operating pressure of the gas in pa, M is the molecular weight of gas in gm/mol, r is the diameter of the pipe in m, L is the length of the pipe in m, and t is the time of release is seconds.

1.19.3 Fireball

A fireball is the rapid turbulent combustion of any fuel. Usually the outcome is in the form of a rising and expanding radiant ball of fire. When a fireball attacks a vessel or a tank containing pressure liquefied gas, the pressure inside the vessel increases and leads to the catastrophic failure of the vessel or the tank. It may lead to the loss of the complete inventory present in the tank. Under boiling liquid expanding vapor explosion (BLEVE) release, the released material is flammable, and may also ignite, which may cause an explosion and thermal radiation hazards. The duration of the heat pulse in BLEVE is about 10–20 seconds, causing high potential damage. The maximum emissive power that results from BLEVE is 270–333 kW/m² in the up/down wind and 278–413 kW/m² in the cross wind.

1.19.4 Flash Fire

A flash fire is the transient fire resulting from the ignition of a gas or vapor cloud. Flash fire is attributed as a special process resulting from a substantial delay between the release of flammable materials and the subsequent ignition. It initially forms a vapor cloud over a larger area and then expands radially. Subsequently, the cloud explodes because of ignition. It is more catastrophic and causes damage to a large area. A wall of flame characterizes flash fire. Similar to fireballs, flash fire can also ignite and remain as a continuous flame. It can also be caused by a delayed ignition and remain for a longer time. The instantaneous effect causes thermal radiation, and the flash fire generates "knock-on" events, such as pool fire, jet fire, and BLEVE. It is important to note that the severity of the flash fire is extremely high.

1.20 STEEL AT ELEVATED TEMPERATURE

Steel is one of the most common and popular construction materials used in industrial and offshore platforms. The behavior of steel at high temperatures is different from its behavior at room temperature, a factor that influences steel design at an elevated temperature. Some of the characteristics of structural steel, such as the modulus of elasticity, stiffness, and yield strength, lessen with increase in temperature, whereas the material ductility increases, showing strength development. Effective yield strength reduces after 400° C in the case of mild carbon steel at 2 percent strain. The decrease in the proportional limit and modulus of elasticity is seen after 100° C, as shown in Figure 1.36. The other major material properties considered in

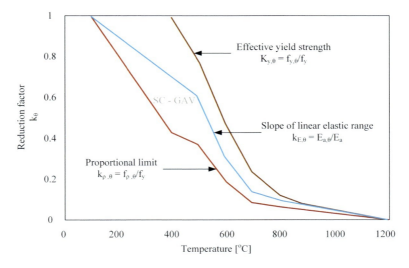

FIGURE 1.36 Material characteristics of carbon steel at high temperature.

the structural response under fire load are thermal conductivity, specific heat, elastic constants, specific weight, thermal expansion, and plasticity. Variations of thermal conductivity, thermal strain, and thermal expansion are shown in Figures 1.37–1.39. It is important to define the thermal and mechanical properties under high temperatures to evaluate the structural response under fire.

1.21 FIRE LOAD ON TOPSIDE

Fire load on the topside depends on the total amount of combustible materials present on the topside. As this has a high resource of hydrocarbon stock in the form of explored crude oil, the topside is vulnerable to hydrocarbon fire explosions. The total amount of combustible material is generally expressed in energy units (either in joules or megajoules). Once this is estimated, then the next step is to assess the behavior of fire over time. It is generally expressed as a heat release rate (HRR) curve or a time-temperature (T-t) curve. The next step is to estimate its consequences on the topside members, including the electro-mechanical equipment. Heat-transfer calculations are made to estimate the temperature rise in structural members due to fire, which helps estimate the fire load. Finally, structural strength to resist fire is estimated in the form of structural fire ratings.

A common question in an engineer's mind is what is a fire load? What contributes to a fire load? Anything capable of burning causes a fire load. In the case of the topside, a large inventory of crude oil and gases, explored from the good form, is a major cause of fire load. It is characterized in terms of energy units (either in joules or in calories). Alternatively, it can also be characterized in energy per unit area (joules/m^2). This is termed fire load energy density (FLED). It represents the total amount of available energy that can be released in the event of a fire. Two major considerations that govern the fire load are the amount of fire load and the geometry of the space.

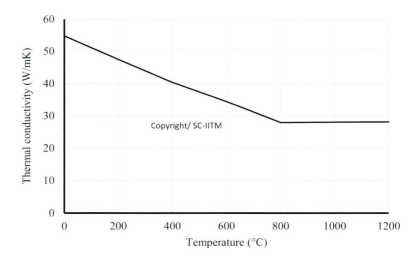

FIGURE 1.37 Thermal conductivity of carbon steel.

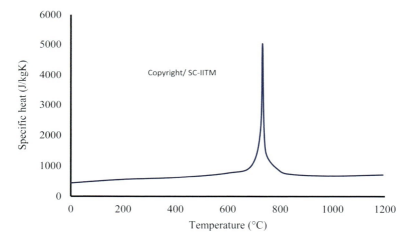

FIGURE 1.38 Specific heat of carbon steel.

In the case of offshore topside, both of these factors contribute to a higher load: fire load is higher due to a large inventory. For example, the combustion of one barrel of petroleum (159 liters) will release approximately 6 gigajoules. Furthermore, the topside space is congested due to the complex layout of machinery and pipelines.

The total energy that can be released during a fire, which is termed as fire load, can be computed using the following relationship (Khan & Gaurav, 2018):

$$E = \sum_{i=1}^{n} m_i C_i \tag{1.57}$$

Materials and Loads on Topside

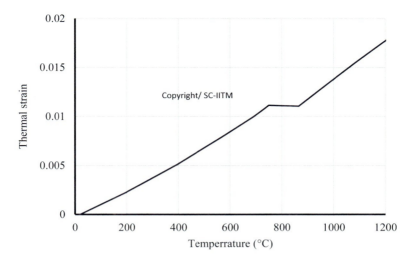

FIGURE 1.39 Thermal strain of carbon steel.

where m_i is the mass of combustible items (kg), and Ci is the calorific value of each combustible item (J/kg or MJ/kg). m_i should be computed based on the inventory of material on the topside, while C_i is available for common materials. For example, see the "Table 9 Calorific Values of Common Materials," reprinted from the National Building Code, Part IV.

Table 9 Calorific Values of Common Materials
(Clause A-1)

Sl No.	Material	Calorific Value ($\times 10^3$ kJ/kg)[1)]	Wood Equivalent (kg/kg)
(1)	(2)	(3)	(4)
i)	**Solid Fuels**		
	a) Anthracite	28.6	1.66
	b) Bituminous coal	30.8	1.75
	c) Charcoal	28.4	1.61
	d) Coke (average)	27.5	1.56
	e) Peat	20.9	1.19
	f) Sub-bituminous coal	22.0	1.25
	g) Woods (hard or softwood)	17.6	1.00
ii)	**Hydrocarbons**		
	a) Benzene	39.6	2.25
	b) Butane	47.1	2.68
	c) Ethane	49.1	2.79
	d) Ethylene	47.7	2.71
	e) Fuel oil	41.6	2.36
	f) Gas oil	42.9	2.44
	g) Hexane	44.9	2.55
	h) Methane (natural gas)	52.8	3.00
	j) Octane	45.3	2.58
	k) Paraffin	39.6-44.0	2.3-2.5
	m) Pentane	46.0	2.61
	n) Propane	47.3	2.69
	p) Propylene	46.2	2.63

Table 9 — (Concluded)

(1)	(2)	(3)	(4)
v)	**Common Solids**		
	a) Asphalt	38.3	2.13
	b) Bitumen	33.4	1.90
	c) Carbon	32.1	1.83
	d) Cotton (dry)	15.8	0.90
	e) Flax	14.3	0.81
	f) Furs and skins	18.7	1.06
	g) Hair (animal)	20.9	1.19
	h) Leather	17.6	1.00
	j) Ozokerite (wax)	43.3	2.46
	k) Paper (average)	15.4	0.88
	m) Paraffin wax	40.9	2.33
	n) Pitch	33.0	1.88
	p) Rubber	37.4	2.13
	q) Straw	13.2	0.75
	r) Tallows	37.6	2.14
	s) Tan bark	20.9	1.19
	t) Tar (bituminous)	35.2	2.00
	u) Wool (raw)	21.6	1.23
	w) Wool (scoured)	19.6	1.11
vi)	**Foodstuffs**		
	a) Barley	14.1	0.80
	b) Bran	11.0	0.63
	c) Bread	9.9	0.56
	d) Butter	29.5	1.68
	e) Cheese (cheddar)	18.1	1.03
	f) Corn meal	14.1	0.80
	e) Flour	14.1	0.80

Let us try to compute the fire load for an example case. Assume that the topside has an inventory of 200 kg of benzene hydrocarbon. Take the C value of benzene as 39.6 MJ/kg. Compute the average heat release rate (HRR) during a fire accident that lasts for about 1 hour.

$$E = 200 \times 39.6 = 7920 \text{ MJ}$$

$$\text{Energy released, } Q = 7920 \text{ MJ} / (60 \times 60 \text{ s}) = 2200 \text{ kW.}$$

If the topside has mixed fuel storage then one can compute the fire load as follows.

Aassume the topside has a stock inventory of 100 kg of benzene (C = 39.6 MJ/kg), 50 kg of butane (C = 47.1 MJ/kg), and 200 kg of ethane (C = 49.1 MJ/kg), then the average energy released during a fire that lasts for 1 hour is estimated as follows

$$E = 100 \times 39.6 + 50 \times 47.1 + 200 \times 49.1 = 16135 \text{ MJ.}$$

$$\text{Energy released, } Q = 16135 \times 10^3 / (60 \times 60) = 4481.9 \text{ kW}$$

The time-temperature (formally called T-t) curve can also quantify the energy released during a fire accident. Figure 1.40 shows a typical T-t curve used in offshore topside design. Growth rate, steady state, and decay state are indicated in the figure.

Let us work out one more example to compute fire load using the T-t curve. Assume that the topside has an inventory of about 100 kg of fuel oil (C = 41.6 MJ/kg). Assuming an ultra-fast fire takes place with a peak HRR of 9 MW, estimate the time of burning.

Total energy content, E = 100 x 41.6 = 4160 MJ
Growth factor for ultra-fast fire, α_g = 0.1874 kW/s²

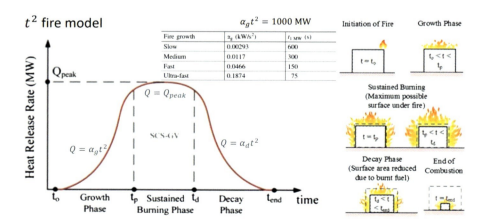

FIGURE 1.40 HRR curve for design. (Courtesy of Gaurav & Chandarsekaran, 2021.)

Time taken to reach peak HRR, $Q_p = t = \sqrt{\dfrac{9000}{0.1874}} = 219$ s

Total energy released in the growth stage, $E_1 = Q_p t_p/3 = 9 \times 219/3 = 657$ MJ $<$ 4160 MJ. As the total energy content available is less than that of the growth stage, fire will grow towards the sustained burning phase.

Total burning time, $t_b = t_p + \dfrac{4160 - 657}{Q_p}$

$$t_p + \dfrac{4160 - 657}{9} = 10.1 \text{ min.}$$

1.21.1 Time-Temperature Behavior

It is necessary to understand the difference between the real and standardized fire, which is clearer in the temperature–time representation of fire load. Fire will have an incipient and starting phase, which is relatively small and lasts about 20 minutes. During this time, possible active fire-fighting mechanisms, such as fire alarms, sprinklers, extinguishers, call to fire stations, and evacuation strategies should be made available without fail to make fire-fighting effective. However, in the topside design of offshore platforms, design is based on the compartmentation of devices and fire-prone areas. Both vertical and horizontal compartmentation are in place in the design to make the topside safe as far as possible in case of any outbreak of fire. It is important to understand that fire-fighting operations should be initiated during this phase of fire growth. This period is designated as the pre-flashover stage, and the temperature remains less than 600° C.

The post-flashover is referred to as the burning stage and instigates the growth of the fire. In the case of hydrocarbon fire, the temperature can even reach as high as 1200° C. During this post-flashover period, the fire grows to a peak temperature, which will take about 90 minutes, and then decays. The decay period will last for about 4–6 hours, and will depend on the unburnt inventory stock left over, air circulation, the efficiency of fire-fighting systems, and other passive fire-fighting methods. It is during this phase that one is more concerned about structural fire safety. To be very precise, the structural system of the platform should not collapse and add more problems to fire-fighting processes. In general, as observed from past accident scenarios, fire accidents under hydrocarbon fire on the topside of the offshore platforms did not show any effective fire-fighting. This is because the intensity of the fire is great, causing a very steep temperature rise. It is represented as the natural fire curve, shown in Figure 1.41. However, structural ratings are derived from the standard fire curve, which starts from the flash-over stage, as shown in the figure. Many codes of practice give standard fire curves (see, for example, IS 3809, ISO 834, ASTM E119).

The T-t curve approach is based on various options derived from the FLED approach. A few empirical relationships are useful in arriving at the time-temperature approach from the FLED approach. Swedish fire curve models (see below) are very useful in estimating the parameters that contribute to the T-t approach. Alternatively, one can also use advanced computational tools like the Fire Dynamics Simulator

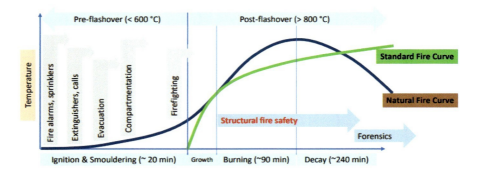

FIGURE 1.41 Fire curve: a comparison. (Gaurav & Chandrasekaran, 2021.)

(FDS)—a free, open-source code—to simulate the T-t fire curve. The following relationship gives the standard fire curve:

$$T = T_0 + 345\log(8t_m + 1) \tag{1.58}$$

Average FLED values are made available by researchers, which helps estimate the FLED for buildings, but no specific data are available for offshore topside design. Table 1.16 shows a comparison of FLED data useful for office building design (Khan & Gaurav, 2018).

1.22 PARAMETRIC FIRE CURVE

The parametric fire curve, seen in Figure 1.42, which is a part of the Euro Code (EN 1992-1-2, Annexure A), is presented here. It considers the effects of FLED, room geometry (shape and size), opening, and wall materials. The parametric fire curve (Figure 1.42) has two phases: the heating and cooling phases. The heating phase follows the standard fire curve, while the cooling phase is modeled as a linear variation, which has its basis in the Swedish fire curve model.

TABLE 1.16
Comparison of FLED Data for Office Buildings

Location	Year	Area surveyed (m²)	Mean FLED (MJ/m²)	Reference
London, UK	1970	2418	330	3
Nationwide, US	1975	7246	130–4805	4
Kanpur, India	1993	11,720	348	5
Wellington, New Zealand	1995	3999	426–947	7
New Zealand Building Code	2001	–	800	16
Ottawa and Gatineau, Canada	2011	935	550–852	13
Ahmedabad, India	2015	938	1334	Present study

Materials and Loads on Topside

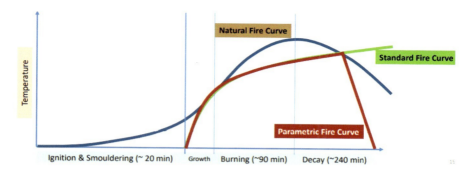

FIGURE 1.42 Parametric fire curve. (Gaurav & Chandrasekaran, 2021.)

The first step in developing the parametric fire curve is to characterize the FLED. The following relationship holds good

$$q_{f,d} = q_{f,k} \cdot m \cdot \delta_{q1} \cdot \delta_{q2} \cdot \delta_n \quad (1.59)$$

where $q_{f,d}$ is the design FLED, and $q_{f,k}$ is the characteristic fire load density per unit floor area (MJ/m²). While it is good to estimate this value based on the surveyed area of machinery and equipment present on the topside, please note that this is an important outcome of the FEED report. For an approximate idea, National Building Code, India, suggests a 25kg/m2 wood equivalent value, which means that FLED is 425 MJ/m² (= 25 × 17MJ/m², which is the C value of wood). In the above equation, m is the combustion factor, which represents the completeness of burning of fuel at source (usually assumed as 0.8), δ_{q1} accounts for the fire activation risk due to the size of the compartment, δ_{q2} accounts for the fire activation risk due to the type of occupancy, and δ_n accounts for the combined effect of all fire-fighting equipment and measures. These factors can be seen from Tables 1.17 and 1.18 (Euro Code 1992-1-2). Typical fire lead densities can be seen from Table 1.19, which can be used if no surveyed data of FLED of the topside are available. However, a good practice is to collect a surveyed data of fire load energy density based on the flammable materials in the site and fit a statistical distribution to arrive at the design fire load. In Table 1.19, taken from Euro Code, the recommended values are arrived at assuming a Gumbel distribution (which is mostly applicable to such surveyed data) and uses the 80 percent fractile value as design FLED (and not the average value). For example, comparing the average value for a typical office building, it is 420 MJ/m² from Table 1.19, whereas a closer value is also recommended by National Building Code, 2005 (425 MJ/m²). Unfortunately, however, due to the densely populated infrastructure, especially in offshore platforms, FLED design values given by Table 1.19 present substantial underestimates compared to more recent studies (see Table 1.16 for details).

The second step is to arrive at the heating phase calculations. The following empirical relationship holds good (Euro Code, EN 1992-1-2):

$$\theta_g = 20 + 1325\left[1 - 0.324\,e^{-0.2t^*} - 0.204\,e^{-1.7t^*} - 0.472\,e^{-19t^*}\right] \quad (1.60)$$

TABLE 1.17
Fire Activation Risk Factors

Compartment Area, at (m2)	Risk Factors		Examples of Occupancies
	Size of Compartment	Type of Occupancy	
25	1.10	0.78	Art gallery, museum, swimming pool.
250	1.50	1.00	Office, residence, hotels, paper industry.
2500	1.90	1.22	Manufacturing heavy machinery and engines.
5000	2.00	1.44	Chemical labs, painting workshops.
10000	2.13	1.66	Manufacturing fireworks or paints.

TABLE 1.18
Factor for Fire-Fighting Equipment and Measures (δ_n)

	Description		Nomenclature	Value
Automatic fire suppression.	Water-extinguisher systems.		δ_{n1}	0.61
	Independent water supply system.		$\delta_{n2.0}$	1.0
			$\delta_{n2.1}$	0.87
			$\delta_{n2.2}$	0.70
Automatic fire detection systems.	Fire detection & alarm.	By smoke.	δ_{n3}	0.87
		By heat.	δ_{n4}	0.73
	Auto-alarm transmission to fire brigade.		δ_{n5}	0.87
Manual fire suppression.	Fire brigade at site.		δ_{n6}	0.61
	Fire brigade off-site.		δ_{n7}	0.78
	Safe access roots available.		δ_{n8}	0.9, 1.0 or 1.5
	Fire-fighting devices available.		δ_{n9}	1.0 or 1.5
	Smoke exhaust system available.		δ_{n10}	1.0 or 1.5

where, θ_g is the gas temperature in the fire compartment (°C). It depends on the time delay, t^* (expressed in hours), which is governed by various factors: opening factor (also called as ventilation factor), area of vertical opening on the walls, and the total area of the enclosure.

$$t^* = t \cdot \Gamma \tag{1.61}$$

where, t is the actual time (in hours), and Γ attributes to other factors, as given below

$$\Gamma = \frac{\left[\dfrac{O}{b}\right]^2}{\left[\dfrac{0.04}{1160}\right]^2} \tag{1.62}$$

TABLE 1.19
Characteristics of Fire Load Density, $q_{f,k}$ (MJ/m²)

Occupancy	Average Value	80% Fractile Recommended for Design
Dwelling	780	948
Hospital room	230	280
Hotel room	310	377
Library	1500	1824
Office	420	511
School classroom	285	347
Shopping center	600	730
Cinema hall	300	365
Public place	100	122

$$\text{opening factor,} \quad O = \frac{A_v \sqrt{h_{eq}}}{A_t} \quad 0.02 \le O \le 0.20 \quad (1.63)$$

A_v is the total area of vertical opening on all walls, A_t is the total area of the enclosure, and h_{eq} is the average height of all window openings on walls (all expressed in m²). The factor (b) in Equation (1.62) depends on material properties and given by the following relationship

$$b = \sqrt{\rho \cdot c \cdot \lambda} \quad 100 \le b \le 2200 \text{ J/m}^2\text{s}^{0.5}\text{K} \quad (1.64)$$

where $(\rho \cdot c \cdot \lambda)$ are density, specific heat, and thermal conductivity of the enclosure material. This product of $(\rho \cdot c \cdot \lambda)$ is also termed as thermal inertia of enclosure material. In case, Γ = unity (as computed from Equation 1.62), then it will result in a standard time-temperature curve. This opening factor governs the fire behavior; in general, all fires are ventilation controlled. If the opening area is very small, the ingress of fresh air will be less. A ventilation-controlled fire will have a lower temperature than open or pool fires, which are quite common in offshore topside. In such cases, due to the free availability of abundant fresh oxygen, the fire will be fuel-controlled and not ventilation-controlled. Still, the fire intensity will be greater and largely depends on the fuel (or flammable materials in stock). It is also interesting to know the duration of the heating phase. On the other hand, it is important to know the limiting time up to which the heating phase will continue. The following relationship gives the time for maximum temperature to be reached. Please note that for Γ = unity, Equation (1.60) will lead to the fire growth of the standard time-temperature curve.

$$t^*_{max} = \text{Max}\left[0.2 \times 10^{-3} \cdot \frac{q_{t,d}}{O}; t_{lim}\right] \quad (1.65)$$

where, t^*_{max} is expressed in hours, $q_{t,d}$ is the design value of fire load density with respect to the total surface area, A_t (m²), and O is the opening factor, as given by Equation (1.63).

$$q_{t,d} = q_{f,d} \frac{A_f}{A_t} \quad 50 \leq q_{t,d} \leq 1000 \text{ MJ/m}^2 \tag{1.66}$$

where, $q_{f,d}$ is the design value of the fire load density computed based only on the surface area of the floor, A_f (m²). The value of limiting time, t_{lim} in Equation (1.65), is taken as 25 minutes, 20 minutes, and 15 minutes in case of slow, medium, and fast fire growths, respectively. It is important to note that the maximum time, t_{max}, is designated by two segments: one part, which is $(0.2 \times 10^{-3} \cdot \frac{q_{t,d}}{O})$ and is termed ventilation-controlled fire; the second part, t_{lim}, is the fuel-controlled fire. Depending upon the Equation (1.65) governing limit, one can identify whether the fire will be ventilation-controlled or fuel-controlled. In the case of hydrocarbon fire, as it occurs on the topside of offshore platforms, one can note that it will always be a fuel-controlled fire. Hence, effective fire-fighting in such cases can only limit the inventory of flammable materials on board. For this reason, one can find a common practice of offloading the explored crude oil from the production platform to the processing platform using large vessels. However, it may increase the cost of production, but might be necessary to ensure fire safety on the production platforms.

The third step is to arrive at the cooling phase temperature calculations. The following relationships hold good

$$\theta_g = \theta_{max} - 625\left(t^* - t^*_{max} \cdot x\right) \quad \text{for } t^*_{max} \leq 0.5 \text{ hrs} \tag{1.67}$$

$$\theta_g = \theta_{max} - 250\left(3 - t^*_{max}\right)\left(t^* - t^*_{max} \cdot x\right) \quad \text{for } 0.5 \leq t^*_{max} \leq 2 \text{ hrs} \tag{1.68}$$

$$\theta_g = \theta_{max} - 250\left(t^* - t^*_{max} \cdot x\right) \quad \text{for } t^*_{max} \geq 2 \text{ hrs} \tag{1.69}$$

$$x = \begin{cases} 1.0 & \text{if } t^*_{max} > t_{lim} \\ \dfrac{t_{lim}\Gamma}{t^*_{max}} & \text{if } t^*_{max} = t_{lim} \end{cases} \tag{1.70}$$

It is important to note that the decay curve is a linear function, as seen from the equation. It decays after (θ_{max}) is reached as a linear decay. The same can also be seen in Figure 1.42 (in the red line).

CREDITS

This chapter is co-authored by Dr. S. Pachaiappan, Assistant Professor, Department of Civil Engineering, Aditya Engineering College, Andhra Pradesh, India, and Dr. Hari Sreenivasan, Assistant Professor, School of Petroleum Technology, Pandit Deendayal Energy University, Gandhinagar, India. The lead authors thank both of them for sharing their research studies and recent updates on the FGMs to offshore applications.

EXERCISE

1. What do you understand by the topside of an offshore platform? What are the facilities that are commonly housed on the topside?
2. List a few advantages of a nonlinear pushover analysis.
3. What do you understand by FORM-dominant design? Explain with an example from an offshore platform.
4. How do the geometric configurations of the topside induce complexities in the analysis?
5. List a few special requirements of the topside which cannot be catered for by X52 steel.
6. What do you understand by functionally graded material (FGM)? Explain its application in various industries.
7. What are critical problems associated with composites that are overcome by FGM?
8. Write a brief note on additive manufacturing, useful in fabricating FGM.
9. List some differences between composites, alloys, and FGM.
10. Compare the mechanical properties of FGM with X52 steel and carbon–manganese steel.
11. Explain the gust factor method used to compute wind load on the topside.
12. How are blast loads idealized for the design of the topside? Why are they idealized?
13. Explain how impact load can be generated from dropped objects? Correlate the importance of the study of dropped objects with accident scenarios on the topside.
14. Explain how the moment–rotation curve plays an important role in pushover analysis.
15. What are the various performance-level indicators to assess a structure under pushover analysis? Explain with reference to international codes of practice.
16. Write a detailed note of FEED.
17. What is EPC? How is it important in an offshore project?
18. Explain what is pre-FEED.
19. Explain the different phases of a FEED report.
20. Explain the basic engineering needs for generating a FEED report.
21. Explain a few factors that govern the design of the topside.
22. Explain what a FEED study should ensure.
23. Explain what is a design review.
24. Write a brief note on HAZOP.
25. Explain the details of activities under the first phase of FEED.
26. Explain the details of activities under the second phase of FEED.
27. Explain the details of activities under the third phase of FEED.
28. Explain the details of activities under the fourth phase of FEED.
29. Explain the details of activities under the fifth phase of FEED.
30. Explain the details of activities under the sixth phase of FEED.

31. Explain how P-M yield interaction is helpful in design.
32. Draw and explain the different sections of the classical P–M interaction curve of reinforced concrete columns.
33. What do you understand by fire load? How is it computed?
34. How is fire classified?
35. What is a pool fire?
36. Write short notes on a flash fire.
37. What do you understand by a fireball?
38. Draw the stress–strain curve of mild steel at elevated temperatures and explain the salient points.
39. What factors influence the mechanical properties of X52 steel at elevated temperature?
40. What are the steps involved in computing fire load on the topside? Explain in detail.
41. Explain the time-temperature behavior of fire with a neat sketch.
42. What do you understand by fire load energy density? How is it computed?
43. Explain the parametric fire curve with a neat sketch.
44. Explain fire activation risk factors in detail.
45. Explain fuel-controlled and ventilation-controlled fires.

2 Basic Design Guidelines

2.1 DESIGN METHODS AND GUIDELINES

Offshore topside structures (Figure 2.1) are typically made of welded tubular members combined with structural steel sections and plated structural elements. The tubular members are typically used as the main vertical load-transfer elements as well as truss elements. Meanwhile, the structural steel sections and plated structural elements are used as framing elements to support the functional areas, instrument cabins, and equipment sections. Substructures support topside structures; either fixed to the sea bottom (for example, jackets and gravity-based platforms), or anchored to the sea bed (for example, compliant towers, tension leg platforms (TLPs), and spar platforms), or completely floating structures like drill ships, floating production storage and offloading (FPSO) units, etc. Topside structures are designed with a high free-board and air gap, safely above wave surface level. In designing the topside, basic environmental loadings and functional loadings are to be considered in the analysis.

Topside structures are designed and installed either in a modular form or in an integrated complex form (AISC, 1989). In a modular form, steel frames are utilized to support multiple production modules, which are subsequently assembled; they ensure a higher structural rigidity to the whole assembly. Each module is usually fabricated at different yards, transported, and assembled on-site at offshore locations. In the end, these modules will be hooked up, commissioned, and tested. Alternatively, an integrated complex topside is chosen for lower center of gravity and weight requirement needs (Amer, 2010). In this case, the topside structure will be fabricated and assembled as a single unit, complete with all instruments and equipment fitted. The disadvantage of this type is the limitation imposed in terms of the capabilities of both the fabrication and lifting equipment facilities. Typical arrangements in the integrated topside are cellar deck (also called the sub-cellar deck, in certain cases), pump access deck, mezzanine deck, main deck, and helideck. Topsides are also equipped with flare boom/vent, rigs, cranes, conductor-support structure, accommodation units, etc.

Before topside design, the overall development plan for the field, physical properties of hydrocarbon in reservoirs, natural conditions, and other environmental factors will be comprehensively considered. To control the weight of the topside, some non-critical elements will be fabricated using aluminum alloys (for example, the helideck). All the structural steel elements are made of primary steel, whether type I or II. In contrast, the secondary and tertiary steel elements are made of type III, IV, and V. Examples of steel elements are tubular members varying in the range of 60–1800 mm as the outer diameter. For tertiary steel, the outer diameter typically does not exceed 60 mm. Rolled steel sections of various shapes and plates of

FIGURE 2.1 Typical view of a topside structure.

different thickness are also used. The typical size of a plate element can be about 3.05 m × 6.1 m, which is modeled as a weightless element in the analysis. For all steel elements, apart from the analysis, design, and scantling details, cutting plans in the form of shop drawings will also be required. Topside structural steel design must comply with the provisions stated in the industry-adopted design codes. Typical international design codes, compliant with the offshore jacket platform topside are as below.

- American Institute of Steel Construction (AISC), Indian Standard (IS) 800-2007, "Design of Steel Structures for Non-Tubular Members," together with American Petroleum Institute, "Recommended Practice for Planning, Designing and Constructing Fixed Offshore Platforms–Working Stress Design," API, RP 2A-WSD, for tubular members.
- BS EN 1993 (EC3) "Design of Steel Structures," NORSOK Standards (N-004), for non-tubular members, together with ISO EN 19902, petroleum and natural gas industries, "Fixed Steel Offshore Structures for Tubular Members."
- DNVGL codes for lifting, transportation, and fatigue analyses.
- AWS D1.1/D1.1M AWS, "Structural Steel Welding Code."
- CAP 437, "Offshore Helicopter Landing Areas—Guidance on Standards."

All primary steel will be of the high-strength category, with BS 7191, Grade 355C, or equivalent. Further, steel grade BS 7191, Grade 355CZ, or Grade 355 EMZ, are used for steel greater than 63 mm. All secondary steel will be of mild steel of BS 7191, Grade 275C and BS 7191, Grade 275CZ, or equivalent, for through-thickness properties. Steel with improved through-thickness properties will be used for all nodes associated with any major lifting attachment and sections under higher through-thickness stress concentration (ATC 40, 1996). Alternative materials of EN10225 and EN10025 are also utilized for the topside. Steel to be used for fabricating topside will meet the specified minimum yield and tensile strength criteria. In the event of difficulty finding stocks in the market, more choices of mild steels from the market could be sourced. Table 2.1 lists the desirable structural properties of steel sections that are commonly used on the topside. For the aluminum helideck, typically, an octagonal-shaped plate with perimeter netting is preferred. It is supported by transition plates, which in turn rest on the helideck support frame. The transition plates will be modified to properly seat and level with the aluminum helideck. Reaction loads at the support points will be transferred to the topside.

2.2 DESIGN LOADS

The planar area of the topside should be designed with enough horizontal and vertical partitioning to support functional areas, instruments, and equipment to operate and maintain a platform successfully. Yet, it also must be light enough to reduce unnecessary loads to the substructure and foundation, considering the high cost of provision, operation, and maintenance of an offshore platform. Design loads on the topside can be reduced using perforated members in the sub-structure. It enables a compact design of the topside as much lesser wave forces are attracted by the supporting legs of the topside (Chandrasekaran & Madhavi, 2015; Chandrasekaran et al., 2015). Force suppression systems are also used in offshore structures to reduce the net design forces on the topside (Chandrasekaran & Merin, 2016). Hence, the major function of a topside deck of an offshore platform is limited for the provision of a few activities, namely well control activities, supporting well work-over equipment and instruments, separation processes for oil, gas, and non-transportable components from the raw hydrocarbon products (for example, water, paraffin wax, and sand), support for pumps and compressors required to transport the product ashore, power generation for the activities on topside, accommodation, and transportation for the operation and maintenance personnel. From the above activities, typical functional loads acting on the topside structure may be classified into the following categories (RP2A-LRFD, 1993).

2.2.1 Dead Loads

Dead loads are the weights of the structure, appurtenances, and any permanent equipment that will not be changed in position during the phases being considered. The following are examples of dead loads in a typical topside structure: weight of the topside structure in air, weight of installation aids, and drilling loads (if relevant).

2.2.2 Live Loads

Live loads may vary in magnitude, position, and direction during the phases being considered. They are not related to accidents or exceptional conditions. The following

TABLE 2.1
Properties of Steel Sections for Topside

Type	Supplementary to PTS 20 207 /BS 7191	BS EN 10225 /BS EN 10025	Description	Thickness Range (mm)**	Minimum Yield Strength (MN/m²)**	Minimum Ultimate Tensile Strength (MN/m²)**	Yield Strength Used in Analysis (MN/m²)
I	355C or 355EM* or API 5L X52 (seamless)	BS EN 10225 S355 G7+N (plate & tubular) S355 G11+N (sections) S355 G14+N (seamless)	High strength (primary applications)	t <= 16 16 < t <= 40 40 < t <= 63 63 < t <= 100	355 345 340 325	460 460 460 460	355 345 340 325
I (450EM)****	450EM	BUN 10225 S460 G1+Q (plate & tubular)	Super high strength (primary applications)	t <= 16 16 < t <= 25 25 < t <= 75	450 430 415	550 550 550	450 430 415
II	355CZ/ 355EMZ (TTP)/ API 5L X52 (seamless)	BS EN 10225 S355 G8+N (plate & tubular) S355 G12 + N (sections) S355 G15 + N (seamless)	High strength (primary applications)	t <= 16 16 < t <= 40 40 < t <= 63 63 < t <= 100	355 345 340 325	460 460 460 460	355 345 340 325
II (450EMZ)****	450EMZ* (TTP)	BS EN 10225: S460 G2+Q (plate & tubular)	Super high strength (primary applications)	t <= 16 16 < t <= 25 25 < t <= 75	450 430 415	550 550 550	450 430 415
III	275C or API 5L B (Seamless)	BS EN 10025: S275 J0+N	Mild steel (secondary applications)	t <= 16 16 < t <= 40	275 265	430 430	248*** 248***
IV	275CZ (TTP) or API 5L B (seamless)	BS EN 10025: S275 J0 + Z35 (TTP)					
V	275C or equivalent	BS EN 10025: S275 J0 + Z35 or equivalent	Mild steel (tertiary applications)	t <= 16 16 < t <= 40	275 265	430 430	248*** 248***

* Refers to modified grade for thickness > 63 mm.
** This refers to PTS 20.207, Amendments/Supplements to BS 7191.
*** 248MPa is applied to cater for another alternative grade (JIS, ASTM A36)

Basic Design Guidelines

are examples of live loads in a typical topside structure: functional activities on the structure, forces exerted on the structure from operations such as drilling and crane usage, stored materials, equipment and liquids, fluid pressure, operations of crane, helideck, and personnel. Table 2.2 summarizes live load on the topside, whose magnitude varies according to the design stages under consideration. Live loads from the topside structure, as computed above, will be taken into account for the following two special cases.

- Maximum compressive load on the foundation for which the substructure will be designed for combination shown in Table 2.3.
- Maximum tension load on the foundation for which the substructure will be designed for combination shown in Table 2.4.

Deformation loads are mainly due to the imposed deformation, which may occur from the differential foundation settlement. It may be applicable under soft-soil conditions, resulting in the differential settlement of foundations of large structures

TABLE 2.2
Typical Live Loads on Topside

Area	Initial Design[c]		Detailed Design[c]	
Handling zone and laydown area.	Blanket load = 15 kN/m².			
Main/drilling deck.	Blanket load = 15 kN/m².	[a]	Functional load = 10. kN/m² or Concentrated load = 30 kN. Equipment + 10.0 kN/m².	[a] [b]
Mezzanine deck.	Blanket load = 10 kN/m².	[a]	Functional load = 7.5. kN/m² or Concentrated load = 15 kN. Equipment + 5 kN/m².	[a] [b]
Cellar/sub-cellar deck.	Blanket load = 10 kN/m².	[a]	Functional load = 7.5. kN/m² or Concentrated load = 15 kN. Equipment + 7.5 kN/m².	[a] [b]
Walkway, access platform, and bridge walkway.	Blanket load = 5 kN/m² Or Concentrated load = 5 kN.	[a]	Functional load = 5. kN/m² or Concentrated load = 5 kN.	[a]
Helideck.	Blanket load = 2 kN/m².		Functional load = 0.5. kN/m².	
Bridges and flare tower.	Blanket load = 10 kN/m².	[a]	Functional load = 7.5. kN/m² or Concentrated load = 15 kN.	[a]
Muster area.	Blanket load = 10 kN/m².	[a]	Functional load = 10. kN/m².	[a]

"Blanket load" includes all non-structural load, i.e., equipment, piping, electrical bulks.
[a] Applicable to the whole flooring area (before equipment installation).
[b] Applicable to the free areas of floor (operating phase).
[c] All trusses and beams on main gridlines will be designed for deadweight and the most severe of either 70 percent of the blanket loading given in the initial design stage or 100 percent of the loading given in the detailed design stage.

TABLE 2.3
Topside Live Load for Maximum Load on Foundation

Item	Location	Uniformly Distributed Area Live Loads
1.	Open area live loads on all normally unoccupied areas (credible area only), including laydown area.	5.0 kN/m²
2.	Walkways, access platforms, and topside areas covered by permanent upper modules, packages, equipment, etc.	0.0 kN/m²

TABLE 2.4
Topside Live Load for Minimum Load on Foundation

Item	Location	Uniformly Distributed Area Live Loads
1.	Open area live loads.	0.0 kN/m²

like gravity-based structure (GBS) platforms (Chandrasekaran, 2014). Accidental loads are generally ill-defined to the intensity and frequency that may occur due to an accident or exceptional situation/ condition. For example, it may arise from dropped objects during operation, as discussed in Chapter 1. Installation loads are imposed on the structure due to loading-out, transportation, lifting, and upending. Environmental loads arise typically for 1-year operating and 100-year storm conditions. These are loads occurring due to environmental actions as follows.

- Wave, anticipating wave-in-deck; reference to be made to API RP 2A or the Kaplan method.
- Wind, which depends on the overall dimension of the topside. The topside will be idealized as a solid cube for the appropriate wind load calculation. No consideration for the details of the area of equipment and piping will be made in the wind load calculation. Wind loads on the drilling rig will be computed where appropriate during the drilling operating condition. Wind speeds used in the design of topside are as follows: 1-min mean, useful for global topside analysis; 3-sec gust, useful for local design; 5-sec gust, useful for topside whose largest horizontal dimension is lesser than 50 m; 15-sec gust, useful for topside whose largest horizontal dimension is greater than 50 m. However, the contribution of wave and current loads on risers and conductors are insignificant and can be omitted in the in-place analysis of the topside.

Dynamic loads are loads imposed on the structure due to the cyclic nature of loads like wind, waves, etc. Future loads are imposed by future appurtenance, and equipment loads will also be considered for topside in-place analysis. Helideck loads are important, and the helideck will be designed to accommodate the appropriate type

Basic Design Guidelines

and size of helicopters. For example, in southeast Asia, typically, wide-bodied Super Puma AS332L1/AS332L2 type helicopters and heavier but smaller Sikorsky type S-92 helicopters are used under normal operating conditions. But obstacle clearance limits required for the Super Puma AS332L1/AS332L2 helicopter should be considered in the design. The markings on the helideck will be imposed as per the S61 helicopter requirements. For a 3-legged platform, the size and weight of a Sikorsky S76 helicopter will be used in the design. The topside will be designed for hydro-test loads for equipment and piping. Bridge reactions are those loads imposed by the bridge onto the support structure, including their self-weight, buildings, equipment, piping, uniformly distributed load (UDL) loads, and wind load onto the bridge. Vent/flare boom reactions are additional loads imposed by the vent or flare boom onto the topside. It includes self-weight, equipment, piping, UDL loads, and wind load on the vent/flare boom.

The topside structure will be designed for appropriate load combinations. This will include environmental load conditions combined with appropriate dead and live loads in the following manner:

Condition I: Extreme storm condition (100-year storm) with maximum topside loads, including the derrick equipment set (DES) and a fill drill string down-hole and helicopter load. Crane is assumed to be in resting mode. All basic loads, i.e., computer-generated loads, dead loads, and operating supplies, are increased by applying the necessary weight contingency factor.

Condition II: Extreme storm condition (100-year storm) with minimum topside loads, without the DES and helicopter load. Crane is assumed to be in resting mode. This load condition is typically prepared for a substructure design.

Condition III: Normal operating condition (100-year storm) with maximum topside loads, including crane operating loads. Load combinations include operating conditions with the DES and heavy helicopter load.

Condition IV: Calm sea condition (1-year operating) with maximum topside loads. The DES load, maximum hook loads, helicopter load, and deck blanket live load are included in these load combinations but excluding environmental loads.

2.3 DESIGN STAGES

The topside structure must be designed to withstand the forces imposed during the construction and installation stages, namely pre-service, load-out, sea transportation, installation, and in-service life. Each of these stages is described in detail as follows.

2.3.1 STATIC IN-PLACE ANALYSIS

Static In-place analysis is applicable for 1-year operating and 100-year storm conditions. This analysis ensures that the platform's topside structure can support the installed facilities in normal operating conditions and extreme storm environmental conditions. Also, care must be given to ensure the adopted design methodologies comply with accepted engineering guidelines and with the authorities/industry regulators/platform owners. The adopted structural design philosophy follows the allowable stress design (ASD) method. The main design guidelines for structural design are API RP 2A-WSD and AISC ASD. The following design procedures are adopted.

  i) Determine appropriate nominal values for the load.
 ii) Determine the external load (applied forces).
iii) Determine load factors to be applied to the load and loads acting in combination.
iv) Compute internal forces due to the action of the load by performing structural analysis.
 v) Confirm the structural strength and stability requirements for members and connections following the adopted design guidelines.

The static in-place analysis can be divided into four major stages, namely:

1. Model generation for stiffness of the structure.
2. Generation of basic loads: gravity, wind load, self-weight, equipment, live loads, and blanket load and applying appropriate load factors.
3. Stiffness analysis of combined topside and part of the substructure.
4. Post-processing of the results: member and joint strength check, deflection check, etc.

The analysis will be performed using computer programs, which generate all functional loads and environmental loads (e.g., wave, wind, and current) from the user input data. In this chapter, the SACS computer program is considered for discussion. The structural model will include the primary structure and sufficient detail of the secondary structure to produce results commensurate with the analysis results required for the design. The full platform model is combined with the topside model to correct jacket stiffness representation during the static in-place analysis. The topside static in-place model will also be used as a basis and modified model in the subsequent load-out, transportation, and lift analyses. Modifications to the basic in-place model for these analyses are described hereafter.

The individual basic load cases considered in this analysis will include computer-generated self-weight, the weight of secondary steel, equipment loads, operating fluid loads, future equipment loads, live loads, drilling rig loads, wind loads, and appurtenance loads. Loading diagrams will be prepared to develop and document the basic load cases during the design process. The dead weights of all topside structural elements are computed by the SACS load module, i.e., the SEA STATE program, using element volume and densities. The weight of un-modeled components, such as gratings, stairs, stiffeners, and appurtenance steel, will be input as joint loads and member loads at appropriate locations on the structure model. The applied load from equipment and bulk materials in dry, operating, and hydro-test conditions will also be considered.

Basic Design Guidelines

The topside will be checked based on the blanket area load as specified in the codes. The actual equipment and piping loads combined with open area load at unoccupied deck area will also be included. If relevant, the topside will also be designed for the derrick equipment set (DES) and hydraulic work-over unit (HWU), which will be used for good maintenance and servicing. However, the above topside main deck is limited to deck loading of 15kPa. If tender assisted drilling (TAD) is available for a four-legged platform, hinged cantilever beams with removable decks beyond the main deck will be provided in the TAD work area. Still, they will be hinged back for better reach of the inner wells by jack-up rigs. The rig load from the drilling operation and wind load will be applied on the skid beams when the rig is in operation at various well locations. The rig load will include both the extreme storm and maximum loads during drilling operations. During the extreme storm, the rig is taken to be located in the middle of well slots with a pedestal crane to be at rest condition. Also, mud weight load will be applied at appropriate load locations on the structure model for drilling load cases. In the drilling deck, all drilling main deck steel members will be adequate to transfer a blow-out preventer (BOP) stack point load located anywhere along the length. This load is not to be considered in combination with UDL.

Impact loads caused by the offshore installation of equipment also will be considered. An impact factor of 2 will be used when assessing impact loads. The topside structure will be designed to resist the loading combination for two environmental conditions, namely normal operating conditions and extreme storm conditions. Drilling loads from the TAD rig, jack-up rigs, and HWU will be considered in both environmental conditions. The topside design will be made for the maximum load cases, suitably combined with the drilling operations. Each member of the deck will be sized by the critical condition which governs for that specific member.

The topside structural truss system will be designed based on the dead weight of structure and appurtenance plus rig load (if relevant) and 70 percent blanket load. However, proper blanket load factors to represent non-structural dead loads are estimated based on the weight control report and applied if the load factors are greater than 70 percent of the blanket load. The static analysis will be performed using SACS PRE and SOLVE processors. SACS POST processor will perform member code checks for the analyzed design conditions. The detail of this is covered in Section 2.6.

2.3.2 LOAD-OUT ANALYSIS

The load-out analysis is essential to ascertain the topside structural integrity during load-out from the fabrication yard onto the transportation barge. The topside will be skidded on skid beams or bogey from land onto the barge. The load-out analysis will consider the most severe and realistic load-out conditions. Typical topsides having four-leg supports will be assumed for loss of support of one leg for each load-out condition. Five support points will be assumed for heavier topsides, and it will be checked for a maximum of two loss-of-supports. Due consideration will be given to scrutinize the load-out supporting conditions if it is deemed that the topside has experienced too extreme load conditions. Three-legged topsides will be supported by three outer legs and two additional temporary supports. These points will be modeled as pinned in all translational directions hence released in rotations. To prevent rigid body movements, the structure will be guided horizontally at the support points. Total loss of support at one time will be considered for load-out analysis of four-legged topside and a maximum of two support losses for three-legged topside. The loss of support is selected based on the support location furthest away from the center of gravity of the topside structure. Tension force at support reaction is considered invalid hence the analysis will be re-run for other cases of support loss. Several lateral restraints will be simulated as required to prevent rigid body movement for the static solution. The applied loads and weight allowances will be based on the information contained in the weight control report. Allowances will be considered for the weight of slings and shackles.

In performing the analysis, the computer model for load-out analysis will be developed from the SACS model used for the static in-place analysis with the following modifications:

a. The jacket model will be removed.
b. All applicable loads, such as future loads, rig loads, wind loads, wind loads, etc., will be excluded.
c. Additional rigging loads (temporary) during load-out will be applied to the topside structure.

Upon completion of the computer analysis, post-processing will be done. Allowable stresses will not exceed those specified in API RP 2A-WSD and AISC codes. No increase in basic allowable stresses is advised. For the tubular joint design, the yield strength will be either the yield strength or 0.80 of the tensile strength (whichever is lower). No increase in basic allowable stresses will be used.

2.3.3 Lifting Analysis

It is essential to ascertain the topside structural integrity to withstand dynamic loads present in the structures during the lifting process as defined by the codes. It is assumed that four points will lift the topside to a single hook. A 60/40 sling load distribution lifting criterion will be adopted. For 60/40 lift criterion, all topside members will be designed for an impact factor of 1.33. In addition to the above condition, topside lift analysis will be carried out with COG shifts specified by the regulatory bodies. A minimum of four locations of a 1-meter COG shift will be considered.

The computer model to perform lift analysis will be based on the model developed for the load-out analysis with the following modifications.

1. Lifting slings will be modeled as a pinned-end rod with appropriate end releases and a minimum sling angle of 60° to the horizontal at the longest sling length. A maximum sling angle of 75° to the horizontal at the shortest sling position is used as a guideline.

Basic Design Guidelines

2. A single hook point will be modeled above the calculated topside lift COG and will be connected to the padeyes by the slings. A couple of forces to account for a 1-meter shift in COG will be performed in a sensitivity study.
3. Eccentricities of padeyes will be modeled with offsets to account for other moments. Padeyes to be designed according to regulatory body specifications.
4. Additional temporary rigging loads during lifting will be applied to the structure.
5. Lateral soft springs to be applied at lower ends of legs.

Two hook points will be created at a common coordinate to simulate the 60/40 sling load distribution. One hook point is simulated to restraint in rotation and translation in all directions, and another hook point is simulated to be free in vertical translation (Figure 2.2). A unit load is to be applied at the later hook point to create unequal sling load distribution. Four slings will be attached to the simulated hook points (2 nos.) diagonally and the topside structure via lifting padeyes (4 nos.). Several lateral restraints at the topside lower ends of the leg joints must be introduced to prevent rigid body movement. A minimum lifting capacity of the anchored heavy-lift vessel (HLV) has to be ascertained and used for the lifting design calculations.

A safety factor of 4 will be applied to the sling-safe working load (SWL) to determine the required sling minimum breaking load (MBL). Shackles will be selected

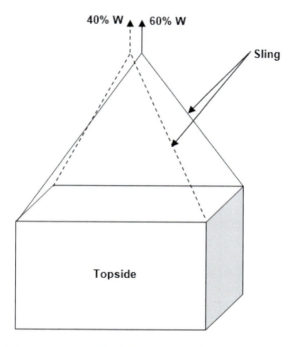

FIGURE 2.2 Typical view of a topside lifting sling configuration.

based on the shackle rating being more than the sling load determined from the factored lift weight and the limitations on bend radius. The ultimate capacity of the shackle will exceed the shackle rating by a factor of at least 4. The load cases to be utilized in this analysis are similar to those used in the load-out analysis with the following exceptions.

- Only dead loads are considered in the analysis.
- Slings are modeled as weightless members with appropriate releases.
- Allowance for rigging is made and input into the computer model.
- A couple of forces to account for the 1-meter shift in the center of gravity will (COG) be performed in a sensitivity study.

Upon completing the computer analysis, post-processing will be done to check that the allowable stresses will not exceed those specified in API RP 2A-WSD and AISC codes. No increase in basic allowable stresses is advised. For the tubular joint design, the yield strength will be either the yield strength or 0.80 of that of the tensile strength, whichever is lower.

2.3.4 Transportation Analysis

This analysis is essential to ascertain the topside structural integrity to withstand inertial forces during sea transportation on a barge. According to the design criteria for barge motions, the analysis will be carried out which can be computed from the actual inertial acceleration. A conservative approach will be used for checking the deck's structural strength. The actual topside sea-fastening and tie-down mechanics may be different. SACS TOW module will generate the gravitational and inertial loads on the structures, the mass of model elements, weights for non-modeled structures, equipment, bulk, etc. The position of the structure relative to the center of motion and the regular and rotational accelerations need to be defined. The module will produce load sets for each defined load case.

The computer model to perform transportation analysis will be based on the model developed for the load-out analysis with the following modifications.

1. The joint coordinates will be transformed in the vertical direction (Z-axis) to represent the correct sea-fastened altitude on the barge. The topside support elevation is 3 meters (minimum) above the barge deck.
2. Additional temporary rigging loads during load-out will be applied to the structure.
3. Wind load is added for sea transportation.

During transportation, the topside will be supported at all translational directions on four-leg nodes for four-legged topside and five leg nodes and two additional temporary support nodes for three-legged topside. These points will be modeled as fixed in the translational direction and released for rotation. During the analysis, mass information for the equipment and other bulk materials will be derived from the weight control report. Appropriate allowances for rigging will be made on the approximate size of slings and shackles for the lifting stage, inclusive of rigging platforms.

During the transportation on the barge, the topside structure is assumed to be located at the extreme position, i.e., the stern of the transportation barge. The topside structure is also assumed to be elevated by 3 meters (minimum) above the deck to reflect the height of the skid beam. The typical barge size utilized in towing analysis is 91.4 m × 27.4 m × 6.1 m. Two different arrangements, namely i) topside positioned longitudinally, and ii) transversely on barge deck, will be analyzed. In performing the analysis, load factors similar to those used for the load-out analysis will be applied to account for the weight growth and estimate accuracy.

The technical notes will specify barge motions from regulatory bodies. For example, for the transportation in East Malaysian South China Sea water, for angular acceleration, pitch about global X & Y axis of the topside is specified as ±8.0° in 5.5 sec resulting in 10.44°/sec². For roll, about global X & Y axis, ±12.5° in 5.0 sec resulting to 19.74°/sec² is specified. Linear acceleration is also required to be computed. For a longer transportation route, the severe conditions will be adopted and checked as a conservative approach. Motion study may be proposed if the results from tow analyses based on the above criteria are too stringent. Conservative wind load will be assumed to act concurrently with the gravitational and inertia loads in the study. The center of rotation for the combined barge and topside structure will be assumed to be at the geometric center of the barge at the water line. A 3-m towing draught will be assumed. A series of load combinations in multi directions will be utilized for respective sides of the barge in transportation analysis (Figure 2.3).

Upon completing the computer analysis, post-processing will be done to check that the allowable stresses will not exceed those specified in API RP 2A-WSD and AISC codes. A one-third increase in basic allowable stresses is advised. For the tubular joint design, the yield strength will be either the yield strength or 0.80 of the tensile strength (whichever is lower).

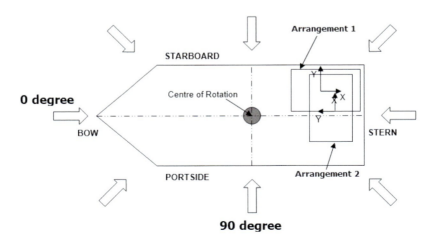

FIGURE 2.3 Transportation analysis configuration.

2.3.5 Analysis of Miscellaneous Items

The topside miscellaneous items will be designed to withstand the forces imposed during pre-service and in-service conditions. The topside miscellaneous design items are as follows.

- Stairways, emergency ladders, landings, and handrails.
- Service platforms.
- Security door.
- Vent/flare boom.
- Crane pedestal and crane boom rest.
- Skidding beams for the drilling rig.
- Drilling deck flooring and drainage.
- Temporary mud protection.
- Drip pans.
- Deck flooring.
- Tender rig requirements—mud return line and air winches supply line.
- Padeyes.
- Platform identification plates, etc.

2.4 WEIGHT CONTROL

Weight control is an important exercise in designing the topside structure. The following weight conditions will be analyzed and monitored in weight control exercise, namely load-out/lift, dry, operating, and hydro-test. Table 2.5 summarizes the weight allowances on the topside weight during analysis.

2.5 NUMERICAL TOOLS

Since the platform model contains hundreds of members, nodes/joints, and elements to be analyzed and designed, computer programs are commonly used in performing

TABLE 2.5
Topside Weight Allowances Provision

		Topside
IAA (item accuracy allowance)	[B] Vendor catalogue.	15%
	[C] Vendor data/quotation.	10%
	[D] Calculated/MTOs.	5%
DCA	(design change allowance).	5%
FCA	(fabrication change allowance).	5%

Weight allowance factor = IAA + DCA + FCA
Factored weight = base weight × (1 + IAA + DCA + FCA)
For primary structural steels and structural appurtenances/miscellaneous items, weight allowance factors of 1.15 and 1.20 are suggested, respectively.

Basic Design Guidelines 81

analyses for the topside structures. The basis for a computer program is the stiffness method of analysis utilizing finite element techniques. The program automatically generates all environmental and functional loading cases from the user-provided input information. This program is extensively explained here, considered the market leader in the industry. In this program, the structural models will be developed, including all the primary structural elements and sufficient detail for the secondary structural elements to produce significant and reliable results for the analysis and design of the topside and substructure. The computer model is believed to correctly model the overall platform stiffness representation during the analysis condition. The topside static in-place model will be used as a basis and modified to suit the stages such as the load-out, transportation, and lift analyses.

In building the topside model, coordinate systems are typically adopted to understand information and design practices better. The global coordinate will be a right-hand system. All elevation will be referenced to mean sea level (MSL), and all horizontal dimensions will be referenced at the center of the jacket. The Z-axis will be vertical with positive upwards; the X-axis will be in the horizontal plane pointing towards platform South. The Y-axis will be in the horizontal plane pointing towards the platform East. The local member coordinate will follow the default system in SACS.

The topside model will comprise all primary structural members, including legs, trusses, deck framing, braces, piles above jacket, etc. (Petronas, 2010). Deck plates will be modeled using the SACS modeling technique. Members will be assumed to be coincident at work points. Brace eccentricities will be introduced into the computer model where actual offsets are greater than 25 percent of the chord diameter. Thickened node barrels and stubs will be modeled for stiffness analysis. Stairs, gratings, ladders, and other miscellaneous items will not be modeled but coded in as loads. The generic topside model will include a vent/flare boom, which is changeable with an inter-platform bridge (if necessary). In addition, the generic topside model will also include a helideck (Norb, 2010). The following are the modules in the SACS suite program.

PRECEDE	Interactive modeling, loading, and plotting.
SEASTATE	Environmental load and mass generation. Combination of basic load cases.
PRE/SOLVE	Static analysis.
POST	Results reporting and member code checking.
JOINT CAN	Tubular joint checks.
DYNPAC	Modal analysis/dynamic characteristic.
FATIGUE	Spectral fatigue analysis.
GAP	Compression or tension only elements.
TOW	Transportation analysis.
SACS IV	Stiffness analysis and internal loads calculation.

The static in-place analysis is performed in SACS-PRE and SOLVE processor modules. The analysis procedure is based on a linear elastic response of the structure under static loading conditions. The structure is idealized as a three-dimensional space frame made of beam-column finite elements with defined boundary conditions. The static analysis will tabulate the joint deflections and member forces and moments for each load condition. Graphical plots of the global deflected shape of the structure could be viewed. SACS-POST processor could perform member code checks, based on the identified international code for the performed analyses and design conditions. SACS derives the member forces and moments from the static in-place analysis and uses the appropriate AISC and API RP 2A-WSD equations to compute member unity check (UC) ratios. Graphical plots of the member UC ratios can be obtained for the individual load conditions or the combined conditions. Overutilized and under-utilized members will be rationalized and resized if necessary.

Deflections in members will comply with the limits set by regulatory/owner specifications. The induced stress levels in the individual structural members will be limited to the basic stresses allowable in international codes such as AISC & API stresses for normal operating conditions. Codes permit a one-third increase in the basic allowable stresses for extreme storm loading. Column buckling effective length factors (K) for all members will be based on the API RP 2A-WSD Section 3.3.1d recommendations. In leg member, conical transitions will be checked following API RP2A-WSD, Section 3.4. Hot-rolled structural steel beam connection checking is not applicable during topside design as this will be performed by the fabricator during the construction stage. Tubular to tubular joint design will be based on API RP 2A-WSD on both the 50 percent effective strength criterion (for major nodes only) and the punching shear criterion for all the members. Joint type classification for punching shear will be based on the load-path method. SACS JOINT CAN processor could perform joint code checks for each of the load conditions. SACS derives member forces and moments from the static in-place analysis and uses the appropriate API RP 2A-WSD equations to compute joint unity check (UC) ratios. It could also perform automatic redesign for an overstressed joint can.

2.6 DESIGN CONSIDERATIONS

The following design considerations are adopted in designing the topside structures.

1) The topside structures will be designed to have sufficient reserve strength to withstand the forces imposed during pre-service and in-service life.
2) The platform topsides will be designed to achieve optimum safety, reliability, cost, and flexibility.
3) The design will also comply with serviceability criteria.
4) The design will comply with the reference codes and standards.
5) The structures will be analyzed for in-service and pre-service conditions.

Structural optimization is considered in terms of member sizing, joint detailing, and to produce an overall economic design about materials utilized and ease of fabrication. However, the optimization will be consistent with the level of analysis and design dictated by schedule and design brief requirements.

Basic Design Guidelines

2.6.1 Design Acceptance Criteria

The following acceptance criteria are specified according to AISC and API RP 2A codes.

Allowable stresses: basic allowable stresses will be following specifications from the international codes. Any increases in basic allowable stresses must be identified.

Slenderness ratio: determination of slenderness ratio, kl/r will be following the provision in the codes. The slenderness ratio for compression members will not be more than 120.

Diameter to thickness ratio: the minimum D/t ratio for tubular members will be 20. The maximum D/t ratio will be 60, except in special circumstances.

Thickness limit: primary braces and beams thickness will not be less than 6 mm. Material thickness exceeding 65 mm will be avoided wherever possible. Generally, the chord thickness of joints will be increased up to 65 mm or a minimum D/T ratio of 20 before considering ring stiffeners.

Allowable deflection: deflections will be maintained within acceptable limits consistent with the required function. The deflection limit will be limited to 1:200. A specific deflection limit will be agreed upon as and when identified as a need.

Joint fatigue life: the minimum fatigue life for joints accessible and inaccessible for inspection is 60 years and 120 years, respectively.

2.7 DESIGN METHODS

Analysis and design can be placed in a close loop. The structural design generally is carried out to control the behavior of the structure under the action of various loads. However, the analysis reflects the response behavior of a designed structure. Hence, it is a common understanding that structural analysis is carried out to check the adequacy of the design. Structural design is more based on experience or understanding of the structural behavior under the action of loads. An analysis is a sophisticated process, which is computer-aided to solve high-end mathematical formulations with the help of various tools. However, the results of any structural analysis need to be checked by a designer to correlate them with the actual behavior of the structure.

Further, as applicable to the topside of an offshore platform, structural design is more focused on the arrangement of functional utilities to obtain a load-balanced geometry. Design of topside should not be seen as sizing of structural members to sustain the encountered load, as it is a larger picture in reality. It is focused on the topside's overall shape and size after accounting for all necessary utilities that form a part of the topside. It is more dependent on practical judgment—lessons learned from the past, with little creativity. The main objective is to minimize weight and maximize its functional ability. Hence, on the topside, functional design preludes structural design. Ideal design philosophy maintains equilibrium between the applied load and the resistance that a structural system can develop. Unlike in the conventional design practices, where resistance is accounted only from the strength of the material used, offshore topside has a paradigm shift. Even the structural form, arrangement of members and their sizing, placement of heavy equipment (termed as vertical and horizontal zoning), space for critical operations such as lifting, loading/unloading, and drilling play a vital role in balancing the load acting on the topside. Hence, they are often termed form-dominant designs. A few examples are as follows.

- TLP in which buoyancy exceeds the weight, by design.
- Triceratops in which topside is partially isolated from the sub-structure using ball joints.
- FPSOs which remain in full-floatation but resistance is derived from their station-keeping characteristics.

A safe design is always termed as when the resistance exceeds the load or capacity exceeds the demand. As the resistance (or capacity) can also arise from the geometric form, form-dominance is carefully exploited to maintain dynamic in-equilibrium between the resistance and the load. Hence, in form-dominant designs, it is very important to note that resistance acclaimed from dynamic positioning of the topside will result in large (excessive) rigid body motion. Most lateral loads caused by waves and wind are counteracted by allowing large deformation (rigid body motion). It is quite commonly misunderstood that large deformation causes plastic deformation. Please note that the deformations being discussed are not at the material level; they are at the member (or structure) level.

Keeping in mind that safe design refers to the fact that capacity should exceed the demand, let us compare a few design methods, namely working stress design (WSD), ultimate load design (ULD), and limit state design (LSD). In the working stress method, demand at working load is ensured to remain lesser than the capacity of the material, while the latter is assigned as the allowable stress limit. Usually, this is kept well within the elastic limit of the material. The ratio of the capacity of the material, which is taken as the yield limit (in this case) to the allowable stresses, is termed as the factor of safety. The margin between the allowable stress limit and the working load demand is termed the safety margin. Hence, in this design method, there exist two-tier safety margins: one for the choice of allowable stress limit and the other for the actual capacity of the material. One should note very carefully that there is still reserve capacity of the material beyond its yield strength, which is completely ignored in this design procedure. In the ultimate load design case, demand at ultimate load is computed by enhancing the demand at working load by the load factor. But still, the ultimate capacity of the material is restricted up to its yield strength. Hence, the safety margin, which was present in the working stress design method, remains unchanged, but the load is enhanced from working load to ultimate load by a factor; the factor of safety is replaced by load factor in this method. Please note that the reserve capacity of the material beyond yield strength remains unexploited even in this method of design. In the limit state design case, the ultimate capacity of the material is reduced by a partial safety factor for material (Υ_m).

In contrast, the demand at the design load is improved (enhanced) from the demand at working load by partial safety factor for the load (Υ_L). But still, material capacity is restricted only up to yield strength, as in other design methods (Lopes & Bernardo, 2003; Mahin et al., 2006). Figure 2.4 explains the comparison.

Basic Design Guidelines

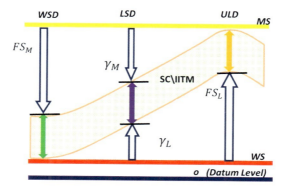

FIGURE 2.4 Design methods: a comparison. (Rupen and Chandrasekaran, 2021.)

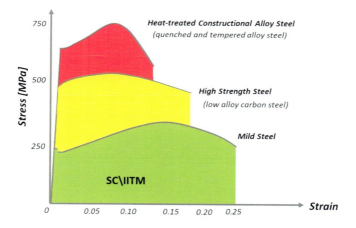

FIGURE 2.5 Structural steel: a comparison. (Rupen and Chandrasekaran, 2021.)

However, reserve energy possessed by structural steel is relatively higher and should be effectively used in the design. Figure 2.5 shows a comparison of the reserve energy of structural steel. It can be observed that structural steel possesses a large, inelastic deformation capacity before it attains a collapsing stage; however, the design procedure should accommodate this plastic state of behavior. Under this specific context, one can compare the design procedures as below.

- Working stress design does not invoke any damage at all (Rustem, 2006).
- Limit state design method invokes nominal damage, but serviceable.
- Ultimate load design invokes damage but no collapse.

Design methods can now be divided into two major domains: force-controlled and displacement-controlled. In the former, failure occurs if the imposed load exceeds yield strength. However, in the latter, no failure occurs even if the imposed displacement exceeds yield displacement, provided the material used for the design is ductile. However, damage occurs in excessive yielding, causing permanent deformation (also termed as plastic deformation). Figure 2.6 explains the concept of plastic design in which the induced force can be even more than the yield strength. It will not cause any damage but results in plastic deformation, which is permanent damage (Wood, 1968). Therefore, the material to be used in the plastic design should possess high post-elastic deformability; steel has this property inadvertently.

While the actual behavior of the steel is shown in Figure 2.7a, its elastic-plastic idealization can be seen in Figure 2.7b.

Based on the idealized behavior, a few assumptions are made to proceed with the plastic design: i) material obeys Hooke's law until it reaches yield, ii) upon further straining, stress remains constant, iii) material remains homogeneous in both elastic and plastic states, iv) steel shows a similar behavior both in tension and compression.

2.8 PLASTIC DESIGN

The plastic analysis permits no shear strain and warping as plane sections are assumed to remain plane and normal to the member's longitudinal axis, even after

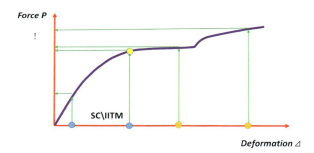

FIGURE 2.6 Plastic design concept. (Rupen and Chandrasekaran, 2021.)

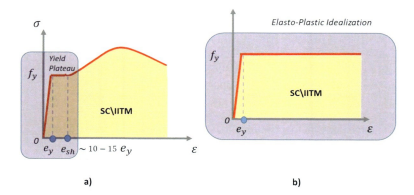

FIGURE 2.7 Characteristics of mild steel a) real behavior, b) idealized behavior. (Rupen and Chandrasekaran, 2021.)

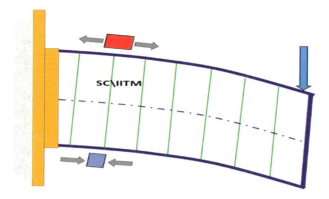

FIGURE 2.8 Plastic analysis: no shear strain and warping. (Rupen and Chandrasekaran, 2021.)

bending (see Figure 2.8). It emphasizes that each fiber in the cross-section is free to deform independently so that successive fibers in the cross-section can yield one after another upon an increase of load.

According to plastic design, the capacity of the member is geometry-dependent and not load-dependent. As discussed earlier, the elastic design procedure restricts the material strength up to the first yield. Steel undergoes local yielding due to the presence of residual stresses. Hence, it is not a good design practice to limit the material capacity at local (first) yield. Static degree of indeterminacy, an index of redundant strength developed from the geometric form, offers additional resistance capacity. In contrast, material reserve in the form of ductility adds substantial input to it. Hence, the idea behind the concept of plastic design is to fully utilize the section's geometric form (shape) and extend the reserve capacity beyond the first yield (which is a falsified indication of yielding).

Concerning the standard stress–strain curve of mild steel, one can understand that when the material fails under load, the following observations are also important: failure can also be due to instability, fatigue, or excessive deformation. If any of these factors does not initiate a failure, the member can carry more loads beyond the elastic limit. Therefore, in plastic design, the gain increases in load-carrying capacity, but at the cost of permanent (plastic) deformation. It is also interesting to note that plastic design is named so even when it is limited until yield value. After yield, the material undergoes (or is permitted to undergo) plastic deformation. Hence, the name "plastic" is affixed to the design method (please refer to the idealized stress–strain curve, Figure 2.7b). In the plastic design, the yielding of the cross-section is assumed to be idealistic, as shown in Figure 2.9.

For simplicity, let us consider a beam with a rectangular cross-section (bxd), as shown in Figure 2.10. Let the material yield strength be denoted as (σ_y). The following relationship gives the section modulus

$$Z_e = \frac{I_z}{y_{max}} = \frac{bd^3}{12} \frac{1}{\frac{d}{2}} = \frac{bd^2}{6} \quad (2.1)$$

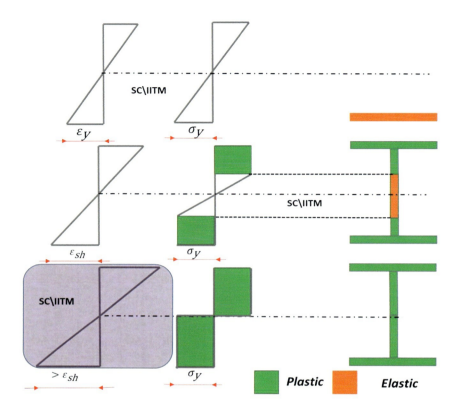

FIGURE 2.9 Idealized yielding progression. (Rupen and Chandrasekaran, 2021.)

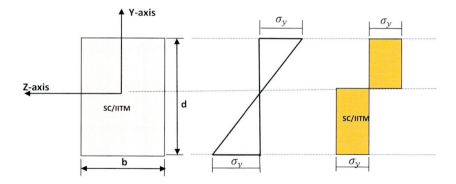

FIGURE 2.10 Rectangular beam (b × d).

Basic Design Guidelines

$$M_z = Z_e \sigma_y \qquad (2.2)$$

$$M_p = \left[\frac{bd}{2} \sigma_y \frac{d}{4} \right] \times 2 = \frac{bd^2}{4} \sigma_y = Z_p \sigma_y \qquad (2.3)$$

where Z_e is termed as elastic section modulus, and σ_y is the yield stress of the material. As seen in Figure 2.10, only the extreme fibers are yielded under the elastic state, while the entire cross-section is yielded under the plastic state. Let us now compare the moment capacity equations of the section under both elastic and plastic states (Equations 2.2 and 2.3). It can be easily inferred that the moment capacity under plastic state (M_p) is greater than that elastic state (M_e). This advantage is the gain achieved only from the section modulus, a geometric property, as the stress in both conditions is the same. As this gain is achieved from the geometric property, we must relate this to the shape of the cross-section. By observing both the moment capacity equations, the ratio of plastic to elastic section modulus (Z_p to Z_e) is termed as shape factor. It is very important to note that the moment capacity is attained only at the section where the member is subjected to maximum bending moment. An additional gain can come from structural redundancy (degree of static indeterminacy).

Further, it is also interesting to note that the maximum stressed section is identified as the critical location. No additional loads can be carried at this section, as the full cross-section (for its entire depth) has already yielded. Such critical sections are termed plastic hinges, and the corresponding section is identified as fully plasticized. Kindly note an important fact that stress level is still at yield only. One should not confuse the term "plastic" with the increase in stress level. It is termed plastic because the section is enabled to undergo plastic deformation at constant stress ordinate (σ_y).

2.8.1 Shape Factor

As discussed in the earlier sections, the plastic design enables a higher moment-carrying capacity, strongly dependent on the cross-section's geometric shape. Let us try to derive the shape factor for any arbitrary cross-section, as shown in Figure 2.11. Let A_1 and A_2 be the equal areas of the cross-section, whose centroid is placed at ($\bar{y}_1, \bar{y}_2$) respectively. The axis, which divides the cross-section into equal areas, is called the equal area axis (EAA). It should not be confused with the neutral axis, which is based on the stress distribution. EAA is based on the geometric shape of the cross-section. If a section has an axis of symmetry to the horizontal plane, it will be the same as EAA.

As the EAA divides the section into equal halves, the total compressive force (C) will be equal to the total tensile force (T), as yield stress in both tension and compression are assumed to be of the same magnitude.

$$A = A_1 + A_2 = A/2 \qquad (2.4)$$

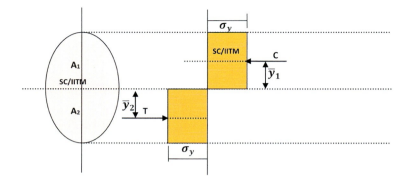

FIGURE 2.11 Shape factor for arbitrary cross-section.

Taking moment about EAA, we get

$$C\bar{y}_1 + T\bar{y}_2 = M \qquad (2.5)$$

$$C = \sigma_y A_1 \qquad (2.6)$$

$$T = \sigma_y A_2 \qquad (2.7)$$

Substituting from the above equations, Equation (2.5) is modified as

$$\sigma_y A_1 \bar{y}_1 + \sigma_y A_2 \bar{y}_2 = M \qquad (2.8)$$

$$\sigma_y \frac{A}{2}(\bar{y}_1 + \bar{y}_2) = M \qquad (2.9)$$

$$\sigma_y Z_p = M_p \qquad (2.10)$$

$$\frac{A}{2}(\bar{y}_1 + \bar{y}_2) = Z_p \qquad (2.11)$$

$$\frac{Z_p}{Z_e} = S \qquad (2.12)$$

where S is called the shape factor.

2.8.2 Depth of Elastic Core

Let us try to understand the depth of elastic core in the context of plastic design. For a plastic design to be effective, the depth of elastic core in a member at any critical section should be zero. Therefore, it is at these sections that plastic hinges are formed. The possible number of plastic hinges depends on the degree of static indeterminacy.

Basic Design Guidelines

If (n) is the degree of static indeterminacy, the (n + 1) plastic hinges are required to convert a beam into a mechanism. More the degree of static indeterminacy, more the possibility of formation of plastic hinges. It will also enhance the moment-carrying capacity of the structure apart from the enhancement that is attained from the shape factor. In simple terms, moment-carrying capacity is achieved in the plastic design in two ways: one by shape factor, which is cross-section dependent, and the other is the degree of indeterminacy, which is form-dependent. Let us consider a rectangular cross-section, as shown in Figure 2.12.

Let the cross-section develops a stress block, which is elastic-plastic, as shown in Figure 2.12. Let (*e*) be the depth of the elastic core. Moment of resistance of the elastic core is given by

$$M_1 = \left[\left(\frac{1}{2}\right)b\left(\frac{e}{2}\right)\left(\frac{2}{3}\right)\left(\frac{e}{2}\right)\right] \times 2\sigma_y \tag{2.13}$$

$$M_1 = \left[\frac{be^2}{6}\right]\sigma_y \tag{2.14}$$

Moment of resistance of the plastic core is given by

$$M_2 = 2\left\{b\left[\frac{h}{2}-\frac{e}{2}\right]\left\{\frac{e}{2}+\left\{\left[\frac{h}{2}-\frac{e}{2}\right]\frac{1}{2}\right\}\right\}\right\}\sigma_y \tag{2.15}$$

$$M_2 = 2\sigma_y \frac{b(h-e)}{2}\left[\frac{e}{4}+\frac{h}{4}\right] \tag{2.16}$$

$$M_2 = 2\sigma_y b \frac{(h-e)}{2} \frac{h+e}{4} \tag{2.17}$$

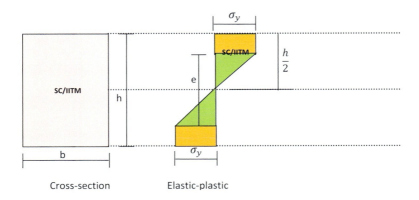

FIGURE 2.12 Elastic-plastic section.

$$M_2 = \sigma_y b \frac{(h^2 - e^2)}{4} \quad (2.18)$$

The total moment of resistance of the elastic-plastic section is given by

$$M = M_1 + M_2 \quad (2.19)$$

$$M = \sigma_y \left\{ \left[\frac{be^2}{6} \right] + \left[\frac{b(h^2 - e^2)}{4} \right] \right\} \quad (2.20)$$

$$= \sigma_y \left[\frac{bh^2}{4} - \frac{be^2}{12} \right] = \frac{\sigma_y bh^2}{4} \left[1 - \frac{e^2}{3h^2} \right]$$

Substituting for Z_p and M_p from Equation (2.3), the above equation can be simplified to the following form

$$M = M_p \left[1 - \frac{e^2}{3h^2} \right] \quad (2.21)$$

The above equation can be used to assess the plastic capacity of any section. For the known cross-section, the plastic section modulus (Z_p) is known. Hence, for any applied moment (M), the depth of the elastic core (e) of the section can be computed from Equation (2.21). If the computed (e) is zero, one can conclude that the section is fully plasticized. Interestingly, the ratio of plastic moment capacity to the elastic moment capacity will also be useful to compute the shape factor. It can be verified as below

$$\frac{M_p}{M_e} = \frac{\sigma_y Z_p}{\sigma_y Z_e} = S \quad (2.22)$$

2.9 SHAPE FACTORS USED IN OFFSHORE TOPSIDE

Let us derive shape factors for a few cross-sections commonly used in offshore topside.

a) Rectangle (Figure 2.13)

$$A_1 = A_2 = \frac{bh}{2} \quad (2.23)$$

$$\bar{y}_1 = \bar{y}_2 = \frac{h}{4} \quad (2.24)$$

$$Z_p = \frac{A}{2}[\bar{y}_1 + \bar{y}_2] = \frac{A}{2}\left[\frac{h}{4} + \frac{h}{4}\right] = \frac{bh^2}{4} \quad (2.25)$$

Basic Design Guidelines

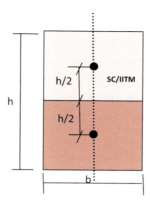

FIGURE 2.13 Rectangular cross-section.

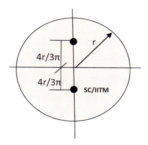

FIGURE 2.14 Circular solid bar.

$$Z_e = \frac{bh^3}{12}\left[\frac{1}{h/2}\right] = \frac{bh^2}{6} \qquad (2.26)$$

$$S = \frac{Z_p}{Z_e} = \frac{bh^2}{4} \times \frac{6}{bh^2} = 1.5 \qquad (2.27)$$

b) Circular, solid bar (Figure 2.14)

$$A_1 = \frac{\pi r^2}{2} = \frac{A}{2} \qquad (2.28)$$

$$\bar{y}_1 = \bar{y}_2 = \frac{4r}{3\pi} \qquad (2.29)$$

$$Z_p = \frac{A}{2}\left[\bar{y}_1 + \bar{y}_2\right] = \frac{\pi r^2}{2}\left[\frac{4r}{3\pi} + \frac{4r}{3\pi}\right] = \frac{4r^3}{3} \qquad (2.30)$$

$$Z_e = \frac{\pi d^4}{64}\left[\frac{1}{d/2}\right] = \frac{\pi d^3}{32} = \frac{\pi (2r)^3}{32} = \frac{\pi r^3}{4} \tag{2.31}$$

$$S = \frac{Z_p}{Z_e} = \frac{4r^3}{3} \times \frac{4}{\pi r^3} = \frac{16}{3\pi} \cong 1.70 \tag{2.32}$$

c) Tubular section

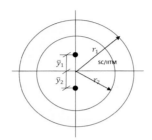

$$A_1 = A_2 = \frac{\left(\pi r_1^2 - \pi r_2^2\right)}{2} = \frac{\pi}{2}\left(r_1^2 - r_2^2\right) \tag{2.33}$$

$$\bar{y}_1 = \bar{y}_2 = \frac{\dfrac{\pi r_1^2}{2}\left(\dfrac{4r_1}{3\pi}\right) - \dfrac{\pi r_2^2}{2}\left(\dfrac{4r_2}{3\pi}\right)}{\dfrac{\pi}{2}\left(r_1^2 - r_2^2\right)} \tag{2.34}$$

$$\bar{y}_1 = \frac{\dfrac{4r_1^3}{3\pi} - \dfrac{4r_2^3}{3\pi}}{\left(r_1^2 - r_2^2\right)} = \frac{4}{3\pi}\frac{r_1^3 - r_2^3}{\left(r_1^2 - r_2^2\right)} \tag{2.35}$$

$$Z_p = \frac{A}{2}\left[\bar{y}_1 + \bar{y}_2\right] = \frac{\pi}{2}\left(r_1^2 - r_2^2\right)\left[\frac{4}{3\pi}\frac{r_1^3 - r_2^3}{\left(r_1^2 - r_2^2\right)}\right]2 = \frac{4}{3}\left(r_1^3 - r_2^3\right) \tag{2.36}$$

$$I_z = \frac{\pi}{64}\left(d_1^4 - d_2^4\right) = \frac{\pi}{64}\left[\left(2r_1\right)^4 - \left(2r_2\right)^4\right] = \frac{\pi}{4}\left[r_1^4 - r_2^4\right] \tag{2.37}$$

$$Z_e = \frac{\pi}{4}\left[r_1^4 - r_2^4\right]\frac{1}{r_1} = \frac{\pi}{4r_1}\left[r_1^4 - r_2^4\right] \tag{2.38}$$

$$S = \frac{Z_p}{Z_e} = \frac{4}{3}\left(r_1^3 - r_2^3\right)\frac{4r_1}{\pi\left[r_1^4 - r_2^4\right]} = \frac{16r_1}{3\pi}\frac{\left(r_1^3 - r_2^3\right)}{\left(r_1^4 - r_2^4\right)} \tag{2.39}$$

Basic Design Guidelines

Let $\dfrac{r_2}{r_1} = k$, then Equation (2.39) takes the following form

$$S = \frac{16 r_1 \left(r_1^3 - r_2^3\right)}{3\pi \left(r_1^4 - r_2^4\right)} = \frac{16 r_1 \left(r_1^3 - k^3 r_1^3\right)}{3\pi \left(r_1^4 - k^4 r_1^4\right)}$$

$$= \frac{16}{3\pi} \frac{\left(1 - k^3\right)}{\left(1 - k^4\right)} \cong 1.7 \frac{\left(1 - k^3\right)}{\left(1 - k^4\right)}$$
(2.40)

By substituting r_2 as zero, Equation (2.40) reduces to the form applicable for a solid circular bar, as given by Equation (2.32). As the shape factor is relatively higher for tubular sections than rectangular, they are highly preferred for topside design apart from other structural advantages (Chandrasekaran, 2020).

e) Channel section

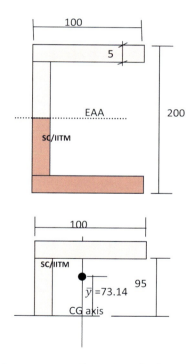

Total area of X-section, $A = (100 \times 5 \times 2) + (190 \times 5) = 1950 \text{ mm}^2$

EAA is located at the center of gravity (CG) by symmetry

$$\bar{y} = \frac{(10 \times 5 \times 97.5) + (95 \times 5 \times 47.5)}{(100 \times 5) + (95 \times 5)} = 73.14 \text{ mm}$$

$$Z_p = \frac{A}{2}\left[\bar{y}_1 + \bar{y}_2\right] = \frac{1950}{2}(73.14 + 73.14)$$

$$= 142623 \text{ mm}^3$$

$$I_z = \left[\frac{100 \times 5^3}{12} + (100 \times 5 \times 97.5^2)\right] \times 2 + \frac{5 \times 190^3}{12}$$

$$= 12.366 \times 10^6 \text{ mm}^3$$

$$Z_e = \frac{I_z}{y_{max}} = \frac{12.366 \times 10^6}{100} = 12.366 \times 10^4 \text{ mm}^3$$

$$S = \frac{Z_p}{Z_e} = \frac{142623}{12.366 \times 10^4} = 1.153$$

f) T-section

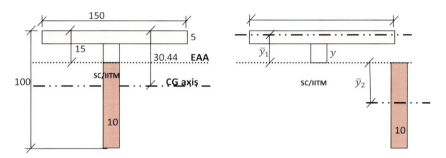

The first step is to locate the EAA, as this section has no horizontal axis of symmetry.

$$\text{Area of the flange} = 150 \times 5 = 750 \text{ mm}^2$$

$$\text{Area of the web} = 95 \times 10 = 950 \text{ mm}^2$$

$$\text{Total area of the X-section} = 1700 \text{ mm}^2$$

Let (y) be the distance, measured from the flange-web junction, as marked in the diagram.

$$\text{Area of the flange} + 10y = \frac{1700}{2} = 850$$

$$750 + 10y = 850;\ y = 10 \text{ mm}$$

$$A_1 = A/2 = \text{Area of the flange} + \text{part of the web}$$

Basic Design Guidelines

$$A_1 = 750 + 10 \times 10 = 850 \text{ mm}^2 = A/2 \text{ (OK)}$$

Now, let us try to locate the CG of the upper area and the lower area $(\bar{y}_1, \bar{y}_2)$.

$$\bar{y}_1 = \frac{(15 \times 5 \times 12.5) + (10 \times 10 \times 5)}{850} = 11.62 \text{ mm}$$

$$\bar{y}_2 = \frac{85}{2} = 42.5 \text{ mm}$$

$$Z_p = \frac{A}{2}[\bar{y}_1 + \bar{y}_2] = \frac{1700}{2}(11.62 + 42.5) = 46002 \text{ mm}^3$$

Now, let us locate the CG of the T-section.

$$\bar{y} = \frac{\Sigma A\bar{y}}{\Sigma A} = \frac{(150 \times 5 \times 2.5) + (95 \times 10 \times 52.5)}{1700} = 30.44 \text{ mm}$$

Moment of inertia (MoI) of the section about the bending plane is obtained as below

$$I_z = \left[\frac{150 \times 5^3}{12} + 150 \times 5 \times (30.44 - 2.5)^2\right]$$

$$+ \left[\frac{10 \times 95^3}{12} + 95 \times 10 \times (52.5 - 30.44)^2\right]$$

$$= 1.767 \times 10^6 \text{ mm}^4$$

$$Z_e = \frac{I_z}{y_{max}} = \frac{1.767 \times 10^6}{(100 - 30.44)} = 25402.5 \text{ mm}^3$$

$$S = \frac{Z_p}{Z_e} = \frac{46002}{25402.5} = 1.81$$

g) L-section

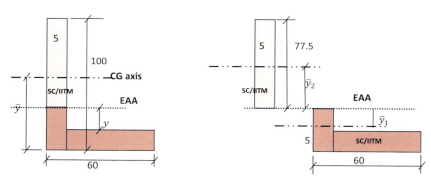

The X-section under consideration is an unequal angle. It has no axis of symmetry.

$$\text{Area of the shorter arm, } A_1 = (60 \times 5) = 300 \text{ mm}^2$$

$$\text{Area of the longer arm, } A_2 = (95 \times 5) = 475 \text{ mm}^2$$

$$\text{Total area of the section} (A_1 + A_2) = 775 \text{ mm}^2$$

We need to locate the EAA.

Let the EAA be at a distance (y) from the intersection phase, as shown in the diagram.

$$(60 \times 5) + (5y) = \frac{775}{2}; y = 17.5 \text{ mm}$$

Check: $A_1 + (17.5 \times 5) = 300 + 87.5 = 387.5 \text{ mm}^2$; OK.

Now, we need to locate the centroid of the upper and lowers areas to compute Z_p.

$$\bar{y}_1 = \frac{(60 \times 5 \times (17.5 + 2.5)) + \left(5 \times 17.5 \times \frac{17.5}{2}\right)}{387.5} = 17.46 \text{ mm}$$

$$\bar{y}_2 = \frac{77.5}{2} = 38.75 \text{ mm}$$

$$Z_p = \frac{A}{2}[\bar{y}_1 + \bar{y}_2] = \frac{775}{2}(17.46 + 38.75) = 21781.38 \text{ mm}^2$$

Now, let us locate the CG of the L-section.

$$\bar{y} = \frac{\Sigma A \bar{y}}{\Sigma A} = \frac{(60 \times 5 \times 2.5) + (95 \times 5 \times 52.5)}{775} = 33.15 \text{ mm}$$

MoI of the section about the bending plane is obtained as below

$$I_z = \left[\frac{60 \times 5^3}{12} + 60 \times 5 \times (33.15 - 2.5)^2\right]$$

$$+ \left[\frac{5 \times 95^3}{12} + 95 \times 5 \times (52.5 - 33.15)^2\right]$$

$$= 8.175 \times 10^5 \text{ mm}^4$$

$$Z_e = \frac{I_z}{y_{max}} = \frac{8.175 \times 10^5}{(100 - 33.15)} = 12228.87 \text{ mm}^3$$

Basic Design Guidelines

$$S = \frac{Z_p}{Z_e} = \frac{21781.38}{12228.87} = 1.78$$

2.10 MOMENT-CURVATURE RELATIONSHIP

Let us consider a simply supported beam of rectangular cross-section (bxd). According to the theory of simple bending, the following relationship holds good (Figure 2.12).

$$\frac{M_z}{I_z} = \frac{E}{R} = \frac{\sigma_{yield}}{y} \quad (2.41)$$

The following relationship gives curvature:

$$\frac{M_z}{E I_z} = \frac{1}{R} = \varphi \quad (2.42)$$

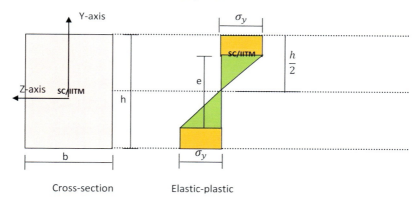

Cross-section   Elastic-plastic

From Equation (2.42), it can be seen that the curvature and the moment are proportional. Any marginal increase in the applied moment caused by the loads will increase the curvature by the corresponding amount (Pfrang et al., 1964). But this linear relationship is valid only up to the limit of proportionality for the elastic material (Pisanty & Regan, 1998, 1993). When the moment reaches (M_{yield}), this relationship is modified as below.

$$\frac{M_{yield}}{E I_z} = \frac{1}{R} = \frac{\sigma_{yield}}{y} = \varphi \quad (2.43)$$

For $y = \frac{h}{2}$, as in the case of the rectangular section, the following relationships hold good.

$$\frac{\sigma_{yield}}{\frac{h}{2}} = \frac{E}{R} \quad (2.44)$$

$$\frac{2\sigma_{yield}}{h} = \frac{E}{R} \tag{2.45}$$

$$\frac{2\sigma_{yield}}{Eh} = \frac{1}{R} \tag{2.46}$$

After reaching yield, if a further moment is applied, the section will get partially plasticized. This state is termed an elastic-plastic state. The depth of the elastic core is identified as (e), as shown in the diagram. For the elastic core, the following expression is valid

$$\frac{2\sigma_{yield}}{Ee} = \frac{1}{R} \tag{2.47}$$

At this stage, the linear relationship between moment and curvature does not hold well. As stated in the plastic design, let us continue to load the beam, which will cause an increase at the moment beyond M_y; the section will continue to carry the load (applied moment) until the section reaches its moment-carrying capacity, M_p. When M_p is reached, the elastic core completely seizes away; hence ($e = o$) condition holds good. From Equation (2.47), one can easily interpret that the curvature tends to become infinity.

$$\frac{1}{R} = \varphi = \frac{M}{EI_z} \tag{2.48}$$

$$\left(\frac{1}{R}\right)_{at\ yield} = \varphi_{yield} = \frac{M_{yield}}{EI_z} \tag{2.49}$$

Then, the following relationship also holds good.

$$\frac{M_p}{M_{yield}} = \frac{\varphi_p}{\varphi_{yield}} \tag{2.50}$$

$$\frac{Z_p \sigma_{yield}}{Z_e \sigma_{yield}} = \frac{\varphi_p}{\varphi_{yield}} = \text{Shape factor}, S \tag{2.51}$$

Referring back to Equation (2.21), which is rewritten below

$$M = M_p \left[1 - \frac{e^2}{3h^2}\right]$$

$$\frac{M}{M_p} = \left[1 - \frac{e^2}{3h^2}\right] \tag{2.52}$$

Substituting for (*e*) from Equation (2.47), we get

$$\frac{2\sigma_{yield}}{Ee} = \frac{1}{R}$$

$$\frac{2R\sigma_{yield}}{E} = e \qquad (2.53)$$

$$\frac{M}{M_p} = 1 - \frac{\left[\frac{2R\sigma_{yield}}{E}\right]^2}{3h^2} = 1 - \frac{4R^2\sigma_y^2}{3E^2h^2} \qquad (2.54)$$

$$\frac{M}{M_p} = 1 - \frac{1}{3}\left[\frac{2\sigma_y}{E}\right]^2 \frac{1}{\left(\frac{h}{R}\right)^2} \qquad (2.55)$$

From Equation (2.46), we know that

$$\frac{2\sigma_{yield}}{E} = \left(\frac{h}{R}\right)_{yield} \qquad (2.56)$$

Substituting the above in Equation (2.55), we get

$$\frac{M}{M_p} = 1 - \frac{1}{3}\left[\frac{\left(\frac{h}{R}\right)_y}{\left(\frac{h}{R}\right)^2}\right] \qquad (2.57)$$

$$\frac{M}{M_p} = 1 - \frac{1}{3}\left[\frac{\left(\frac{h}{R}\right)_y}{\frac{h}{R}}\right] \qquad (2.58)$$

2.11   LOAD FACTOR

Load factor, Q, is defined as the ratio of collapse load to working load.

$$Q = \frac{W_c}{W_w} \qquad (2.59)$$

We also know that the moment, M, is proportional to the applied load. Hence,

$$M \propto W \tag{2.60}$$

$$M_w = k W_w \tag{2.61}$$

$$M_p = k W_c \tag{2.62}$$

$$Q = \frac{W_c}{W_w} = \frac{M_p}{M_w} \tag{2.63}$$

Further,

$$M_p = \sigma_y Z_p \tag{2.64}$$

$$M_w = \sigma_{all} Z_e \tag{2.65}$$

Substituting Equations (2.64–2.65) in Equation (2.63), we get

$$Q = \frac{M_p}{M_w} = \frac{\sigma_y Z_p}{\sigma_{all} Z_e} = \left(\frac{Z_p}{Z_e}\right)\left(\frac{\sigma_y}{\sigma_{all}}\right) = S \times FOS \tag{2.66}$$

Hence, in plastic design, factor of safety (FOS) is enhanced by the shape factor, ensuring that the design procedure is safe. Plastic design claims one more advantage: without changing the material, one can improve the moment capacity by simply choosing geometry with a better shape factor. Citing the above examples solved for shape factor, one can observe that closed, rolled sections of even welded sections possess higher shape factor. Due to this fact, tubular sections and closed, hot-rolled sections are used on the topside.

2.12 STABILITY OF THE STRUCTURAL SYSTEM

The stability of structural systems, in general, and steel structures, in particular, is extremely important in offshore structures, as most of the compliant structures alleviate the encountered environmental loads by their geometric form and not by strength. The configuration of any structural system is posed to a challenge under the given loads and boundary conditions. However, if the structural configuration satisfies the conditions of static equilibrium, compatibility, and force-displacement relationships, it is stable (Priestley et al., 2007). An unstable condition is often referred to as a failed state. Stability, therefore, demands an understanding of failure.

Stability is affected significantly under compressive forces. Further, geometric stability is more important than material stability, as the former may challenge the functional requirements of the structural member. Therefore, stability refers to a stable state of equilibrium and is defined as the ability of any structural system to remain (or continue to remain) in its geometric form, which can perform the intended function even if the geometric position is disturbed by external forces. By this definition, compliant offshore structures are stable as they can perform their intended function at the disturbed geometric position. For example, a TLP is said to be in a disturbing position under the combined effect of offset and set-down. As long as this change in geometric position does not affect its load-disbursing capacity, TLP is said to be in a stable condition. Therefore, structures don't need to remain (or continue to remain) in their original geometric form to classify them as stable. They may continue to remain stable even under the deformed geometric position if they perform their intended function successfully.

The three criteria of checking stability are i) Euler's static criterion, ii) Lyapunov's dynamic criterion, and iii) the potential energy stability criterion. Euler's static criterion is applicable under the non-trivial equilibrium state. It evaluates the stability of a structural system by examining the optimum geometric configuration of the system other than the original (initially straight) configuration at which the structural system can still disburse the applied load (P_{cr}). Under the given boundary conditions and initially perfect straight geometry, structures are examined. Euler's criterion evaluates whether the structure is capable of carrying the load (maybe in lesser magnitude, which is P_{cr} where $P > P_{cr}$) instead of remaining in a state where it is unable to carry any load at all. It is interesting to note that the load-carrying capacity of the structural member is reduced from P to P_{cr}. Still, the important fact is that the structural member can carry at least P_{cr} even at a changed geometric form, which is relatively weak from that of the initial form. This value of load with lesser magnitude compared to that of the originally intended load (P) is termed as critical load or buckling load. It can be easily seen from the standard literature that P_{cr} is easily computed from the boundary conditions, cross-sectional properties of the member, and slenderness ratio. However, instability occurs when two or more adjacent equilibrium positions correspond to different mode shapes.

For assessing the stability of offshore compliant structures, it is reasonably simple to disagree with Euler's criterion, as the boundary conditions of the member do not permit examining Euler's stability criterion. They are either completely floating under hydrodynamic stability, or compliant, and hence stability is dependent on the high-magnitude pretension of tethers. For example, in the case of TLPs, high pretension imposed on tethers ensures stability and re-centering of the platform under lateral loads. Functional working of the platform is not lost even under the deformed position of the platform under wave loads. Please note that this condition is true even though the offset values are quite large (about 10 percent of that of the initial pretension of tethers), causing large deformation. Lyapunov's condition examines the stability of the structural system under dynamic excitations. If a member is subjected to a continuously varying disturbing force, it is necessary to examine whether equilibrium under the dynamic forces is satisfied. As explained in the literature (Chandrasekaran, 2015, 2016, 2017), stability can be influenced by both varying amplitudes of the exciting force and its period of excitation. While the former can influence the member's load-carrying capacity, which is the strength-dependent criterion, the latter can result in unconditional response at the near-resonance state of vibration. Hence, Lyapunov's assessment of stability is focused on the dynamic response behavior of the structural system instead of purely assessing its load-carrying capacity, as in the case of Euler's criterion.

Lyapunov's condition is more significant for structures designed to perform their intended function under varying external forces. The most challenging element of stability is that the structure will be assuming a different geometric position with time. Structural systems like ships, offshore compliant platforms, and floating production platforms fall under this category of stability check. Stability calculations of ships focus on estimating the center of gravity, buoyancy, meta-center of vessels, and interaction. The saving part of such systems is that they are designed to remain hydro-statically stable at any instant of time. In case of structures that are permitted to undergo large displacements, as in the case of TLPs, then the geometric design ensures proper re-centering, which means that the structural system continuously tries to regain its original geometric position with the help of dedicated members present within the system. In the case of TLPs, the design is tethered. A potential energy stability criterion applies to structural systems for which the system's potential energy ceases to be the minimum, which is more relevant to conservative systems.

2.13 EULER'S CRITICAL LOAD

Consider an ideal column, as shown in Figure 2.15. The column is assumed to be initially straight and compressed by a concentric load, P. The column is pinned at both the supports and is uniformly slender. Further, it is assumed to be laterally restrained in position at both the supports. The column section is assumed to be of negligible weight and perfectly elastic. Stresses developed by the axial forces are assumed to be within the proportional limit of the column material. If the applied force, P, is lesser than the critical value, the column will continue to remain straight and undergo only axial compression. Under this state, the column is said to be in *stable equilibrium*. Under such conditions, if, say, a lateral load is applied at any point, it will result in lateral deflection at the mid-height of the column. Importantly, however, the column will return to its original position in terms of geometry, shape, and size in relation to its cross-section.

FIGURE 2.15 Euler's column.

Basic Design Guidelines

However, upon a continuous (steady rate) increase of the axial load, P, the straight form of equilibrium tends to become gradually unstable. Under this condition, even the lateral of a very small magnitude can cause deflection, which will not disappear upon removal of the lateral load. This is not the case when P is lesser than the axial capacity of the member, as discussed earlier. Based on these two sets of explanations, one can define a *critical load*. The critical load is the axial load necessary to maintain (or continue to maintain) the member in its initial straight position (Stephen & Gere, 1961). This critical load can be computed based on the elastic curve equation (Livesley & Chandler, 1956).

$$\frac{d^2 y}{d^2 x} = \frac{M}{EI} \tag{2.67}$$

where M is bending moment, I is the moment of inertia, and E is the modulus of elasticity. Figure 2.16 shows the free-body diagram of the column member under the applied load.

With reference to Figure 2.16, the equilibrium of the free-body diagram is written as below

$$EI \frac{d^2 y}{d^2 x} = M = -Py \tag{2.68}$$

$$EI \frac{d^2 y}{d^2 x} + Py = 0 \tag{2.69}$$

$$y = A \sin\left[\frac{\alpha x}{L}\right] + B \cos\left[\frac{\alpha x}{L}\right] \tag{2.70a}$$

where

$$\alpha = L\sqrt{\frac{P}{EI}} \tag{2.70b}$$

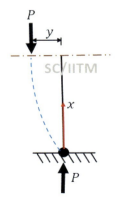

FIGURE 2.16 Free-body diagram of the column member.

For the boundary condition: @ $x = 0$; $y = 0$, $B = 0$; hence, Equation (2.70a) becomes as below

$$y = A\sin\left[\frac{\alpha x}{L}\right] \qquad (2.71)$$

Further, @ $x = L$, $y = 0$. Applying this boundary condition, Equation (2.71) becomes as below

$$A\sin(\alpha) = 0 \qquad (2.78)$$

which means that either $A = 0$ or $\sin(\alpha) = 0$. If $A = 0$, there will no lateral deflection. Therefore, setting $\sin(\alpha) = 0$

$$\alpha = n\pi \quad \text{for } n = 0,1,2,3,\ldots \qquad (2.79)$$

Substituting Equation (2.79), we get

$$n\pi = L\sqrt{\frac{P}{EI}} \qquad (2.80)$$

Squaring,

$$n^2\pi^2 = L^2\frac{P}{EI} \qquad (2.81)$$

$$P = \frac{n^2\pi^2 EI}{L^2} \quad \text{for } n = 1,2,3,4,\ldots \qquad (2.82)$$

$n = 0$ is meaningless as this will cause no axial load ($P = 0$).

In Equation (2.82), P is called a Euler critical load.

$$P_E = \frac{n^2\pi^2 EI}{L^2} \quad \text{for } n = 1,2,3,4,\ldots \qquad (2.83)$$

2.14 STANDARD BEAM ELEMENT, NEGLECTING AXIAL DEFORMATION

A beam element, shown in Figure 2.17, is one of the basic elements used in stability analysis. A few sign conventions must be followed before deriving the stiffness matrix of the standard beam element.

- The end moment, joint rotation, and joint moments, which are anti-clockwise in nature, are considered to be positive.
- Upward force (or displacement) of the joint is considered a positive value.
- Force or axial displacement towards the right of the joint is also considered as positive.

Basic Design Guidelines

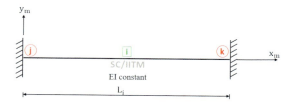

FIGURE 2.17 Standard beam element.

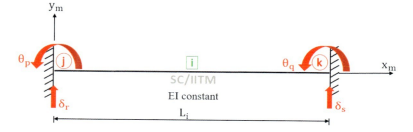

FIGURE 2.18 Rotational and translational moments in the standard beam.

- Upward end shear at the ends of the beam is positive.
- Right direction force at the ends of the beam is positive.

Consider a fixed beam undergoing deformation due to bending, neglecting the axial deformation. The standard fixed beam is shown in Figure 2.17. End nodes of the beam are designated as j and k end, while the length of the member is designated as L_i; subscript "i" refers to the member index. The beam has a constant EI over its entire length. The axes (x_m, y_m) are the local axes of the member. It is very important to note the axis system. The axis system is such that it has an origin at the jth end; x_m is directed towards a kth end; y_m is counter-clockwise 90° to the x_m axis. Therefore, (x_m, y_m) plane defines the plane of bending the beam element.

Neglecting the axial deformation, one should identify both the translational and rotational displacements at each end of the beam, as shown in Figure 2.18. Suitable subscripts are used for denoting the rotational and translational displacements, as marked in the figure. The displacements at the jth end and k^{th} end are (θ_p, δ_r) and (θ_q, δ_s), respectively. All these displacements happen in x_m, y_m plane, and there is no out-of-plane bending. By classical definition, the stiffness coefficient, k_{ij}, is the force in the ith degree-of-freedom by imposing unit displacement (either translational or rotational) in the jth degree-of-freedom by keeping all other degrees-of-freedom restrained. As seen in Figure 2.18, there are four degrees of freedom (2 rotations and 2 translations). One should give unit displacement (or rotation) in each degree-of-freedom to find the forces (or moments) in the respective degrees-of-freedom by keeping the remaining degrees-of-freedom restrained. Imposing unit displacement represents $\delta_r = 1$ or $\delta_s = 1$ and that of unit rotation implies $\theta_p = 1$ or $\theta_q = 1$.

Let us apply unit rotation at the jth end, keeping all other degrees of freedom restrained, as shown in Figure 2.19. It will invoke the members with the end forces, $k^i_{pp}, k^i_{qp}, k^i_{rp}, k^i_{sp}$ as seen in the figure. k^i_{pp} is the force in p^{th} degree-of-freedom by giving unit displacement in p^{th} degree-of-freedom; superscript "i" refers to the ith member. Similarly, k^i_{qp} is the force in q^{th} degree-of-freedom by giving unit displacement in p^{th} degree-of-freedom in the ith member. The second subscript in all the notations is common, which is "p," indicating that the unit displacement (in this case, it is unit rotation since p is a rotational degree-of-freedom) is applied at p^{th} degree. The stiffness coefficients derived column-wise correspond to the first column of the stiffness matrix. Similarly, applying unit rotation at the kth end of the ith member, as shown in Figure 2.20, develops the stiffness coefficients $(k^i_{pq}, k^i_{qq}, k^i_{rq}, k^i_{sq})$.

The stiffness coefficients are obtained by applying unit displacements at the jth end and k^{th} end, as shown in Figures 2.21 and 2.20, respectively, which yields the stiffness coefficients at the jth end as $(k^i_{pr}, k^i_{qr}, k^i_{rr}, k^i_{sr})$ and at the kth end as $(k^i_{ps}, k^i_{qs}, k^i_{rs}, k^i_{ss})$, respectively. As shown in Figures 2.21 and 2.22, a tangent can be drawn by connecting the deflected position of the beam at which the unit rotation is applied and the initial position of the beam at the other end. It can be easily inferred that the beam has undergone a rotation of $(1/L_i)$, where L_i is the length of the ith beam element. The rotation at the ends of the beam is equal to $(1/L_i)$.

The corresponding end reactions (moment, shear) for the ith beam element are required to be estimated under the arbitrary end displacements $\left(\theta_p, \delta_r\right)$ and $\left(\theta_q, \delta_s\right)$

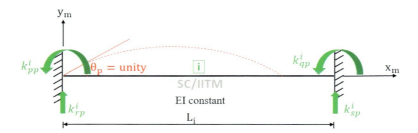

FIGURE 2.19 Unit rotation at the jth end.

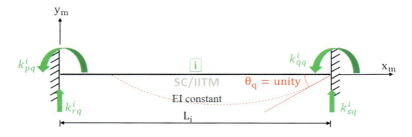

FIGURE 2.20 Unit rotation at the jth end.

Basic Design Guidelines

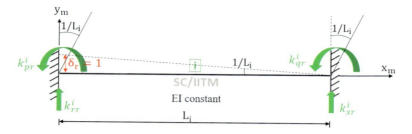

FIGURE 2.21 Unit displacement at the jth end.

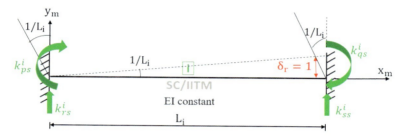

FIGURE 2.22 Unit displacement at the kth end.

at the jth and k^{th} ends of the beam. They are estimated by maintaining the equilibrium of the restrained member. The governing equations are as follows

$$m_p^i = k_{pp}^i \theta_p + k_{pq}^i \theta_q + k_{pr}^i \delta_r + k_{ps}^i \delta_s \tag{2.84}$$

$$m_q^i = k_{qp}^i \theta_p + k_{qq}^i \theta_q + k_{qr}^i \delta_r + k_{qs}^i \delta_s \tag{2.85}$$

$$p_r^i = k_{rp}^i \theta_p + k_{rq}^i \theta_q + k_{rr}^i \delta_r + k_{rs}^i \delta_s \tag{2.86}$$

$$p_s^i = k_{sp}^i \theta_p + k_{sq}^i \theta_q + k_{sr}^i \delta_r + k_{ss}^i \delta_s \tag{2.87}$$

It can be seen from the above equations that the first subscript corresponds to the end at which the unit rotation (or displacement) is applied. The above set of equations gives the end moments and end shear forces for arbitrary displacements $(\theta_p, \theta_q, \delta_r, \delta_s)$, which are unity at respective degrees of freedom. These equations can be generalized as follows

$$\{m_i\} = [k]_i \{\delta_i\} \tag{2.88}$$

$$\{m_i\} = \begin{Bmatrix} m_p \\ m_q \\ p_r \\ p_s \end{Bmatrix} \tag{2.89}$$

$$\{\delta_i\} = \begin{Bmatrix} \theta_p \\ \theta_q \\ \delta_r \\ \delta_s \end{Bmatrix} \quad (2.90)$$

$$[k]_i = \begin{bmatrix} k_{pp} & k_{pq} & k_{pr} & k_{ps} \\ k_{qp} & k_{qq} & k_{qr} & k_{qs} \\ k_{rp} & k_{rq} & k_{rr} & k_{rs} \\ k_{sp} & k_{sq} & k_{sr} & k_{ss} \end{bmatrix} \quad (2.91)$$

Figure 2.23 shows the forces at both the ends of the standard beam member under unit rotation applied at the jth end of the member.

In the standard fixed beam element with unit rotation at the jth end, moments developed at the ends to control the applied unit rotation are (k^i_{pp}, k^i_{qp}), respectively. It results in the development of an anti-clockwise moment ($k^i_{pp} + k^i_{qp}$), which a coupled shear should counteract. The shear forces at the ends of the beam are determined as $\left[\dfrac{k^i_{pp} + k^i_{qp}}{L_i} \right]$. At the jth and k^{th} end of the beam, the magnitude of the shear will be the same, but it will be in the opposite direction at the jth end of the member. It can be seen that the second subscript in the stiffness coefficients indicates the end at which the unit rotation is applied, and the first subscript indicates the forces in the respective degrees of freedom. With reference to Figure 2.23, it is clear that one needs to evaluate only a set of rotational coefficients in Equation (2.91). These rotational coefficients are ($k^i_{pp}, k^i_{pq}, k^i_{qp}, k^i_{qq}$). Knowing the rotational coefficients, end shear can be expressed as follows

$$k^i_{rp} = \frac{k^i_{pp} + k^i_{qp}}{L_i} \quad (2.92)$$

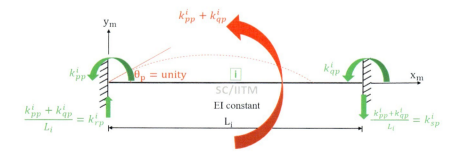

FIGURE 2.23 Rotation coefficients of the standard beam.

Basic Design Guidelines

$$k_{sp}^i = -\frac{k_{pp}^i + k_{qp}^i}{L_i} \tag{2.93}$$

The negative sign in Equation (2.93) is because the direction of k_{sp}^i is opposite to that of the end shear developed by the restraining moments, as shown in Figure 2.23. For the unit rotation applied at the kth end, end shear, in terms of the rotational coefficients is given by

$$k_{rq}^i = \frac{k_{pq}^i + k_{qq}^i}{L_i} \tag{2.94}$$

$$k_{sq}^i = -\frac{k_{pq}^i + k_{qq}^i}{L_i} \tag{2.95}$$

By applying unit displacement at the jth end of the beam element, the stiffness coefficients can be expressed as

$$k_{pr}^i = \frac{k_{pp}^i + k_{pq}^i}{L_i} \tag{2.96}$$

$$k_{qr}^i = -\frac{k_{qp}^i + k_{qq}^i}{L_i} \tag{2.97}$$

$$k_{rr}^i = \frac{k_{pr}^i + k_{qr}^i}{L_i} = \left[\frac{k_{pp}^i + k_{pq}^i}{(L_i)^2}\right] + \left[\frac{k_{qp}^i + k_{qq}^i}{(L_i)^2}\right] \tag{2.98}$$

$$k_{rr}^i = \frac{k_{pp}^i + k_{pq}^i + k_{qp}^i + k_{qq}^i}{(L_i)^2} \tag{2.99}$$

$$k_{sr}^i = -\frac{k_{pp}^i + k_{pq}^i + k_{qp}^i + k_{qq}^i}{(L_i)^2} \tag{2.100}$$

By imposing unit displacement at the kth end of the beam element, the stiffness coefficients are expressed as

$$k_{ps}^i = -\frac{k_{pp}^i + k_{pq}^i}{L_i} \tag{2.101}$$

$$k_{qs}^i = -\frac{k_{qp}^i + k_{qq}^i}{L_i} \tag{2.102}$$

$$k_{rs}^i = -\frac{k_{ps}^i + k_{qs}^i}{L_i} = -\frac{k_{pp}^i + k_{pq}^i + k_{qp}^i + k_{qq}^i}{(L_i)^2} \quad (2.103)$$

$$k_{ss}^i = \frac{k_{pp}^i + k_{pq}^i + k_{qp}^i + k_{qq}^i}{(L_i)^2} \quad (2.104)$$

It can be seen from the above expression that out of the sixteen coefficients of the stiffness matrix given in Equation (2.91), one needs to evaluate only the four rotational coefficients ($k_{pp}^i, k_{pq}^i, k_{qp}^i, k_{qq}^i$). The remaining coefficients can be expressed as a function of these rotational coefficients. The complete stiffness matrix is given below

$$[k] = \begin{bmatrix} k_{pp} & k_{pq} & \dfrac{k_{pp}+k_{pq}}{L} & -\left(\dfrac{k_{pp}+k_{pq}}{L}\right) \\ k_{qp} & k_{qq} & \dfrac{k_{qp}+k_{qq}}{L} & -\left(\dfrac{k_{qp}+k_{qq}}{L}\right) \\ \dfrac{k_{pp}+k_{pq}}{L} & \dfrac{k_{pq}+k_{qq}}{L} & \dfrac{k_{pp}+k_{pq}+k_{qp}+k_{qq}}{L^2} & \dfrac{k_{pp}+k_{pq}+k_{qp}+k_{qq}}{L^2} \\ -\left(\dfrac{k_{pp}+k_{pq}}{L}\right) & -\left(\dfrac{k_{pq}+k_{qq}}{L}\right) & -\left(\dfrac{k_{pp}+k_{pq}+k_{qp}+k_{qq}}{L^2}\right) & -\left(\dfrac{k_{pp}+k_{pq}+k_{qp}+k_{qq}}{L^2}\right) \end{bmatrix}$$

(2.105)

2.14.1 Rotational Coefficients

Consider a simply supported beam as shown in Figures 2.24 and 2.25. Unit rotation is applied at the (j,k) ends of the beam to obtain the flexibility coefficients, respectively. The flexibility coefficients $\left(\delta_{jj}^i, \delta_{kj}^i\right)$ define rotations at end j and k, respectively of the ith member, caused due to unit moment applied at the jth end. Similarly, the flexibility coefficients $\left(\delta_{jk}^i, \delta_{kk}^i\right)$ define rotations at jth and kth ends of the ith member due to unit moment applied at the kth end.

Let us now consider a beam fixed at the kth end, imposed by unit rotation at the jth end (Figure 2.26) and unit rotation at the kth end as shown in Figure 2.27. The

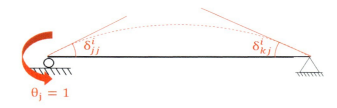

FIGURE 2.24 Unit rotation at the jth end of simply supported beam.

Basic Design Guidelines

FIGURE 2.25 Unit rotation at the kth end of simply supported beam.

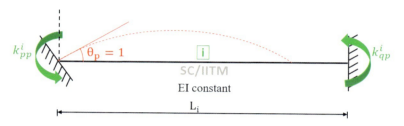

FIGURE 2.26 Unit rotation at the jth end of fixed beam.

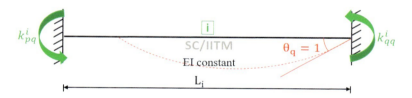

FIGURE 2.27 Unit rotation at the kth end of fixed beam.

stiffness coefficients $\left(k_{pp}^i, k_{qp}^i\right)$ define end moments required at j^{th} and k^{th} ends to maintain equilibrium when the jth end is subjected to unit rotation while the kth end is restrained. Similarly, the stiffness coefficients $\left(k_{pq}^i, k_{qq}^i\right)$ define end moments required at the jth and the kth ends to maintain equilibrium, when the kth end is subjected to unit rotation, and the jth end is restrained.

Thus,

$$\begin{bmatrix} \delta_{ij} & \delta_{jk} \\ \delta_{kj} & \delta_{kk} \end{bmatrix} \cdot \begin{bmatrix} k_{pp} & k_{pq} \\ k_{qp} & k_{qq} \end{bmatrix} = \begin{bmatrix} 1 & 0 \\ 0 & 1 \end{bmatrix} \tag{2.106}$$

Expanding the above equation, we get

$$k_{pp}^i \delta_{jj}^i + k_{qp}^i \delta_{jk}^i = 1 \tag{2.107a}$$

$$k_{pp}^i \delta_{kj}^i + k_{qp}^i \delta_{kk}^i = 1 \tag{2.107b}$$

$$k^i_{pq}\delta^i_{jj} + k^i_{qq}\delta^i_{jk} = 1 \tag{2.107c}$$

$$k^i_{pq}\delta^i_{kj} + k^i_{qq}\delta^i_{kk} = 1 \tag{2.107d}$$

Let us denote the flexibility matrix as $[D_r]$ and stiffness matrix as $[k_r]$. The subscript r refers to the rotational degrees of freedom. To estimate the flexibility matrix for the beam element, assume the simply supported beam with the unit moment at the jth end as shown in Figure 2.28. The anticlockwise moment is balanced by the clockwise couple created by the forces. The bending moment diagram is also shown in the figure with tension at the top and compression at the bottom.

Let us replace the loading diagram with a conjugate beam, as shown in Figure 2.29. Taking a moment about A,

$$V_B = \left[\left\{\frac{1}{2}L_i\left(\frac{1}{EI}\right)\right\}\frac{1}{3}L_i\right]\frac{1}{L_i} = \frac{L_i}{6EI} \text{ (downward)} \tag{2.108a}$$

$$V_A = \left\{\frac{1}{2}L_i\left(\frac{1}{EI}\right)\right\} - \frac{L_i}{6EI} = \frac{L_i}{3EI} \text{ (upward)} \tag{2.108b}$$

The same procedure is followed for the other case to derive the following flexibility matrix.

$$D_r = \begin{bmatrix} \dfrac{L}{3EI} & -\dfrac{L}{6EI} \\ -\dfrac{L}{6EI} & \dfrac{L}{3EI} \end{bmatrix} \tag{2.109}$$

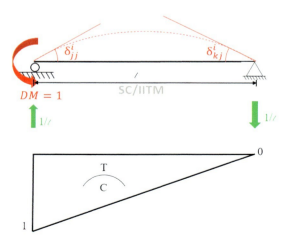

FIGURE 2.28 Simply supported beam with unit moment at the jth end.

Basic Design Guidelines

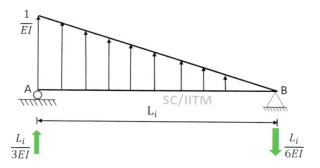

FIGURE 2.29 Conjugate beam.

$$k_r = [D_r]^{-1} = \frac{12(EI)^2}{L^2} \begin{bmatrix} \frac{L}{3EI} & \frac{L}{6EI} \\ \frac{L}{6EI} & \frac{L}{3EI} \end{bmatrix} = \begin{bmatrix} \frac{4EI}{L} & \frac{2EI}{L} \\ \frac{2EI}{L} & \frac{4EI}{L} \end{bmatrix} \quad (2.110)$$

Thus, from the above four rotational coefficients, the whole stiffness matrix can be derived below

$$K_i = \begin{bmatrix} \frac{4EI}{l} & \frac{2EI}{l} & \frac{6EI}{l^2} & -\frac{6EI}{l^2} \\ \frac{2EI}{l} & \frac{4EI}{l} & \frac{6EI}{l^2} & -\frac{6EI}{l^2} \\ \frac{6EI}{l^2} & \frac{6EI}{l^2} & \frac{12EI}{l^3} & -\frac{12EI}{l^3} \\ -\frac{6EI}{l^2} & -\frac{6EI}{l^2} & -\frac{12EI}{l^3} & \frac{12EI}{l^3} \end{bmatrix} \quad (2.111)$$

2.15 STABILITY FUNCTIONS UNDER AXIAL COMPRESSION

Consider a beam element, both ends fixed, as shown in Figure 2.30. It is important to note that the beam element is subjected to axial compressive load, P_a, as shown in the figure.

2.15.1 ROTATION FUNCTIONS

Let us now apply unit rotation at the jth end of the member to obtain the stiffness coefficients. Figure 2.31 shows the rotations and shear developed at both the member ends due to the unit rotation applied at the jth end of the member.

From Figure 2.31, and based on the expressions derived in the earlier sections, the following relationship holds good:

$$\frac{(k_{pp} + k_{qp})}{L_i} = k_{rp} = -k_{sp} \quad (2.112)$$

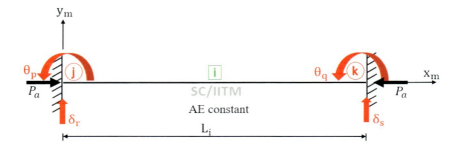

FIGURE 2.30 Fixed beam under axial compressive load.

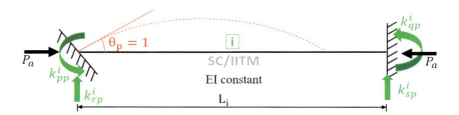

FIGURE 2.31 Unit rotation at the jth end of fixed beam.

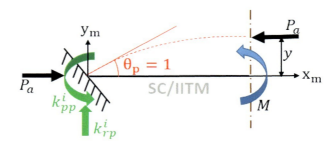

FIGURE 2.32 Free-body diagram under axial load and unit rotation at the jth end.

A free-body diagram, under the influence of the applied unit rotation and the axial compressive load, is shown in Figure 2.32.

With reference to Figure 2.32, and applying the differential equation, we get

$$EI \frac{d^2 y}{d^2 x} = M \qquad (2.113)$$

$$= -P_a(y) - k_{pp} + k_{rp}(x) \qquad (2.114)$$

Substituting Equation (2.112) in Equation (2.114), we get

Basic Design Guidelines

$$EI\frac{d^2y}{d^2x} = -P_a(y) - k_{pp} + \frac{(k_{pp} + k_{qp})}{L_i}(x) \tag{2.115}$$

In Equation (2.115), let us express the axial load (P_a as a function of Euler load P_E), given below

$$P_a = \varphi_i P_E \tag{2.116}$$

Substituting Equation (2.83) in Equation (2.116), we get

$$P_a = \frac{\pi^2 \varphi_i EI}{L^2} \text{ for } n = 1 \tag{2.117}$$

It is important to note that buckling is happening in the plane where unit rotation is applied. Therefore, Equation (2.115) is modified as below

$$EI\frac{d^2y}{d^2x} = -\frac{\pi^2 \varphi_i EI}{L_i^2}(y) - k_{pp} + \frac{(k_{pp} + k_{qp})}{L_i}(x) \tag{2.118}$$

Dividing by (EI), and rearranging the terms, we get

$$\frac{d^2y}{d^2x} + \frac{\pi^2 \varphi_i}{L_i^2}(y) = \frac{1}{EI}\left[(k_{pp} + k_{qp})\frac{x}{L_i} - (k_{pp})\right] \tag{2.119}$$

The general solution of Equation (2.119) is given by

$$y = A\sin\left[\frac{\alpha_i x}{L_i}\right] + B\cos\left[\frac{\alpha_i x}{L_i}\right] + \frac{L_i^2}{\alpha_i^2 EI}\left[(k_{pp} + k_{qp})\frac{x}{L_i} - (k_{pp})\right] \tag{2.120}$$

$$\alpha_i = \pi\sqrt{\varphi_i} \tag{2.121}$$

Applying the boundary conditions: y = 0 at x = 0, we get

$$B = \frac{L_i^2}{\alpha_i^2 EI} k_{pp} \tag{2.122}$$

Substituting another boundary condition y = 0 @ x = L in Equation (2.120), we get

$$0 = A\sin(\alpha_i) + B\cos(\alpha_i) + \frac{L_i^2}{\alpha_i^2 EI} k_{qp} \tag{2.123}$$

$$-\left[\frac{L_i^2}{\alpha_i^2 EI}\right](k_{qp}) = A\sin(\alpha_i) + B\cos(\alpha_i) \tag{2.124}$$

$$-\left[\frac{L_i^2}{\alpha_i^2 EI}\right](k_{qp}) - \left[\frac{L_i^2}{\alpha_i^2 EI}\right](k_{pp})\cos(\alpha_i) = A\sin(\alpha_i) \qquad (2.125)$$

$$-\left[\frac{L_i^2}{\alpha_i^2 EI}\right]\left[\frac{(k_{qp}) + (k_{pp})\cos(\alpha_i)}{\sin(\alpha_i)}\right] = A \qquad (2.126)$$

$$-\left[\frac{L_i^2}{\alpha_i^2 EI}\right]\left[k_{qp}\operatorname{cosec}(\alpha_i) + k_{pp}\cot(\alpha_i)\right] = A\sqrt{2} \qquad (2.127)$$

Substituting Equation (2.127) and Equation (2.122) in Equation (2.120), we get

$$\frac{\alpha_i^2 EI}{L_i^2} y = -\left[k_{pp}\cot(\alpha_i) + k_{qp}\operatorname{cosec}(\alpha_i)\right]\sin\left[\frac{\alpha_i x}{L_i}\right]$$
$$+ k_{pp}\cos\left[\frac{\alpha_i x}{L_i}\right] + (k_{pp} + k_{qp})\frac{x}{L_i} - k_{pp} \qquad (2.128)$$

Differentiating once, we get

$$\frac{\alpha_i^2 EI}{L_i^2}\frac{dy}{dx} = -\frac{\alpha_i}{L_i}\left[k_{pp}\cot(\alpha_i) + k_{qp}\operatorname{cosec}(\alpha_i)\right]\cos\left[\frac{\alpha_i x}{L_i}\right]$$
$$-\frac{\alpha_i}{L_i}k_{pp}\sin\left[\frac{\alpha_i x}{L_i}\right] + (k_{pp} + k_{qp})\frac{1}{L_i} \qquad (2.129)$$

$$\frac{\alpha_i^2 EI}{L_i}\frac{dy}{dx} = k_{pp}\left[1 - \alpha_i\sin\left[\frac{\alpha_i x}{L_i}\right] - \alpha_i\cot(\alpha_i)\cos\left[\frac{\alpha_i x}{L_i}\right]\right]$$
$$+ k_{qp}\left[1 - \alpha_i\operatorname{cosec}(\alpha_i)\cos\left[\frac{\alpha_i x}{L_i}\right]\right] \qquad (2.130)$$

Note: At x = 0, slope (dy/dx) is equal to θ_p, which is equal to unity in the present case
Applying the boundary condition (@ x = L, dy/dx = 0), we get

$$0 = k_{pp}\left[1 - \alpha_i\sin(\alpha_i) - \alpha_i\cot(\alpha_i)\cos(\alpha_i)\right]$$
$$+ k_{qp}\left[1 - \alpha_i\operatorname{cosec}(\alpha_i)\cos(\alpha_i)\right] \qquad (2.131)$$

$$0 = k_{pp} - k_{pp}\alpha_i\left[\sin(\alpha_i) + \frac{\cos(\alpha_i)^2}{\sin(\alpha_i)}\right] + k_{qp} - k_{qp}\alpha_i\cot(\alpha_i) \qquad (2.132)$$

The above equation is simplified to obtain the stiffness coefficient as below

Basic Design Guidelines

$$k_{qp} = \left[\frac{\alpha_i - \sin(\alpha_i)}{\sin(\alpha_i) - \alpha_i \cos(\alpha_i)} \right] k_{pp} \qquad (2.133)$$

Similarly, at x = 0, the slope is unity, which implies the following relationship

$$k_{pp} = \left[\frac{\alpha_i (\sin(\alpha_i) - \alpha_i \cos(\alpha_i))}{2(1 - \cos(\alpha_i)) - \alpha_i \sin(\alpha_i)} \right] \frac{EI}{L_i} \qquad (2.134)$$

By expressing the stiffness coefficients as a function of rotation functions (r_i and c_i), we get the following set of equations:

$$r_i = \frac{\alpha_i (\sin(\alpha_i) - \alpha_i \cos(\alpha_i))}{2(1 - \cos(\alpha_i)) - \alpha_i \sin(\alpha_i)} \qquad (2.135)$$

$$c_i = \frac{\alpha_i - \sin(\alpha_i)}{\sin(\alpha_i) - \alpha_i \cos(\alpha_i)} \qquad (2.136)$$

The above equations are termed *rotation functions* for compressive axial load cases.

2.15.2 ROTATION FUNCTIONS UNDER ZERO AXIAL LOAD

For a special case of *zero axial loads*, for which (α_i) becomes zero, one needs to apply L'Hospital's rule to obtain the limit of the quotient. L'Hospital's rule is briefly explained for the benefit of the readers.

Suppose $f(x)$ and $g(x)$ are differentiable functions and $g'(x) \neq 0$ on an open interval I, which contains (a) {except @ a}, then the following conditions apply:

Suppose, $\lim_{x \to a} f(x) = 0$; $\lim_{x \to a} g(x) = 0$ (or) $\lim_{x \to a} f(x) = \pm \infty$; $\lim_{x \to a} g(x) = \pm \infty$ then, it may reduce to a form (0/0) or (∞ / ∞). In such cases, the following equations hold good $\lim_{x \to 0} \frac{f(x)}{g(x)} = \lim_{x \to a} \frac{f'(x)}{g'(x)}$ if the limit of RHS exists. For example,

$$\lim_{x \to 0} \frac{e^x - 1}{x^2 + x} = \lim_{x \to 0} \frac{\frac{d}{dx}(e^x - 1)}{\frac{d}{dx}(x^2 + x)} = \lim_{x \to 0} \frac{e^x}{(2x + 1)} = 1 \qquad (2.137)$$

L'Hospital's rule uses derivatives to evaluate the limits involving indeterminate forms. It states that for indeterminate functions (or forms), where the unity tends to a form (0/0) or (∞ / ∞), the limit of that form is equal to the limit of the derivatives. L'Hospital's rule may be applied as many times as necessary until the function does not reduce the form (0/0) or (∞ / ∞).

Now, let us consider Equations (2.135–2.136). As (α_i) approaches zero, both $\left[\dfrac{f(\alpha_i)}{g(\alpha_i)}\right]$ approaches zero. Then, one can apply L'Hospital's rule as explained earlier, which will yield the following results:

$$r_{i@\,\varphi_i=0} = 4 \tag{2.138}$$

$$c_{i@\,\varphi_i=0} = 0.5 \tag{2.139}$$

Hence, at zero axial loads, the stiffness coefficients reduce to the conventional carry-over factors of the beam. Substituting Equations (2.135–2.136) in Equation (2.134–2.135), we get

$$k_{pp} = r_i \frac{EI}{L_i} \tag{2.140}$$

$$k_{qp} = c_i\, k_{pp} \tag{2.141}$$

$$k_{rp} = \frac{k_{pp} + k_{qp}}{L_i} = r_i \frac{EI}{L_i^2}(1 + c_i) \tag{2.142}$$

$$k_{sp} = -k_{rp} \tag{2.143}$$

By applying unit rotation at the kth end (please refer to Figure 2.33), another set of stiffness coefficients can be derived as follows

$$k_{pq} = c_i\, r_i \frac{EI}{L_i} \tag{2.144}$$

$$k_{qq} = r_i \frac{EI}{L_i} \tag{2.145}$$

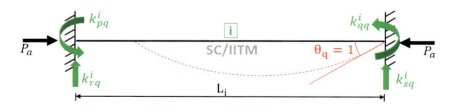

FIGURE 2.33 Unit rotation at the kth end under axial load.

Basic Design Guidelines

2.15.3 ROTATION FUNCTIONS UNDER AXIAL TENSILE LOAD

If the beam member is subjected to axial tensile load, then (φ_i) becomes negative. In that case, the following condition holds good.

$$\beta_i = \pi\sqrt{-\varphi_i} = i\pi\sqrt{\varphi_i} = i\alpha_i \quad (2.146)$$

Further,

$$\sin(\beta_i) = \frac{e^{i\beta_i} - e^{-i\beta_i}}{2i} \quad (2.147)$$

$$\cos(\beta_i) = \frac{e^{i\beta_i} + e^{-i\beta_i}}{2} \quad (2.148)$$

Substituting Equation (2.146) in Equations (2.147–2.148), we get

$$\sin(\beta_i) = \frac{e^{i^2\alpha_i} - e^{-i^2\alpha_i}}{2i} = \frac{e^{-\alpha_i} - e^{\alpha_i}}{2i} \quad (2.149)$$

$$\cos(\beta_i) = \frac{e^{i^2\alpha_i} + e^{-i^2\alpha_i}}{2} = \frac{e^{-\alpha_i} + e^{\alpha_i}}{2} \quad (2.150)$$

Rotation constants at the jth end can be obtained by substituting Equations (2.149–2.150) along with Equation (2.146), given below

$$r_i = \frac{i\alpha_i\left\{\left[\dfrac{e^{-\alpha_i} - e^{\alpha_i}}{2i}\right] - i\alpha_i\left[\dfrac{e^{-\alpha_i} + e^{\alpha_i}}{2}\right]\right\}}{2\left[1 - \dfrac{e^{-\alpha_i} + e^{\alpha_i}}{2}\right] - i\alpha_i\left[\dfrac{e^{-\alpha_i} - e^{\alpha_i}}{2i}\right]} \quad (2.151)$$

$$c_i = \frac{i\alpha_i - \left[\dfrac{e^{-\alpha_i} - e^{\alpha_i}}{2i}\right]}{\dfrac{e^{-\alpha_i} - e^{\alpha_i}}{2i} - i\alpha_i\left[\dfrac{e^{-\alpha_i} + e^{\alpha_i}}{2}\right]} \quad (2.152)$$

The above equations can be further simplified as follows

$$r_i = \frac{\alpha_i\left\{\left[\dfrac{e^{-\alpha_i} - e^{\alpha_i}}{2}\right] + \alpha_i\left[\dfrac{e^{-\alpha_i} + e^{\alpha_i}}{2}\right]\right\}}{2\left[1 - \dfrac{e^{-\alpha_i} + e^{\alpha_i}}{2}\right] - \alpha_i\left[\dfrac{e^{-\alpha_i} - e^{\alpha_i}}{2}\right]} \quad (2.153)$$

$$c_i = \frac{\alpha_{i+}\left[\dfrac{e^{-\alpha_i}-e^{\alpha_i}}{2}\right]}{-\left[\dfrac{e^{-\alpha_i}-e^{\alpha_i}}{2}\right]-\alpha_i\left[\dfrac{e^{-\alpha_i}+e^{\alpha_i}}{2}\right]} \qquad (2.154)$$

Using the following hyperbolic functions given below

$$\sinh \alpha_i = \frac{e^{\alpha_i}-e^{\alpha_i}}{2} \qquad (2.155)$$

$$\cosh \alpha_i = \frac{e^{\alpha_i}+e^{\alpha_i}}{2} \qquad (2.156)$$

By considering the absolute values of φ_i, Equations (2.153–2.154) will reduce to the following form

$$r_i = \frac{\alpha_i(\alpha_i \cosh \alpha_i - \sinh \alpha_i)}{2(1-\cosh \alpha_i)+\alpha_i \sinh \alpha_i} \qquad (2.157)$$

$$c_i = \frac{\alpha_i - \sinh \alpha_i}{\sinh \alpha_i - \alpha_i \cosh \alpha_i} \qquad (2.158)$$

Rotation constants of the member, under compressive and tensile axial loads, as derived above, are plotted for a wide range of values of φ. Table 2.6 also gives these values at closer intervals, useful in calculating the critical buckling load.

2.15.4 Translation Function under Axial Compressive Load

Figure 2.34 shows unit translation at the jth end of the beam under axial compressive load.

By taking moments of all forces about the kth end of the member, we get

$$P_a(1) - k_{pr} + k_{rr}(L_i) - k_{qr} = 0 \qquad (2.159)$$

$$k_{rr} = \frac{P_a(1) - (k_{pr} + k_{qr})}{L_i} \qquad (2.160)$$

$$k_{sr} = -k_{rr} \qquad (2.161)$$

With reference to Figure 2.34, the following expressions can be written. Due to the end rotations, we get the end moments as below

$$k_{pr} = r_i\left(\frac{EI}{L_i}\right)\left(\frac{1}{L_i}\right) + r_i c_i\left(\frac{EI}{L_i}\right)\left(\frac{1}{L_i}\right) \qquad (2.162)$$

Basic Design Guidelines

TABLE 2.6
Stability Functions (negative sign indicates tensile axial load)

Phi	r	c	t	Phi	r	c	t
−10	11.1864	0.1118	4.9678	−6	9.0436	0.1483	3.8512
−9.9	11.1382	0.1124	4.9429	−5.9	8.9830	0.1497	3.8192
−9.8	11.0897	0.1131	4.9179	−5.8	8.9220	0.1511	3.7869
−9.7	11.0410	0.1137	4.8928	−5.7	8.8605	0.1526	3.7544
−9.6	10.9921	0.1144	4.8675	−5.6	8.7986	0.1541	3.7216
−9.5	10.9430	0.1150	4.8421	−5.5	8.7362	0.1556	3.6885
−9.4	10.8936	0.1157	4.8166	−5.4	8.6733	0.1572	3.6551
−9.3	10.8440	0.1164	4.7909	−5.3	8.6100	0.1588	3.6215
−9.2	10.7941	0.1171	4.7652	−5.2	8.5461	0.1604	3.5875
−9.1	10.7440	0.1178	4.7392	−5.1	8.4818	0.1621	3.5532
−9	10.6937	0.1185	4.7131	−5	8.4169	0.1639	3.5187
−8.9	10.6431	0.1193	4.6869	−4.9	8.3515	0.1657	3.4838
−8.8	10.5922	0.1200	4.6606	−4.8	8.2855	0.1676	3.4485
−8.7	10.5411	0.1208	4.6341	−4.7	8.2190	0.1695	3.4129
−8.6	10.4897	0.1215	4.6074	−4.6	8.1520	0.1715	3.3770
−8.5	10.4380	0.1223	4.5806	−4.5	8.0843	0.1735	3.3407
−8.4	10.3860	0.1231	4.5536	−4.4	8.0161	0.1757	3.3040
−8.3	10.3338	0.1239	4.5265	−4.3	7.9472	0.1778	3.2669
−8.2	10.2813	0.1248	4.4992	−4.2	7.8777	0.1801	3.2295
−8.1	10.2285	0.1256	4.4717	−4.1	7.8075	0.1824	3.1916
−8	10.1754	0.1265	4.4441	−4	7.7367	0.1848	3.1533
−7.9	10.1220	0.1274	4.4163	−3.9	7.6652	0.1873	3.1146
−7.8	10.0683	0.1283	4.3884	−3.8	7.5930	0.1899	3.0755
−7.7	10.0143	0.1292	4.3602	−3.7	7.5201	0.1926	3.0359
−7.6	9.9600	0.1301	4.3319	−3.6	7.4465	0.1954	2.9958
−7.5	9.9054	0.1311	4.3034	−3.5	7.3721	0.1983	2.9552
−7.4	9.8504	0.1321	4.2747	−3.4	7.2969	0.2013	2.9141
−7.3	9.7951	0.1331	4.2458	−3.3	7.2209	0.2044	2.8725
−7.2	9.7395	0.1341	4.2167	−3.2	7.1441	0.2076	2.8304
−7.1	9.6835	0.1351	4.1875	−3.1	7.0664	0.2110	2.7877
−7	9.6272	0.1362	4.1580	−3	6.9878	0.2145	2.7444
−6.9	9.5706	0.1373	4.1283	−2.9	6.9084	0.2182	2.7005
−6.8	9.5136	0.1384	4.0984	−2.8	6.8280	0.2220	2.6560
−6.7	9.4562	0.1395	4.0683	−2.7	6.7466	0.2260	2.6108
−6.6	9.3984	0.1407	4.0380	−2.6	6.6642	0.2302	2.5650
−6.5	9.3403	0.1419	4.0074	−2.5	6.5808	0.2346	2.5185
−6.4	9.2817	0.1431	3.9766	−2.4	6.4963	0.2392	2.4712
−6.3	9.2228	0.1444	3.9456	−2.3	6.4107	0.2440	2.4232
−6.2	9.1635	0.1457	3.9144	−2.2	6.3239	0.2491	2.3744
−6.1	9.1037	0.1470	3.8829	−2.1	6.2360	0.2544	2.3248
−2	6.1468	0.2600	2.2743	0.2	3.7297	0.5550	0.8298

(Continued)

TABLE 2.6 (CONTINUED)
Stability Functions (negative sign indicates tensile axial load)

Phi	r	c	t	Phi	r	c	t
−1.9	6.0564	0.2659	2.2230	0.21	3.7158	0.5581	0.8210
−1.8	5.9645	0.2721	2.1706	0.22	3.7019	0.5612	0.8122
−1.7	5.8714	0.2787	2.1174	0.23	3.6879	0.5644	0.8033
−1.6	5.7767	0.2857	2.0631	0.24	3.6739	0.5676	0.7943
−1.5	5.6806	0.2931	2.0077	0.25	3.6598	0.5708	0.7854
−1.4	5.5828	0.3010	1.9512	0.26	3.6457	0.5741	0.7764
−1.3	5.4835	0.3094	1.8935	0.27	3.6315	0.5774	0.7674
−1.2	5.3824	0.3183	1.8346	0.28	3.6174	0.5807	0.7584
−1.1	5.2795	0.3279	1.7743	0.29	3.6031	0.5841	0.7493
−1	5.1748	0.3381	1.7127	0.3	3.5889	0.5875	0.7402
−0.9	5.0681	0.3490	1.6496	0.31	3.5746	0.5910	0.7310
−0.8	4.9593	0.3608	1.5850	0.32	3.5602	0.5945	0.7218
−0.7	4.8483	0.3735	1.5187	0.33	3.5458	0.5981	0.7126
−0.6	4.7351	0.3872	1.4508	0.34	3.5314	0.6017	0.7034
−0.5	4.6194	0.4021	1.3809	0.35	3.5169	0.6053	0.6941
−0.4	4.5013	0.4183	1.3092	0.36	3.5024	0.6090	0.6848
−0.3	4.3804	0.4360	1.2354	0.37	3.4878	0.6127	0.6754
−0.2	4.2567	0.4553	1.1593	0.38	3.4732	0.6165	0.6660
−0.1	4.1299	0.4765	1.0809	0.39	3.4586	0.6203	0.6566
0	4.0000	0.5000	1.0000	0.4	3.4439	0.6242	0.6471
0.01	3.9868	0.5025	0.9918	0.41	3.4292	0.6281	0.6376
0.02	3.9736	0.5050	0.9835	0.42	3.4144	0.6321	0.6281
0.03	3.9604	0.5075	0.9752	0.43	3.3995	0.6361	0.6185
0.04	3.9471	0.5101	0.9669	0.44	3.3847	0.6402	0.6089
0.05	3.9338	0.5127	0.9585	0.45	3.3698	0.6443	0.5992
0.06	3.9204	0.5153	0.9502	0.46	3.3548	0.6485	0.5895
0.07	3.9070	0.5179	0.9418	0.47	3.3398	0.6528	0.5798
0.08	3.8936	0.5206	0.9333	0.48	3.3247	0.6571	0.5701
0.09	3.8802	0.5233	0.9249	0.49	3.3096	0.6614	0.5603
0.1	3.8667	0.5260	0.9164	0.5	3.2945	0.6659	0.5504
0.11	3.8531	0.5288	0.9078	0.51	3.2793	0.6703	0.5405
0.12	3.8396	0.5316	0.8993	0.52	3.2640	0.6749	0.5306
0.13	3.8260	0.5344	0.8907	0.53	3.2487	0.6795	0.5206
0.14	3.8123	0.5372	0.8821	0.54	3.2334	0.6841	0.5106
0.15	3.7987	0.5401	0.8735	0.55	3.2180	0.6889	0.5006
0.16	3.7849	0.5430	0.8648	0.56	3.2025	0.6937	0.4905
0.17	3.7712	0.5460	0.8561	0.57	3.1870	0.6985	0.4804
0.18	3.7574	0.5490	0.8474	0.58	3.1715	0.7035	0.4702
0.19	3.7436	0.5520	0.8386	0.59	3.1559	0.7085	0.4600
0.6	3.1403	0.7136	0.4498	1	2.4674	1.0000	0.0000
0.61	3.1246	0.7187	0.4395	1.01	2.4493	1.0101	−0.0124

(Continued)

Basic Design Guidelines

TABLE 2.6 (CONTINUED)
Stability Functions (negative sign indicates tensile axial load)

Phi	r	c	t	Phi	r	c	t
0.62	3.1088	0.7239	0.4291	**1.02**	2.4311	1.0204	−0.0248
0.63	3.0930	0.7292	0.4187	**1.03**	2.4128	1.0309	−0.0373
0.64	3.0771	0.7346	0.4083	**1.04**	2.3944	1.0416	−0.0498
0.65	3.0612	0.7401	0.3978	**1.05**	2.3760	1.0526	−0.0625
0.66	3.0453	0.7456	0.3873	**1.06**	2.3575	1.0638	−0.0752
0.67	3.0293	0.7513	0.3768	**1.07**	2.3389	1.0752	−0.0879
0.68	3.0132	0.7570	0.3661	**1.08**	2.3202	1.0868	−0.1007
0.69	2.9971	0.7628	0.3555	**1.09**	2.3015	1.0987	−0.1136
0.7	2.9809	0.7687	0.3448	**1.1**	2.2827	1.1109	−0.1266
0.71	2.9646	0.7746	0.3340	**1.11**	2.2638	1.1233	−0.1396
0.72	2.9484	0.7807	0.3233	**1.12**	2.2448	1.1360	−0.1527
0.73	2.9320	0.7869	0.3124	**1.13**	2.2258	1.1490	−0.1658
0.74	2.9156	0.7932	0.3015	**1.14**	2.2066	1.1623	−0.1790
0.75	2.8991	0.7995	0.2906	**1.15**	2.1874	1.1759	−0.1923
0.76	2.8826	0.8060	0.2796	**1.16**	2.1681	1.1898	−0.2057
0.77	2.8660	0.8126	0.2686	**1.17**	2.1487	1.2040	−0.2192
0.78	2.8494	0.8193	0.2575	**1.18**	2.1293	1.2185	−0.2327
0.79	2.8327	0.8261	0.2463	**1.19**	2.1097	1.2335	−0.2463
0.8	2.8159	0.8330	0.2351	**1.2**	2.0901	1.2487	−0.2599
0.81	2.7991	0.8400	0.2239	**1.21**	2.0704	1.2644	−0.2737
0.82	2.7822	0.8472	0.2126	**1.22**	2.0506	1.2804	−0.2875
0.83	2.7653	0.8544	0.2013	**1.23**	2.0307	1.2968	−0.3014
0.84	2.7483	0.8618	0.1899	**1.24**	2.0107	1.3137	−0.3153
0.85	2.7312	0.8693	0.1784	**1.25**	1.9906	1.3309	−0.3294
0.86	2.7141	0.8770	0.1669	**1.26**	1.9705	1.3487	−0.3435
0.87	2.6969	0.8848	0.1554	**1.27**	1.9502	1.3669	−0.3577
0.88	2.6797	0.8927	0.1438	**1.28**	1.9299	1.3855	−0.3720
0.89	2.6623	0.9008	0.1321	**1.29**	1.9094	1.4047	−0.3864
0.9	2.6450	0.9090	0.1204	**1.3**	1.8889	1.4244	−0.4009
0.91	2.6275	0.9173	0.1086	**1.31**	1.8683	1.4447	−0.4154
0.92	2.6100	0.9258	0.0968	**1.32**	1.8476	1.4655	−0.4300
0.93	2.5924	0.9345	0.0849	**1.33**	1.8267	1.4869	−0.4447
0.94	2.5748	0.9433	0.0729	**1.34**	1.8058	1.5089	−0.4595
0.95	2.5570	0.9523	0.0609	**1.35**	1.7848	1.5316	−0.4744
0.96	2.5392	0.9615	0.0489	**1.36**	1.7637	1.5549	−0.4894
0.97	2.5214	0.9709	0.0367	**1.37**	1.7425	1.5790	−0.5044
0.98	2.5035	0.9804	0.0246	**1.38**	1.7212	1.6038	−0.5196
0.99	2.4855	0.9901	0.0123	**1.39**	1.6997	1.6293	−0.5348
1.4	1.6782	1.6557	−0.5502	**1.8**	0.7170	4.4969	−1.2537
1.41	1.6566	1.6828	−0.5656	**1.81**	0.6900	4.6924	−1.2739
1.42	1.6348	1.7109	−0.5811	**1.82**	0.6629	4.9051	−1.2944

(*Continued*)

TABLE 2.6 (CONTINUED)
Stability Functions (negative sign indicates tensile axial load)

Phi	r	c	t	Phi	r	c	t
1.43	1.6130	1.7399	−0.5967	1.83	0.6356	5.1377	−1.3149
1.44	1.5910	1.7699	−0.6125	1.84	0.6081	5.3929	−1.3357
1.45	1.5690	1.8009	−0.6283	1.85	0.5804	5.6742	−1.3566
1.46	1.5468	1.8329	−0.6442	1.86	0.5526	5.9859	−1.3776
1.47	1.5245	1.8661	−0.6602	1.87	0.5246	6.3331	−1.3989
1.48	1.5021	1.9005	−0.6763	1.88	0.4964	6.7223	−1.4202
1.49	1.4796	1.9361	−0.6925	1.89	0.4680	7.1616	−1.4418
1.5	1.4570	1.9731	−0.7089	1.9	0.4394	7.6612	−1.4635
1.51	1.4342	2.0114	−0.7253	1.91	0.4107	8.2345	−1.4855
1.52	1.4114	2.0512	−0.7418	1.92	0.3817	8.8990	−1.5076
1.53	1.3884	2.0926	−0.7585	1.93	0.3526	9.6785	−1.5298
1.54	1.3653	2.1356	−0.7752	1.94	0.3232	10.6056	−1.5523
1.55	1.3420	2.1804	−0.7921	1.95	0.2937	11.7264	−1.5749
1.56	1.3187	2.2271	−0.8090	1.96	0.2639	13.1087	−1.5978
1.57	1.2952	2.2757	−0.8261	1.97	0.2339	14.8562	−1.6208
1.58	1.2716	2.3264	−0.8433	1.98	0.2038	17.1355	−1.6440
1.59	1.2479	2.3794	−0.8606	1.99	0.1734	20.2327	−1.6674
1.6	1.2240	2.4348	−0.8781	2	0.1428	24.6841	−1.6910
1.61	1.2000	2.4927	−0.8956	2.01	0.1120	31.6264	−1.7148
1.62	1.1759	2.5534	−0.9133	2.02	0.0809	43.9616	−1.7388
1.63	1.1516	2.6170	−0.9311	2.03	0.0497	71.9627	−1.7631
1.64	1.1272	2.6838	−0.9490	2.04	0.0182	197.3863	−1.7875
1.65	1.1027	2.7540	−0.9670	2.05	−0.0135	−267.2161	−1.8122
1.66	1.0780	2.8278	−0.9852	2.06	−0.0455	−79.8138	−1.8371
1.67	1.0532	2.9056	−1.0035	2.07	−0.0777	−46.9612	−1.8622
1.68	1.0282	2.9877	−1.0219	2.08	−0.1101	−33.2921	−1.8875
1.69	1.0031	3.0744	−1.0405	2.09	−0.1428	−25.8013	−1.9130
1.7	0.9779	3.1662	−1.0592	2.1	−0.1757	−21.0722	−1.9388
1.71	0.9525	3.2635	−1.0780	2.11	−0.2089	−17.8154	−1.9648
1.72	0.9270	3.3667	−1.0969	2.12	−0.2423	−15.4361	−1.9911
1.73	0.9013	3.4766	−1.1160	2.13	−0.2760	−13.6217	−2.0176
1.74	0.8754	3.5936	−1.1353	2.14	−0.3099	−12.1925	−2.0444
1.75	0.8494	3.7187	−1.1546	2.15	−0.3441	−11.0376	−2.0714
1.76	0.8233	3.8524	−1.1741	2.16	−0.3786	−10.0850	−2.0986
1.77	0.7969	3.9960	−1.1938	2.17	−0.4134	−9.2858	−2.1261
1.78	0.7705	4.1504	−1.2136	2.18	−0.4485	−8.6059	−2.1539
1.79	0.7438	4.3169	−1.2336	2.19	−0.4838	−8.0203	−2.1820
2.2	−0.5194	−7.5107	−2.2103	2.6	−2.2490	−2.2312	−3.6335
2.21	−0.5553	−7.0632	−2.2389	2.61	−2.3020	−2.1959	−3.6785
2.22	−0.5916	−6.6673	−2.2678	2.62	−2.3557	−2.1618	−3.7241
2.23	−0.6281	−6.3143	−2.2970	2.63	−2.4100	−2.1289	−3.7704

(Continued)

TABLE 2.6 (CONTINUED)
Stability Functions (negative sign indicates tensile axial load)

Phi	r	c	t	Phi	r	c	t
2.24	−0.6649	−5.9978	−2.3264	2.64	−2.4650	−2.0971	−3.8172
2.25	−0.7020	−5.7124	−2.3562	2.65	−2.5206	−2.0665	−3.8647
2.26	−0.7395	−5.4537	−2.3863	2.66	−2.5769	−2.0369	−3.9128
2.27	−0.7773	−5.2181	−2.4166	2.67	−2.6339	−2.0082	−3.9616
2.28	−0.8154	−5.0027	−2.4473	2.68	−2.6915	−1.9805	−4.0111
2.29	−0.8538	−4.8050	−2.4783	2.69	−2.7499	−1.9538	−4.0613
2.3	−0.8926	−4.6230	−2.5096	2.7	−2.8091	−1.9278	−4.1122
2.31	−0.9318	−4.4547	−2.5413	2.71	−2.8690	−1.9027	−4.1639
2.32	−0.9713	−4.2988	−2.5733	2.72	−2.9296	−1.8784	−4.2162
2.33	−1.0111	−4.1540	−2.6056	2.73	−2.9911	−1.8548	−4.2694
2.34	−1.0513	−4.0190	−2.6383	2.74	−3.0533	−1.8319	−4.3233
2.35	−1.0919	−3.8930	−2.6713	2.75	−3.1164	−1.8097	−4.3781
2.36	−1.1328	−3.7750	−2.7047	2.76	−3.1803	−1.7882	−4.4336
2.37	−1.1742	−3.6644	−2.7384	2.77	−3.2451	−1.7673	−4.4900
2.38	−1.2159	−3.5604	−2.7725	2.78	−3.3108	−1.7470	−4.5473
2.39	−1.2580	−3.4626	−2.8070	2.79	−3.3774	−1.7273	−4.6055
2.4	−1.3006	−3.3703	−2.8419	2.8	−3.4449	−1.7081	−4.6645
2.41	−1.3435	−3.2831	−2.8772	2.81	−3.5133	−1.6895	−4.7245
2.42	−1.3869	−3.2006	−2.9129	2.82	−3.5828	−1.6714	−4.7854
2.43	−1.4307	−3.1225	−2.9490	2.83	−3.6532	−1.6538	−4.8474
2.44	−1.4749	−3.0484	−2.9855	2.84	−3.7246	−1.6366	−4.9103
2.45	−1.5196	−2.9780	−3.0224	2.85	−3.7972	−1.6200	−4.9742
2.46	−1.5647	−2.9111	−3.0598	2.86	−3.8707	−1.6038	−5.0392
2.47	−1.6103	−2.8473	−3.0976	2.87	−3.9454	−1.5880	−5.1053
2.48	−1.6563	−2.7865	−3.1359	2.88	−4.0213	−1.5726	−5.1725
2.49	−1.7028	−2.7286	−3.1746	2.89	−4.0983	−1.5576	−5.2409
2.5	−1.7499	−2.6732	−3.2138	2.9	−4.1765	−1.5430	−5.3104
2.51	−1.7974	−2.6202	−3.2534	2.91	−4.2559	−1.5288	−5.3811
2.52	−1.8454	−2.5695	−3.2936	2.92	−4.3366	−1.5149	−5.4531
2.53	−1.8939	−2.5210	−3.3342	2.93	−4.4186	−1.5014	−5.5263
2.54	−1.9430	−2.4744	−3.3754	2.94	−4.5019	−1.4882	−5.6008
2.55	−1.9926	−2.4298	−3.4170	2.95	−4.5866	−1.4754	−5.6767
2.56	−2.0427	−2.3869	−3.4592	2.96	−4.6727	−1.4628	−5.7540
2.57	−2.0934	−2.3457	−3.5020	2.97	−4.7602	−1.4506	−5.8327
2.58	−2.1447	−2.3061	−3.5452	2.98	−4.8492	−1.4387	−5.9129
2.59	−2.1965	−2.2680	−3.5891	2.99	−4.9398	−1.4270	−5.9946
3	−5.0320	−1.4157	−6.0778	3.4	−10.9082	−1.1223	−11.5754
3.01	−5.1258	−1.4046	−6.1627	3.41	−11.1485	−1.1179	−11.8055
3.02	−5.2212	−1.3937	−6.2491	3.42	−11.3965	−1.1135	−12.0435
3.03	−5.3184	−1.3832	−6.3373	3.43	−11.6528	−1.1093	−12.2897
3.04	−5.4174	−1.3728	−6.4273	3.44	−11.9178	−1.1052	−12.5445

(*Continued*)

TABLE 2.6 (CONTINUED)
Stability Functions (negative sign indicates tensile axial load)

Phi	r	c	t	Phi	r	c	t
3.05	−5.5182	−1.3628	−6.5191	3.45	−12.1919	−1.1011	−12.8084
3.06	−5.6209	−1.3529	−6.6127	3.46	−12.4757	−1.0972	−13.0820
3.07	−5.7256	−1.3433	−6.7083	3.47	−12.7697	−1.0933	−13.3657
3.08	−5.8323	−1.3339	−6.8058	3.48	−13.0745	−1.0896	−13.6602
3.09	−5.9410	−1.3247	−6.9055	3.49	−13.3907	−1.0859	−13.9661
3.1	−6.0519	−1.3157	−7.0072	3.5	−13.7190	−1.0824	−14.2840
3.11	−6.1651	−1.3069	−7.1111	3.51	−14.0601	−1.0789	−14.6148
3.12	−6.2805	−1.2983	−7.2174	3.52	−14.4149	−1.0755	−14.9591
3.13	−6.3984	−1.2899	−7.3259	3.53	−14.7842	−1.0722	−15.3180
3.14	−6.5186	−1.2817	−7.4369	3.54	−15.1689	−1.0690	−15.6922
3.15	−6.6415	−1.2737	−7.5505	3.55	−15.5702	−1.0659	−16.0829
3.16	−6.7669	−1.2659	−7.6666	3.56	−15.9890	−1.0628	−16.4912
3.17	−6.8951	−1.2582	−7.7854	3.57	−16.4267	−1.0599	−16.9183
3.18	−7.0262	−1.2508	−7.9071	3.58	−16.8845	−1.0570	−17.3655
3.19	−7.1601	−1.2434	−8.0316	3.59	−17.3640	−1.0542	−17.8343
3.2	−7.2971	−1.2363	−8.1592	3.6	−17.8668	−1.0514	−18.3264
3.21	−7.4373	−1.2293	−8.2899	3.61	−18.3946	−1.0488	−18.8435
3.22	−7.5807	−1.2224	−8.4238	3.62	−18.9494	−1.0462	−19.3875
3.23	−7.7276	−1.2157	−8.5611	3.63	−19.5335	−1.0438	−19.9608
3.24	−7.8779	−1.2092	−8.7019	3.64	−20.1492	−1.0413	−20.5657
3.25	−8.0320	−1.2028	−8.8464	3.65	−20.7993	−1.0390	−21.2049
3.26	−8.1899	−1.1965	−8.9947	3.66	−21.4868	−1.0367	−21.8814
3.27	−8.3518	−1.1904	−9.1469	3.67	−22.2150	−1.0345	−22.5987
3.28	−8.5178	−1.1844	−9.3032	3.68	−22.9879	−1.0324	−23.3606
3.29	−8.6881	−1.1786	−9.4639	3.69	−23.8096	−1.0304	−24.1713
3.3	−8.8629	−1.1729	−9.6290	3.7	−24.6852	−1.0284	−25.0358
3.31	−9.0425	−1.1673	−9.7988	3.71	−25.6201	−1.0265	−25.9597
3.32	−9.2269	−1.1618	−9.9734	3.72	−26.6208	−1.0247	−26.9492
3.33	−9.4165	−1.1565	−10.1532	3.73	−27.6945	−1.0229	−28.0117
3.34	−9.6114	−1.1512	−10.3382	3.74	−28.8496	−1.0212	−29.1556
3.35	−9.8119	−1.1461	−10.5289	3.75	−30.0960	−1.0196	−30.3907
3.36	−10.0183	−1.1412	−10.7253	3.76	−31.4449	−1.0180	−31.7284
3.37	−10.2308	−1.1363	−10.9279	3.77	−32.9098	−1.0165	−33.1820
3.38	−10.4497	−1.1315	−11.1369	3.78	−34.5066	−1.0151	−34.7673
3.39	−10.6755	−1.1269	−11.3526	3.79	−36.2539	−1.0138	−36.5033
3.8	−38.1745	−1.0125	−38.4124				
3.81	−40.2956	−1.0112	−40.5220				
3.82	−42.6506	−1.0101	−42.8655				
3.83	−45.2809	−1.0090	−45.4842				
3.84	−48.2381	−1.0079	−48.4298				
3.85	−51.5874	−1.0070	−51.7674				

(Continued)

Basic Design Guidelines

TABLE 2.6 (CONTINUED)
Stability Functions (negative sign indicates tensile axial load)

Phi	r	c	t
3.86	−55.4130	−1.0061	−55.5813
3.87	−59.8247	−1.0052	−59.9813
3.88	−64.9691	−1.0045	−65.1139
3.89	−71.0459	−1.0037	−71.1789
3.9	−78.3349	−1.0031	−78.4560
3.91	−87.2400	−1.0025	−87.3492
3.92	−98.3675	−1.0020	−98.4647
3.93	−112.6696	−1.0015	−112.7548
3.94	−131.7337	−1.0011	−131.8069
3.95	−158.4168	−1.0008	−158.4779
3.96	−198.4334	−1.0005	−198.4823
3.97	−265.1166	−1.0003	−265.1534
3.98	−398.4666	−1.0001	−398.4912
3.99	−798.4833	−1.0000	−798.4956
4	65535.0000	−1.0000	INF

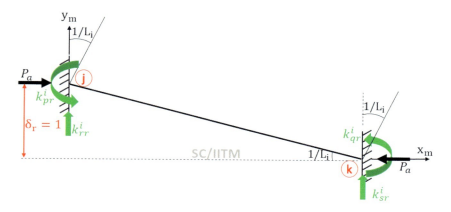

FIGURE 2.34 Unit translation at the jth end of fixed beam under axial load.

$$k_{qr} = r_i c_i \left(\frac{EI}{L_i}\right)\left(\frac{1}{L_i}\right) + r_i \left(\frac{EI}{L_i}\right)\left(\frac{1}{L_i}\right) \quad (2.163)$$

$$k_{qr} = k_{pr} = r_i(1+c_i)\left(\frac{EI}{L_i^2}\right) \quad (2.164)$$

Substituting Equations (2.162–2.164) in Equation (2.160), we get the expressions for end shear as below

$$k_{rr} = -k_{sr} = 2r_i(1+c_i)\left(\frac{EI}{L_i^3}\right) - \frac{P_a}{L_i} \qquad (2.165)$$

For zero axial load, the equation for the end shear reduces to the following form

$$k_{rr} = -k_{sr} = 2r_i(1+c_i)\left(\frac{EI}{L_i^3}\right) \qquad (2.166)$$

Hence, to make a general form, in both the presence and absence of axial load, the following function holds good to express end shear

$$k_{rr} = -k_{sr} = t_i\left[2r_i(1+c_i)\left(\frac{EI}{L_i^3}\right)\right] \qquad (2.167)$$

where t_i is known as the translation function. To obtain the translation function, let us now equate the axial load function as discussed below

We know the following relationship

$$P_a = \frac{\pi^2 \varphi_i EI}{L_i^2} \text{ for } n=1$$

Now, equating (2.167) and (2.165), we get

$$t_i\left[2r_i(1+c_i)\left(\frac{EI}{L_i^3}\right)\right] = 2r_i(1+c_i)\left(\frac{EI}{L_i^3}\right) - \frac{P_a}{L_i} \qquad (2.168)$$

$$t_i = 1 - \left\{\left[\frac{P_a}{2r_i(1+c_i)}\right]\left[\frac{L_i^2}{EI}\right]\right\} \qquad (2.169)$$

Substituting for P_a, we get

$$t_i = 1 - \frac{\pi^2 \varphi_i}{2r_i(1+c_i)} \qquad (2.170)$$

Similarly, by inducing unit rotation at the kth end of the beam, other coefficients, (k_{ps}, k_{qs}, k_{rs}, k_{ss}) will be generated, as given below

$$k_{ps} = k_{rs} = -r_i(1+c_i)\left(\frac{EI}{L_i^2}\right) \qquad (2.171)$$

$$k_{rs} = -k_{ss} = -2t_i\, r_i(1+c_i)\left(\frac{EI}{L_i^3}\right) \qquad (2.172)$$

Basic Design Guidelines

By neglecting the axial deformation, the member stiffness matrix of a fixed beam under axial load (either compressive or tensile) is given by the following form

$$[K_i] = EI \begin{bmatrix} \dfrac{r_i}{L_i} & \dfrac{c_i r_i}{L_i} & \dfrac{r_i(1+c_i)}{L_i^2} & -\dfrac{r_i(1+c_i)}{L_i^2} \\ \dfrac{c_i r_i}{L_i} & \dfrac{r_i}{L_i} & \dfrac{r_i(1+c_i)}{L_i^2} & -\dfrac{r_i(1+c_i)}{L_i^2} \\ \dfrac{r_i(1+c_i)}{L_i^2} & \dfrac{r_i(1+c_i)}{L_i^2} & \dfrac{2t_i\, r_i(1+c_i)}{L_i^3} & -\dfrac{2t_i\, r_i(1+c_i)}{L_i^3} \\ -\dfrac{r_i(1+c_i)}{L_i^2} & -\dfrac{r_i(1+c_i)}{L_i^2} & -\dfrac{2t_i\, r_i(1+c_i)}{L_i^3} & \dfrac{2t_i\, r_i(1+c_i)}{L_i^3} \end{bmatrix} \quad (2.173)$$

2.16 LATERAL LOAD FUNCTIONS UNDER UNIFORMLY DISTRIBUTED LOAD

In this section, we will derive the lateral load's functions for the fixed beam loaded under uniformly distributed load in the presence of the axial load. Consider a fixed beam, loaded as shown in Figure 2.35. Here, M_p does not refer to the plastic moment of resistance but indicates the moment at the jth end under the rotation of θ_p. Alternatively, M_p also refers to the fixed end moment (FeM$_p$).

Consider the free-body diagram, as shown in Figure 2.36.

Under the fixed beam subjected to uniformly distributed load, end rotation and shear are developed as shown in Figures 2.35 and 2.36. The following equation holds good.

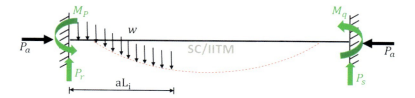

FIGURE 2.35 Fixed beam under uniformly distributed load and axial compressive load.

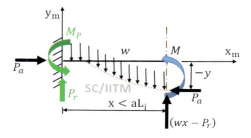

FIGURE 2.36 Free-body diagram ($x < aL_i$).

$$EI\frac{d^2y}{dx^2} = M \quad \text{(for } x < aL_i\text{)} \tag{2.174}$$

$$= -\left[M_p - P_a(-y) - P_r(x) + (-w)\frac{x^2}{2}\right] \tag{2.175}$$

The general solution to Equation (2.175) is given by

$$y = A\sin\left(\frac{\alpha_i x}{L_i}\right) + B\cos\left(\frac{\alpha_i x}{L_i}\right) - \frac{L_i^2}{\alpha_i^2 EI}\left[M_p + w\left(\frac{L_i^2}{\alpha_i^2}\right) - \frac{wx}{2}\left(\frac{2P_r}{w} + x\right)\right] \tag{2.176}$$

Differentiating once, we get

$$\frac{dy}{dx} = \frac{\alpha_i}{L_i}A\cos\left(\frac{\alpha_i x}{L_i}\right) - \frac{\alpha_i}{L_i}B\sin\left(\frac{\alpha_i x}{L_i}\right) + \frac{L_i^2}{\alpha_i^2 EI}(P_r + wx) \tag{2.177}$$

Applying the Boundary conditions:

$$@\ x = 0, \begin{cases} y = 0 \\ \dfrac{dy}{dx} = 0 \end{cases} \tag{2.178}$$

$$B = \frac{L_i^2}{\alpha_i^2 EI}\left[M_p + w\left(\frac{L_i^2}{\alpha_i^2}\right)\right] \tag{2.179}$$

$$A = -\left[\frac{L_i^3}{\alpha_i^3 EI}P_r\right] \tag{2.180}$$

Substituting Equations (2.179–2.180) in Equation (2.176), we get

$$\left(\frac{\alpha_i^2 EI}{L_i^2}\right)y = -\left[\frac{L_i}{\alpha_i}P_r\right]\sin\left(\frac{\alpha_i x}{L_i}\right) + \left[M_p + w\left(\frac{L_i^2}{\alpha_i^2}\right)\right]\cos\left(\frac{\alpha_i x}{L_i}\right)$$
$$-M_p - w\left(\frac{L_i^2}{\alpha_i^2}\right) + \frac{wx}{2}\left(\frac{2P_r}{w} + x\right) \quad \text{for } 0 \le x \le aL_i \tag{2.181}$$

Substituting Equation (2.180) in Equation (2.177), we get

$$\left(\frac{\alpha_i^2 EI}{L_i^2}\right)\frac{dy}{dx} = -\left(\frac{\alpha_i}{L_i}\right)P_r\cos\left(\frac{\alpha_i x}{L_i}\right) - \frac{\alpha_i}{L_i}\left[M_p + w\left(\frac{L_i^2}{\alpha_i^2}\right)\right]\sin\left(\frac{\alpha_i x}{L_i}\right)$$
$$+ (P_r + wx) \quad \text{for } 0 \le x \le aL_i \tag{2.182}$$

For the position of the uniformly distributed load ($x > aL_i$), Equations (2.175–2.176) reduces to the following form (see Figure 2.37):

Basic Design Guidelines

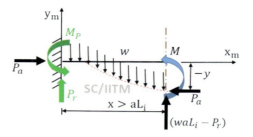

FIGURE 2.37 Free-body diagram ($x > aL_i$).

$$EI\frac{d^2y}{dx^2} = M = -\left[M_p - P_a(-y) - P_r(x) + (-waL_i)\left(x - \frac{aL_i}{2}\right)\right] \quad (2.183)$$

The general solution to Equation (2.183) is given by

$$y = A\sin\left(\frac{\alpha_i x}{L_i}\right) + B\cos\left(\frac{\alpha_i x}{L_i}\right) - \frac{L_i^2}{\alpha_i^2 EI}\left[M_p + \frac{wa^2 L_i^2}{2} - (P_r + waL_i)x\right] \quad (2.184)$$

Differentiating once, we get the following form

$$\frac{dy}{dx} = \frac{\alpha_i}{L_i}A\cos\left(\frac{\alpha_i x}{L_i}\right) - \frac{\alpha_i}{L_i}B\sin\left(\frac{\alpha_i x}{L_i}\right) + \frac{L_i^2}{\alpha_i^2 EI}(P_r + waL_i) \quad (2.185)$$

Applying the Boundary conditions:

$$@\ x = L_i, \begin{cases} y = 0 \\ \dfrac{dy}{dx} = 0 \end{cases} \quad (2.186)$$

$$A = \frac{L_i^2}{\alpha_i^2 EI}\left[\left(M_p + \frac{wa^2 L_i^2}{2}\right)\sin(\alpha_i) - L_i(P_r + waL_i)\left(\sin(\alpha_i) + \frac{\cos(\alpha_i)}{\alpha_i}\right)\right] \quad (2.187)$$

$$B = \frac{L_i^2}{\alpha_i^2 EI}\left[\left(M_p + \frac{wa^2 L_i^2}{2}\right)\cos(\alpha_i) - L_i(P_r + waL_i)\left(\cos(\alpha_i) - \frac{\sin(\alpha_i)}{\alpha_i}\right)\right] \quad (2.188)$$

Substituting the above constants in Equation (2.184, 2.185), we get

$$\frac{\alpha_i^2 EI}{L_i^2}y = \left[\left(M_p + \frac{wa^2 L_i^2}{2}\right)\sin(\alpha_i) - L_i(P_r + waL_i)\left(\sin(\alpha_i) + \frac{\cos(\alpha_i)}{\alpha_i}\right)\right]\sin\left(\frac{\alpha_i x}{L_i}\right)$$

$$+ \left[\left(M_p + \frac{wa^2 L_i^2}{2}\right)\cos(\alpha_i) - L_i(P_r + waL_i)\left(\cos(\alpha_i) - \frac{\sin(\alpha_i)}{\alpha_i}\right)\right]\cos\left(\frac{\alpha_i x}{L_i}\right) \quad (2.189)$$

$$- \left[M_p + \frac{wa^2 L_i^2}{2} - (P_r + waL_i)x\right] \text{ for } aL_i \le x \le L_i$$

$$\frac{\alpha_i^2 EI}{L_i^2}\frac{dy}{dx} = \frac{\alpha_i}{L_i}\left[\left(M_p + \frac{wa^2L_i^2}{2}\right)\sin(\alpha_i) - L_i(P_r + waL_i)\left(\sin(\alpha_i) + \frac{\cos(\alpha_i)}{\alpha_i}\right)\right]\cos\left(\frac{\alpha_i x}{L_i}\right)$$

$$-\frac{\alpha_i}{L_i}\left[\left(M_p + \frac{wa^2L_i^2}{2}\right)\cos(\alpha_i) - L_i(P_r + waL_i)\left(\cos(\alpha_i) - \frac{\sin(\alpha_i)}{\alpha_i}\right)\right]\sin\left(\frac{\alpha_i x}{L_i}\right)$$

$$+(P_r + waL_i) \text{ for } aL_i \leq x \leq L_i$$

(2.190)

Now, for equating the displacement function (y) of both the segment lengths, one needs to equate Equation (2.181) and Equation (2.189). For equating the slope function (dy/dx) of both the segment lengths, one needs to equate Equation (2.182) and Equation (2.190). Equating as above, and simplifying, we get the following

$$P_r = waL_i\left(\frac{a}{2} - 1\right) + \frac{M_p + M_q}{L_i} \quad (2.191)$$

$$M_p = -wL_i^2\left\{\left[\frac{\frac{a^2}{2}(1-\cos(\alpha_i)) + \left(\frac{\sin(a\alpha_i)}{\alpha_i} - a\right)\left(\frac{\sin(\alpha_i)}{\alpha_i} - \cos(\alpha_i)\right)}{2(1-\cos(\alpha_i)) - \alpha_i\sin(\alpha_i)}\right]\right.$$

$$\left. + \left[\frac{\frac{1-\cos(a\alpha_i)}{\alpha_i}\left[\frac{1-\cos(\alpha_i)}{\alpha_i} - \sin(\alpha_i)\right]}{2(1-\cos(\alpha_i)) - \alpha_i\sin(\alpha_i)}\right]\right\}$$

(2.192)

$$M_q = wL_i^2\left\{\left[\frac{a\left(1-\frac{a}{2}\right)(1-\cos(\alpha_i)) + \left(\frac{\sin(a\alpha_i)}{\alpha_i} - a\right)\left(\frac{\sin(\alpha_i)}{\alpha_i} - \cos(\alpha_i)\right)}{2(1-\cos(\alpha_i)) - \alpha_i\sin(\alpha_i)}\right]\right.$$

$$\left. + \left[\frac{\frac{1-\cos(\alpha_i)}{\alpha_i}\left[\frac{1-\cos(a\alpha_i)}{\alpha_i} - \sin(a\alpha_i)\right]}{2(1-\cos(\alpha_i)) - \alpha_i\sin(\alpha_i)}\right]\right\}$$

(2.193)

For the case of *zero axial loads*, one should apply L'Hospital's rule to Equations (2.191–2.193), to obtain the following results

$$M_p = -wa^2L_i^2\left(\frac{1}{2} - \frac{2}{3}a + \frac{1}{4}a^2\right) \quad (2.194)$$

Basic Design Guidelines

$$M_q = wa^2 L_i^2 \left(\frac{1}{3} - \frac{1}{4}a \right) \qquad (2.195)$$

For the uniformly distributed load applied to the complete span of the beam, end moments will take the following form

$$M_p = -M_q = \frac{wL_i^2}{\alpha_i^2} \left[1 - \frac{\alpha_i \sin(\alpha_i)}{2(1-\cos(\alpha_i))} \right] \qquad (2.196)$$

Alternatively, they can also be expressed as

$$M_p = -M_q = m_i \frac{wL_i^2}{\alpha_i^2} \qquad (2.197)$$

$$m_i = \frac{12}{\alpha_i^2} \left[1 - \frac{\alpha_i \sin(\alpha_i)}{2(1-\cos(\alpha_i))} \right] \qquad (2.198)$$

Where m_i is termed as load function, which is to be multiplied to the end moments of a fixed beam under uniformly distributed load acting over the entire length of the member to account for the effect of the axial load. In the absence of axial compressive load ($\varphi_i = 0$), L'Hospital's rule is applied to Equation (2.196). It reduces to the following form, a standard expression for a fixed beam under uniformly distributed load acting upon its entire length.

$$(M_p)_{@\,\varphi_i=0} = -(M_q)_{@\,\varphi_i=0} = -\frac{wL_i^2}{12} \qquad (2.199)$$

2.17 FIXED BEAM UNDER TENSILE AXIAL LOAD

Under tensile axial loads, the load function and end moments will take a different form. For a special case where (φ_i) is negative, Equations (2.192, 2.193) will reduce to the following form.

$$M_p = -wL_i^2 \left\{ \left[\frac{\frac{a^2}{2}(1-\cosh(\alpha_i)) + \left(\frac{\sinh(a\alpha_i)}{\alpha_i} - a \right)\left(\frac{\sinh(\alpha_i)}{\alpha_i} - \cosh(\alpha_i) \right)}{2(1-\cosh(\alpha_i)) - \alpha_i \sinh(\alpha_i)} \right] \right.$$

$$\left. - \left[\frac{\frac{1-\cosh(a\alpha_i)}{\alpha_i}\left[\frac{1-\cosh(\alpha_i)}{\alpha_i} - \sinh(\alpha_i) \right]}{2(1-\cosh(\alpha_i)) - \alpha_i \sinh(\alpha_i)} \right] \right\} \qquad (2.200)$$

$$M_q = wL_i^2 \left\{ \left[\frac{a\left(1-\dfrac{a}{2}\right)(1-\cosh(\alpha_i)) + \left(\dfrac{\sinh(a\alpha_i)}{\alpha_i} - a\right)\left(\dfrac{\sinh(\alpha_i)}{\alpha_i} - \cosh(\alpha_i)\right)}{2(1-\cosh(\alpha_i)) - \alpha_i \sinh(\alpha_i)} \right] \right.$$

$$\left. - \left[\frac{\dfrac{1-\cosh(\alpha_i)}{\alpha_i}\left[\dfrac{1-\cosh(a\alpha_i)}{\alpha_i} - \sinh(a\alpha_i)\right]}{2(1-\cosh(\alpha_i)) - \alpha_i \sinh(\alpha_i)} \right] \right\}$$

(2.201)

For the uniformly distributed load applied to the complete span of the beam and in the presence of tensile axial load, end moments will take the following form

$$M_p = -M_q = \frac{wL_i^2}{\alpha_i^2}\left[1 + \frac{\alpha_i \sinh(\alpha_i)}{2(1-\cosh(\alpha_i))}\right] \qquad (2.202)$$

Alternatively, they can also be expressed as

$$M_p = -M_q = m_i \frac{wL_i^2}{\alpha_i^2} \qquad (2.203)$$

$$m_i = -\frac{12}{\alpha_i^2}\left[1 - \frac{\alpha_i \sinh(\alpha_i)}{2(1-\cosh(\alpha_i))}\right] \qquad (2.204)$$

2.18 LATERAL LOAD FUNCTIONS FOR CONCENTRATED LOAD

In this section, we will derive the lateral load's functions for the fixed beam loaded under concentrated load in the presence of the axial load. Consider a fixed beam, loaded as shown in Figure 2.38.

Consider the free-body diagram, where ($x < aL_i$), as shown in Figure 2.39.

Under the fixed beam subjected to concentrated load, end rotations and shear are developed. The following equation holds good.

$$EI\frac{d^2y}{dx^2} = M \text{ (for } x < aL_i) \qquad (2.205)$$

$$= -\left[M_p - P_a(-y) - P_r(x)\right] \qquad (2.206)$$

Basic Design Guidelines

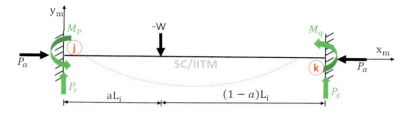

FIGURE 2.38 Fixed beam under concentrated load and axial compressive load.

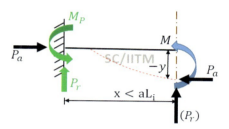

FIGURE 2.39 Free-body diagram ($x < aL_i$).

The general solution to Equation (2.206) is given by

$$y = A\sin\left(\frac{\alpha_i x}{L_i}\right) + B\cos\left(\frac{\alpha_i x}{L_i}\right) - \frac{L_i^2}{\alpha_i^2 EI}\left[M_p - P_r(x)\right] \quad (2.207)$$

Differentiating once, we get

$$\frac{dy}{dx} = \frac{\alpha_i}{L_i} A\cos\left(\frac{\alpha_i x}{L_i}\right) - \frac{\alpha_i}{L_i} B\sin\left(\frac{\alpha_i x}{L_i}\right) + \frac{L_i^2}{\alpha_i^2 EI}(P_r) \quad (2.208)$$

Applying the Boundary conditions

$$@\ x = 0, \begin{cases} y = 0 \\ \dfrac{dy}{dx} = 0 \end{cases} \quad (2.209)$$

$$B = \frac{L_i^2}{\alpha_i^2 EI}\left[M_p\right] \quad (2.210)$$

$$A = -\left[\frac{L_i^3}{\alpha_i^3 EI} P_r\right] \quad (2.211)$$

Substituting Equations (2.209–2.211) in Equation (2.207), we get

$$\left(\frac{\alpha_i^2 EI}{L_i^2}\right) y = -\left[\frac{L_i}{\alpha_i} P_r\right] \sin\left(\frac{\alpha_i x}{L_i}\right) + M_p \cos\left(\frac{\alpha_i x}{L_i}\right) - \left[M_p - P_r(x)\right] \quad \text{for } 0 \le x \le aL_i$$

(2.212)

Substituting Equation (2.211) in Equation (2.208), we get

$$\left(\frac{\alpha_i^2 EI}{L_i^2}\right)\frac{dy}{dx} = -\left(\frac{\alpha_i}{L_i}\right) P_r \cos\left(\frac{\alpha_i x}{L_i}\right) - \frac{\alpha_i}{L_i}\left[M_p\right] \sin\left(\frac{\alpha_i x}{L_i}\right) + P_r \quad \text{for } 0 \le x \le aL_i \quad (2.213)$$

For the position of the concentrated load ($x > aL_i$), Equation (2.205) reduces to the following form (see Figure 2.40)

$$EI \frac{d^2 y}{dx^2} = M = -\left[M_p - P_a(-y) - P_r(x) + (-W)(x - aL_i)\right] \quad (2.214)$$

The general solution to Equation (2.214) is given by

$$y = A \sin\left(\frac{\alpha_i x}{L_i}\right) + B \cos\left(\frac{\alpha_i x}{L_i}\right) - \frac{L_i^2}{\alpha_i^2 EI}\left[M_p + WaL_i - (P_r + W)x\right] \quad (2.215)$$

Differentiating once, we get the following form

$$\frac{dy}{dx} = \frac{\alpha_i}{L_i} A \cos\left(\frac{\alpha_i x}{L_i}\right) - \frac{\alpha_i}{L_i} B \sin\left(\frac{\alpha_i x}{L_i}\right) + \frac{L_i^2}{\alpha_i^2 EI}(P_r + W) \quad (2.216)$$

We can evaluate the constants by equating the expressions for deflection and slope for two segments (aL_i) and ($1-a$)L_i. Now, equating Equation (2.212) and Equation (2.215) for the deflection, and Equation (2.213) and Equation (2.216) for the slope, respectively, we get the following

$$A = -\frac{L_i^3}{\alpha_i^3 EI}\left[P_r + W \cos(a\alpha_i)\right] \quad (2.217)$$

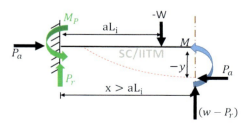

FIGURE 2.40 Free-body diagram ($x > aL_i$).

Basic Design Guidelines

$$B = \frac{L_i^2}{\alpha_i^2 EI}\left[M_p + \frac{L_i W}{\alpha_I}\sin(a\alpha_i)\right] \quad (2.218)$$

Substituting the above constants in Equation (2.215 and 2.216), we get

$$\frac{\alpha_i^2 EI}{L_i^2} y = -\frac{L_i}{\alpha_i}\left[P_r + W\cos(a\alpha_i)\right]\sin\left(\frac{\alpha_i x}{L_i}\right)$$

$$+ \left[M_p + \frac{L_i W}{\alpha_I}\sin(a\alpha_i)\right]\cos\left(\frac{\alpha_i x}{L_i}\right) \quad (2.219)$$

$$- \left[M_p + WaL_i - (P_r + W)x\right] \quad \text{for } aL_i \leq x \leq L_i$$

$$\frac{\alpha_i^2 EI}{L_i^2}\frac{dy}{dx} = -\left[P_r + W\cos(a\alpha_i)\right]\cos\left(\frac{\alpha_i x}{L_i}\right)$$

$$- \frac{\alpha_i}{L_i}\left[M_p + \frac{L_i W}{\alpha_I}\sin(a\alpha_i)\right]\sin\left(\frac{\alpha_i x}{L_i}\right) \quad (2.220)$$

$$+ \left[P_r + W\right] \quad \text{for } aL_i \leq x \leq L_i$$

Applying the following boundary conditions

$$@\, x = L, \begin{cases} y = 0 \\ \dfrac{dy}{dx} = 0 \end{cases} \quad (2.221)$$

Eliminating P_r from Equation (2.219) and Equation (2.220) after applying the above boundary conditions, the end moment assumes the following form

$$M_p = -WL_i \left\{ \begin{array}{c} \dfrac{\left[a(1-\cos(\alpha_i))\right] - \left[(1-\cos(a\alpha_i))\sin\left(\dfrac{\alpha_i x}{L_i}\right)\right] - \cos(\alpha_i)}{2(1-\cos(\alpha_i)) - \alpha_i \sin(\alpha_i)} \\ \\ + \dfrac{\left[\dfrac{1-\cos(\alpha_i)}{\alpha_i} - \sin(\alpha_i)\right]\sin(a\alpha_i)}{2(1-\cos(\alpha_i)) - \alpha_i \sin(\alpha_i)} \end{array} \right\} \quad (2.222)$$

For zero axial loads, one can estimate the end moment by applying L'Hospital's rule.

$$M_p = WaL_i\left(a^2 - 1\right) \quad (2.223)$$

Under tensile axial loads, which is a special case where (φ_i) is negative, Equation (2.223) will reduce to the following form

$$M_p = -WL_i \left\{ \left[\frac{a(1-\cosh(\alpha_i)) - (1-\cosh(a\alpha_i))\left(\frac{\sinh(\alpha_i)}{\alpha_i} - \cosh(\alpha_i)\right)}{2(1-\cosh(\alpha_i)) + \alpha_i \sinh(\alpha_i)} \right] \right.$$

$$\left. + \left[\frac{\left(\frac{1-\cosh(\alpha_i)}{\alpha_i} + \sinh(\alpha_i)\right)\sinh(a\alpha_i)}{2(1-\cosh(\alpha_i)) + \alpha_i \sinh(\alpha_i)} \right] \right\}$$

(2.224)

2.19 EXERCISE PROBLEMS ON STABILITY ANALYSIS USING MATLAB®

Exercise problem 1

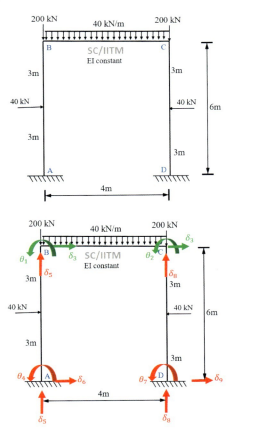

MATLAB® PROGRAM (WITHOUT AXIAL LOAD)

```matlab
%% This MATLAB code is for analysis of structures without
considering axial load
% Re-type the following code in MATLAB new script and run the
file to get the output.
%% Input
clc;
clear;
n = 3; % number of members
I = [1 1 1]*8.333e-6; %Moment of inertia in m4
E = [1 1 1]*2.1e11; % Young's modulus
L = [6 4 6]; % length in m
uu = 3; % Number of unrestrained degrees of freedom
ur = 6; % Number of restrained degrees of freedom
uul = [1 2 3]; % global labels of unrestrained dof
url = [4 5 6 7 8 9]; % global labels of restrained dof
l1 = [1 4 3 6]; % Global labels for member 1
l2 = [1 2 5 8]; % Global labels for member 2
l3 = [2 7 3 9]; % Global labels for member 3
l= [l1; l2; l3];
dof = uu+ur;
Ktotal = zeros (dof);
fem1= [-30 30 -20 -20]; % Local Fixed end moments of member 1
fem2= [53.333 -53.333 280 280]; % Local Fixed end moments of
member 2
fem3= [30 -30 20 20]; % Local Fixed end moments of member 3
 %% rotation coefficients for each member
rc1 = 4.*E.*I./L;
rc2 = 2.*E.*I./L;
%% stiffness matrix 4 by 4 (axial deformation neglected)
for i = 1:n
   Knew = zeros (dof);
   k1 = [rc1(i); rc2(i); (rc1(i)+rc2(i))/L(i);
       (-(rc1(i)+rc2(i))/L(i))];
   k2 = [rc2(i); rc1(i); (rc1(i)+rc2(i))/L(i);
       (-(rc1(i)+rc2(i))/L(i))];
   k3 = [(rc1(i)+rc2(i))/L(i); (rc1(i)+rc2(i))/L(i);
       (2*(rc1(i)+rc2(i))/(L(i)^2)); (-2*(rc1(i)+rc2(i))/
       (L(i)^2))];
   k4 = -k3;
   K = [k1 k2 k3 k4];
   fprintf ('Member Number =');
   disp (i);
   fprintf ('Local Stiffness matrix of member, [K] = \n');
   disp (K);
   for p = 1:4
     for q = 1:4
       Knew((l(i,p)),(l(i,q))) =K(p,q);
     end
   end
   Ktotal = Ktotal + Knew;
```

```
    if i == 1
      Kg1=K;
    elseif i == 2
      Kg2 = K;
    else
      Kg3=K;
    end
end
fprintf ('Stiffness Matrix of complete structure, [Ktotal] = \n');
disp (Ktotal);
Kunr = zeros(uu);
for x=1:uu
  for y=1:uu
    Kunr(x,y)= Ktotal(x,y);
  end
end
fprintf ('Unrestrained Stiffness sub-matix, [Kuu] = \n');
disp (Kunr);
KuuInv= inv(Kunr);
fprintf ('Inverse of Unrestrained Stiffness sub-matix, [KuuInverse] = \n');
disp (KuuInv);
 %% Creation of joint load vector
jl= [-23.333; 23.333; 0; -30; -280; 20; 30; -280; -20]; % values given in kN or kNm
jlu = jl(1:uu,1); % load vector in unrestrained dof
delu = KuuInv*jlu;
fprintf ('Joint Load vector, [Jl] = \n');
disp (jl');
fprintf ('Unrestrained displacements, [DelU] = \n');
disp (delu');
delr = zeros (ur,1);
del = [delu; delr];
deli= zeros (4,1);
for i = 1:n
  for p = 1:4
    deli(p,1) = del((l(i,p)),1) ;
  end
  if i == 1
     delbar1 = deli;
     mbar1= (Kg1 * delbar1)+fem1';
     fprintf ('Member Number =');
     disp (i);
     fprintf ('Global displacement matrix [DeltaBar] = \n');
     disp (delbar1');
     fprintf ('Global End moment matrix [MBar] = \n');
     disp (mbar1');
  elseif i == 2
     delbar2 = deli;
```

Basic Design Guidelines

```
            mbar2= (Kg2 * delbar2)+fem2';
            fprintf ('Member Number =');
            disp (i);
            fprintf ('Global displacement matrix [DeltaBar] = \n');
            disp (delbar2');
            fprintf ('Global End moment matrix [MBar] = \n');
            disp (mbar2');
        else
            delbar3 = deli;
            mbar3= (Kg3 * delbar3)+fem3';
            fprintf ('Member Number =');
            disp (i);
            fprintf ('Global displacement matrix [DeltaBar] = \n');
            disp (delbar3');
            fprintf ('Global End moment matrix [MBar] = \n');
            disp (mbar3');
        end
end
%% check
mbar = [mbar1'; mbar2'; mbar3'];
jf = zeros(dof,1);
for a=1:n
    for b=1:4 % size of k matrix
        d = l(a,b);
        jfnew = zeros(dof,1);
        jfnew(d,1)=mbar(a,b);
        jf=jf+jfnew;
    end
end
fprintf ('Joint forces = \n');
disp (jf');
```

OUTPUT:

```
Member Number =    1
Local Stiffness matrix of member, [K] =
  1.0e+06 *
  1.1666   0.5833   0.2917  -0.2917
  0.5833   1.1666   0.2917  -0.2917
  0.2917   0.2917   0.0972  -0.0972
 -0.2917  -0.2917  -0.0972   0.0972
Member Number =    2
Local Stiffness matrix of member, [K] =
  1.0e+06 *
  1.7499   0.8750   0.6562  -0.6562
  0.8750   1.7499   0.6562  -0.6562
  0.6562   0.6562   0.3281  -0.3281
 -0.6562  -0.6562  -0.3281   0.3281
Member Number =    3
Local Stiffness matrix of member, [K] =
```

```
   1.0e+06 *
   1.1666    0.5833    0.2917   -0.2917
   0.5833    1.1666    0.2917   -0.2917
   0.2917    0.2917    0.0972   -0.0972
  -0.2917   -0.2917   -0.0972    0.0972
Stiffness Matrix of complete structure, [Ktotal] =
   1.0e+06*
   2.9166   0.8750   0.2917   0.5833   0.6562  -0.2917        0  -0.6562        0
   0.8750   2.9166   0.2917        0   0.6562        0   0.5833  -0.6562  -0.2917
   0.2917   0.2917   0.1944   0.2917        0  -0.0972   0.2917        0  -0.0972
   0.5833        0   0.2917   1.1666        0  -0.2917        0        0        0
   0.6562   0.6562        0        0   0.3281        0        0  -0.3281        0
  -0.2917        0  -0.0972  -0.2917        0   0.0972        0        0        0
        0   0.5833   0.2917        0        0        0   1.1666        0  -0.2917
  -0.6562  -0.6562        0        0  -0.3281        0        0   0.3281        0
        0  -0.2917  -0.0972        0        0        0  -0.2917        0   0.0972
Unrestrained Stiffness sub-matix, [Kuu] =
   1.0e+06 *
   2.9166   0.8750   0.2917
   0.8750   2.9166   0.2917
   0.2917   0.2917   0.1944
Inverse of Unrestrained Stiffness sub-matix, [KuuInverse] =
   1.0e-05 *
   0.0416  -0.0073  -0.0514
  -0.0073   0.0416  -0.0514
  -0.0514  -0.0514   0.6686
Joint Load vector, [Jl] =
  -23.3330  23.3330         0  -30.0000  -280.0000   20.0000   30.0000
  -280.0000 -20.0000
Unrestrained displacements, [DelU] =
   1.0e-04 *
  -0.1143    0.1143   -0.0000
Member Number =  1
Global displacement matrix [DeltaBar] =
   1.0e-04 *
  -0.1143        0  -0.0000        0
Global End moment matrix [MBar] =
  -43.3331  23.3334  -23.3333  -16.6667
Member Number =  2
Global displacement matrix [DeltaBar] =
   1.0e-04 *
  -0.1143    0.1143        0        0
Global End moment matrix [MBar] =
   43.3331  -43.3331  280.0000  280.0000
Member Number =  3
Global displacement matrix [DeltaBar] =
   1.0e-04 *
   0.1143         0   -0.0000        0
Global End moment matrix [MBar] =
   43.3331  -23.3334   23.3333   16.6667
Joint forces =
        0  -0.0000        0  23.3334  280.0000  -16.6667  -23.3334  280.0000  16.6667
>>
```

Basic Design Guidelines

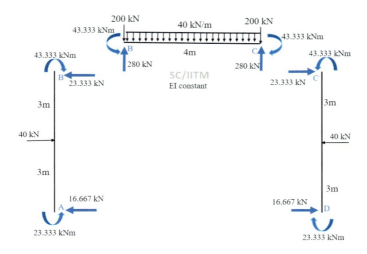

MATLAB® PROGRAM (WITH AXIAL LOAD)

```
%% This MATLAB code is for stability analysis of structures,
which includes the effect of axial load on the end moments and
shear
% Re-type the following code in MATLAB new script and run the
file to get the output.
%% Input
clc;
clear;
n = 3; % number of members
I = [1 1 1]*8.333e-6; %Moment of inertia in m4
E = [1 1 1]*2.1e11; % Young's modulus in N/m^2
L = [6 4 6]; % length in m
uu = 3; % Number of unrestrained degrees of freedom
ur = 6; % Number of restrained degrees of freedom
uul = [1 2 3]; % global labels of unrestrained dof
url = [4 5 6 7 8 9]; % global labels of restrained dof
l1 = [1 4 3 6]; % Global labels for member 1
l2 = [1 2 5 8]; % Global labels for member 2
l3 = [2 7 3 9]; % Global labels for member 3
l= [l1; l2; l3];
dof = uu+ur;
Ktotal = zeros (dof);
fem1= [-30 30 -20 -20]; % Local Fixed end moments of member 1
fem2= [53.333 -53.333 280 280]; % Local Fixed end moments of
member 2
fem3= [30 -30 20 20]; % Local Fixed end moments of member 3
pa = [280 23.333 280]*1000; %Axial load in N
load = [1 1 1]; % 0-zero load, 1-compression, 2-tension
%% Load and angle calculation
pe = pi*pi.*E.*I./(L.*L); % Euler's load in N
phi = pa./pe;
alrad = pi.*sqrt(phi);
```

```
al = radtodeg (alrad);
r = zeros(1,n);
c = zeros(1,n);
t = zeros(1,n);
for i = 1:n
  if load(i)==1
    % rotation coefficients only for compression loads
    r(i) = (alrad(i).*((sind(al(i))-(alrad(i).*cosd(al(i))))))./
         ((2.*(1-cosd(al(i))))-(alrad(i).*(sind(al(i))))); %
         rotation function
    c(i) = (alrad(i)-sind(al(i)))./(sind(al(i))-(alrad(i).*co
sd(al(i))))); % rotation function
    t(i) = 1-((pi*pi*phi(i))./(2.*r(i).*(1+c(i)))); %
Translation function
  elseif load(i)==2
    % rotation coefficients only for tension loads
    r(i)= (alrad(i)*((alrad(i)*cosh(alrad(i)))-sinh(alrad(i)
)))/((2*(1-cosh(alrad(i))))+(alrad(i)*sinh(alrad(i))));
    c(i) = (alrad(i)-sinh(alrad(i)))/(sinh(alrad(i))-(alrad
(i)*cosh(alrad(i))));
    t(i) = 1-((pi*pi*(-phi(i)))./(2.*r(i).*(1+c(i))));
  else
    r(i) = 4;
    c(i) = 0.5;
    t(i) = 1;
  end
end
 %% rotation coefficients for each member
rc1 = r.*E.*I./L;
rc2 = c.*E.*I.*r./L;
 %% stiffness matrix 4 by 4 (axial deformation neglected)
for i = 1:n
   Knew = zeros (dof);
   k1 = [rc1(i); rc2(i); (rc1(i)+rc2(i))/L(i); 
(-(rc1(i)+rc2(i))/L(i))];
   k2 = [rc2(i); rc1(i); (rc1(i)+rc2(i))/L(i); 
(-(rc1(i)+rc2(i))/L(i))];
   k3 = [(rc1(i)+rc2(i))/L(i); (rc1(i)+rc2(i))/L(i); 
(2*t(i)*(rc1(i)+rc2(i))/(L(i)^2)); (-2*t(i)*(rc1(i)+rc2(i))/
(L(i)^2))];
   k4 = -k3;
   K = [k1 k2 k3 k4];
   fprintf ('Member Number =');
   disp (i);
   fprintf ('Local Stiffness matrix of member, [K] = \n');
   disp (K);
   for p = 1:4
     for q = 1:4
       Knew((l(i,p)),(l(i,q))) =K(p,q);
     end
   end
```

```
    Ktotal = Ktotal + Knew;
    if i == 1
      Kg1=K;
    elseif i == 2
      Kg2 = K;
    else
      Kg3=K;
    end
end
fprintf ('Stiffness Matrix of complete structure, [Ktotal] = \n');
disp (Ktotal);
Kunr = zeros(uu);
for x=1:uu
  for y=1:uu
    Kunr(x,y)= Ktotal(x,y);
  end
end
fprintf ('Unrestrained Stiffness sub-matix, [Kuu] = \n');
disp (Kunr);
KuuInv= inv(Kunr);
fprintf ('Inverse of Unrestrained Stiffness sub-matix,
[KuuInverse] = \n');
disp (KuuInv);
%% Creation of joint load vector
jl= [-23.333; 23.333; 0; -30; -280; 20; 30; -280; -20]; %
values given in kN or kNm
jlu = jl(1:uu,1); % load vector in unrestrained dof
delu = KuuInv*jlu;
fprintf ('Joint Load vector, [Jl] = \n');
disp (jl');
fprintf ('Unrestrained displacements, [DelU] = \n');
disp (delu');
delr = zeros (ur,1);
del = [delu; delr];
deli= zeros (4,1);
for i = 1:n
  for p = 1:4
    deli(p,1) = del((l(i,p)),1) ;
  end
  if i == 1
     delbar1 = deli;
     mbar1= (Kg1 * delbar1)+fem1';
     fprintf ('Member Number =');
     disp (i);
     fprintf ('Global displacement matrix [DeltaBar] = \n');
     disp (delbar1');
     fprintf ('Global End moment matrix [MBar] = \n');
     disp (mbar1');
  elseif i == 2
     delbar2 = deli;
     mbar2= (Kg2 * delbar2)+fem2';
```

```
      fprintf ('Member Number =');
      disp (i);
      fprintf ('Global displacement matrix [DeltaBar] = \n');
      disp (delbar2');
      fprintf ('Global End moment matrix [MBar] = \n');
      disp (mbar2');
    else
      delbar3 = deli;
      mbar3= (Kg3 * delbar3)+fem3';
      fprintf ('Member Number =');
      disp (i);
      fprintf ('Global displacement matrix [DeltaBar] = \n');
      disp (delbar3');
      fprintf ('Global End moment matrix [MBar] = \n');
      disp (mbar3');
    end
end
%% check
mbar = [mbar1'; mbar2'; mbar3'];
jf = zeros(dof,1);
for a=1:n
  for b=1:4 % size of k matrix
    d = l(a,b);
    jfnew = zeros(dof,1);
    jfnew(d,1)=mbar(a,b);
    jf=jf+jfnew;
  end
end
fprintf ('Joint forces = \n');
disp (jf');
```

OUTPUT:

```
Member Number =    1
Local Stiffness matrix of member, [K] =
  1.0e+05 *
  9.2333  6.5121  2.6242  -2.6242
  6.5121  9.2333  2.6242  -2.6242
  2.6242  2.6242  0.4081  -0.4081
 -2.6242 -2.6242 -0.4081   0.4081
Member Number =    2
Local Stiffness matrix of member, [K] =
  1.0e+06 *
  1.7375  0.8781  0.6539  -0.6539
  0.8781  1.7375  0.6539  -0.6539
  0.6539  0.6539  0.3211  -0.3211
 -0.6539 -0.6539 -0.3211   0.3211
Member Number =    3
Local Stiffness matrix of member, [K] =
  1.0e+05 *
```

```
     9.2333   6.5121    2.6242    -2.6242
     6.5121   9.2333    2.6242    -2.6242
     2.6242   2.6242    0.4081    -0.4081
    -2.6242  -2.6242   -0.4081     0.4081
  Stiffness Matrix of complete structure, [Ktotal] =
    1.0e+06 *
    2.6608  0.8781  0.2624  0.6512  0.6539 -0.2624       0 -0.6539       0
    0.8781  2.6608  0.2624       0  0.6539       0  0.6512 -0.6539 -0.2624
    0.2624  0.2624  0.0816  0.2624       0 -0.0408  0.2624       0 -0.0408
    0.6512       0  0.2624  0.9233       0 -0.2624       0       0       0
    0.6539  0.6539       0       0  0.3211       0       0 -0.3211       0
   -0.2624       0 -0.0408 -0.2624       0  0.0408       0       0       0
         0  0.6512  0.2624       0       0       0  0.9233       0 -0.2624
   -0.6539 -0.6539       0       0 -0.3211       0       0  0.3211       0
         0 -0.2624 -0.0408       0       0       0 -0.2624       0  0.0408
  Unrestrained Stiffness sub-matix, [Kuu] =
    1.0e+06 *
    2.6608   0.8781    0.2624
    0.8781   2.6608    0.2624
    0.2624   0.2624    0.0816
  Inverse of Unrestrained Stiffness sub-matix, [KuuInverse] =
    1.0e-04 *
    0.0055 -0.0001 -0.0174
   -0.0001  0.0055 -0.0174
   -0.0174 -0.0174  0.2342
  Joint Load vector, [Jl] =
   -23.3330  23.3330       0 -30.0000 -280.0000  20.0000  30.0000
  -280.0000 -20.0000
  Unrestrained displacements, [DelU] =
    1.0e-04 *
    -0.1309  0.1309       0
  Member Number =   1
  Global displacement matrix [DeltaBar] =
    1.0e-04 *
    -0.1309       0       0       0
  Global End moment matrix [MBar] =
   -42.0852 21.4765 -23.4348 -16.5652
  Member Number =   2
  Global displacement matrix [DeltaBar] =
    1.0e-04 *
    -0.1309  0.1309       0       0
  Global End moment matrix [MBar] =
    42.0852 -42.0852 280.0000 280.0000
  Member Number =   3
  Global displacement matrix [DeltaBar] =
    1.0e-04 *
     0.1309       0       0       0
  Global End moment matrix [MBar] =
    42.0852 -21.4765 23.4348 16.5652
  Joint forces =
    0.0000 -0.0000       0 21.4765 280.0000 -16.5652 -21.4765 280.0000 16.5652
```

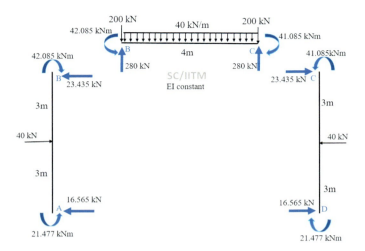

Exercise problem 2

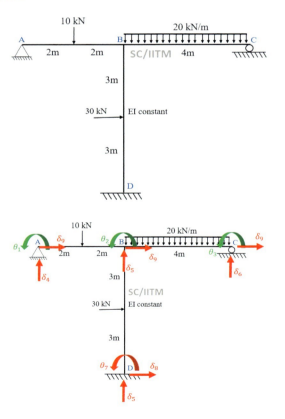

MATLAB® PROGRAM INPUT (WITHOUT AXIAL LOAD):

```
n = 3; % number of members
I = [1 1 1]*8.333e-6; %Moment of inertia in m4
E = [1 1 1]*2.1e11; % Young's modulus in N/m^2
L = [3 4 6]; % length in m
uu = 3; % Number of unrestrained degrees of freedom
ur = 6; % Number of restrained degrees of freedom
uul = [1 2 3]; % global labels of unrestained dof
url = [4 5 6 7 8 9]; % global labels of restained dof
l1 = [1 2 4 5]; % Global labels for member 1
l2 = [2 3 5 6]; % Global labels for member 2
l3 = [2 7 9 8]; % Global labels for member 3
fem1= [3.75 -3.75 5 5]; % Local Fixed end moments of member 1
fem2= [26.667 -26.667 40 40]; % Local Fixed end moments of member 2
fem3= [-22.5 22.5 -15 -15]; % Local Fixed end moments of member 3
jl= [-3.75; -0.417; 26.667; -5; -45; -40; -22.5; 15; 15]; % values given in kN or kNm
```

OUTPUT:

```
Member Number =    1
Local Stiffness matrix of member, [K] =
  1.0e+06 *
  2.3332  1.1666  1.1666  -1.1666
  1.1666  2.3332  1.1666  -1.1666
  1.1666  1.1666  0.7777  -0.7777
 -1.1666 -1.1666 -0.7777   0.7777
Member Number =    2
Local Stiffness matrix of member, [K] =
  1.0e+06 *
  1.7499  0.8750  0.6562  -0.6562
  0.8750  1.7499  0.6562  -0.6562
  0.6562  0.6562  0.3281  -0.3281
 -0.6562 -0.6562 -0.3281   0.3281
Member Number =    3
Local Stiffness matrix of member, [K] =
  1.0e+06 *
  1.1666  0.5833  0.2917  -0.2917
  0.5833  1.1666  0.2917  -0.2917
  0.2917  0.2917  0.0972  -0.0972
 -0.2917 -0.2917 -0.0972   0.0972
Stiffness Matrix of complete structure, [Ktotal] =
  1.0e+06 *
  2.3332  1.1666   0  1.1666 -1.1666   0   0   0
```

```
   1.1666   5.2498   0.8750   1.1666  -0.5104  -0.6562   0.5833  -0.2917   0.2917
        0   0.8750   1.7499        0   0.6562  -0.6562        0        0        0
   1.1666   1.1666        0   0.7777  -0.7777        0        0        0        0
  -1.1666  -0.5104   0.6562  -0.7777   1.1059  -0.3281        0        0        0
        0  -0.6562  -0.6562        0  -0.3281   0.3281        0        0        0
        0   0.5833        0        0        0        0   1.1666  -0.2917   0.2917
        0  -0.2917        0        0        0        0  -0.2917   0.0972  -0.0972
        0   0.2917        0        0        0        0   0.2917  -0.0972   0.0972
Unrestrained Stiffness sub-matix, [Kuu] =
     2333240     1166620           0
     1166620     5249790      874965
           0      874965     1749930
Inverse of Unrestrained Stiffness sub-matix, [KuuInverse] =
  1.0e-06 *
   0.4877  -0.1182   0.0591
  -0.1182   0.2365  -0.1182
   0.0591  -0.1182   0.6306
Joint Load vector, [Jl] =
  -3.7500   -0.4170   26.6670   -5.0000  -45.0000  -40.0000  -22.5000
  15.0000   15.0000
Unrestrained displacements, [DelU] =
  1.0e-04 *
  -0.0020  -0.0281   0.1664
Member Number =    1
Global displacement matrix [DeltaBar] =
  1.0e-05 *
  -0.0203  -0.2808        0        0
Global End moment matrix [MBar] =
        0 -10.5390   1.4870   8.5130
Member Number =    2
Global displacement matrix [DeltaBar] =
  1.0e-04 *
  -0.0281   0.1664        0        0
Global End moment matrix [MBar] =
  36.3150        0  49.0787  30.9213
Member Number =    3
Global displacement matrix [DeltaBar] =
  1.0e-05 *
  -0.2808        0        0        0
Global End moment matrix [MBar] =
 -25.7760  20.8620 -15.8190 -14.1810
Joint forces =
        0  -0.0000        0   1.4870  57.5917  30.9213
  20.8620 -14.1810 -15.8190
>>
```

Basic Design Guidelines

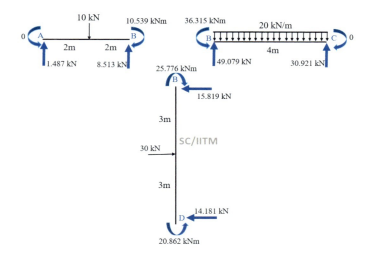

MATLAB® PROGRAM ADDITIONAL INPUT (WITH AXIAL LOAD):

pa = [15 15 45]*1000; %Axial load in N
load = [2 1 1]; % 1-compression, 2-tension

OUTPUT:

```
Member Number =    1
Local Stiffness matrix of member, [K] =
  1.0e+06 *
  2.3392  1.1651  1.1681  -1.1681
  1.1651  2.3392  1.1681  -1.1681
  1.1681  1.1681  0.7837  -0.7837
 -1.1681 -1.1681 -0.7837   0.7837
Member Number =    2
Local Stiffness matrix of member, [K] =
  1.0e+06 *
  1.7419  0.8770  0.6547  -0.6547
  0.8770  1.7419  0.6547  -0.6547
  0.6547  0.6547  0.3236  -0.3236
 -0.6547 -0.6547 -0.3236   0.3236
Member Number =    3
Local Stiffness matrix of member, [K] =
  1.0e+06 *
  1.1302  0.5926  0.2871  -0.2871
  0.5926  1.1302  0.2871  -0.2871
  0.2871  0.2871  0.0882  -0.0882
 -0.2871 -0.2871 -0.0882   0.0882
```

```
Stiffness Matrix of complete structure, [Ktotal] =
  1.0e+06 *
  2.3392   1.1651      0    1.1681 -1.1681    0       0       0       0
  1.1651   5.2113   0.8770   1.1681 -0.5134 -0.6547  0.5926 -0.2871  0.2871
     0    0.8770   1.7419      0    0.6547 -0.6547    0       0       0
  1.1681   1.1681      0    0.7837 -0.7837    0       0       0       0
 -1.1681  -0.5134   0.6547  -0.7837  1.1074 -0.3236    0       0       0
     0   -0.6547  -0.6547      0   -0.3236  0.3236    0       0       0
     0    0.5926      0       0       0       0    1.1302 -0.2871  0.2871
     0   -0.2871      0       0       0       0   -0.2871  0.0882 -0.0882
     0    0.2871      0       0       0       0    0.2871 -0.0882  0.0882
Unrestrained Stiffness sub-matix, [Kuu] =
  1.0e+06 *
  2.3392   1.1651      0
  1.1651   5.2113   0.8770
     0    0.8770   1.7419
Inverse of Unrestrained Stiffness sub-matix, [KuuInverse] =
  1.0e-06 *
  0.4867  -0.1189   0.0599
 -0.1189   0.2387  -0.1202
  0.0599  -0.1202   0.6346
Joint Load vector, [Jl] =
 -3.7500  -0.4170  26.6670  -5.0000 -45.0000 -40.0000 -22.5000
 15.0000  15.0000
Unrestrained displacements, [DelU] =
  1.0e-04 *
 -0.0018  -0.0286   0.1675
Member Number =    1
Global displacement matrix [DeltaBar] =
  1.0e-05 *
 -0.0179  -0.2858      0       0
Global End moment matrix [MBar] =
  0.0000 -10.6453   1.4516   8.5484
Member Number =    2
Global displacement matrix [DeltaBar] =
  1.0e-04 *
 -0.0286   0.1675      0       0
Global End moment matrix [MBar] =
 36.3757       0   49.0939  30.9061
Member Number =    3
Global displacement matrix [DeltaBar] =
  1.0e-05 *
 -0.2858       0      0       0
Global End moment matrix [MBar] =
-25.7304  20.8062 -15.8207 -14.1793
Joint forces =
  0.0000   0.0000      0    1.4516  57.6423  30.9061
 20.8062 -14.1793 -15.8207
>>
```

Basic Design Guidelines

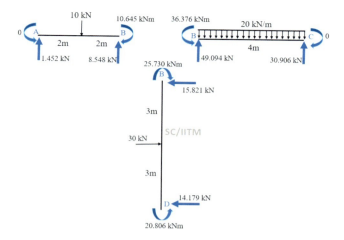

Exercise problem 3

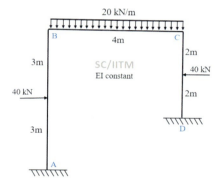

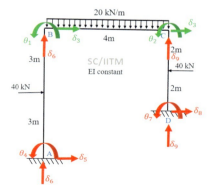

MATLAB® PROGRAM INPUT (WITHOUT AXIAL LOAD):

```
n = 3; % number of members
I = [1 1 1]*8.333e-6; %Moment of inertis in m4
E = [1 1 1]*2.1e11; % Young's modulus in N/m^2
L = [6 4 4]; % length in m
uu = 3; % Number of unrestrained degrees of freedom
ur = 6; % Number of restrained degrees of freedom
uul = [1 2 3]; % global labels of unrestained dof
url = [4 5 6 7 8 9]; % global labels of restained dof
l1 = [1 4 3 5]; % Global labels for member 1
l2 = [1 2 6 9]; % Global labels for member 2
l3 = [2 7 3 8]; % Global labels for member 3
l= [l1; l2; l3];
dof = uu+ur;
Ktotal = zeros (dof);
fem1= [-30 30 -20 -20]; % Local Fixed end moments of member 1
fem2= [26.667 -26.667 40 40]; % Local Fixed end moments of member 2
fem3= [20 -20 20 20]; % Local Fixed end moments of member 3
jl= [3.333; 6.667; 0; -30; 20; -40; 20; -20; -40]; % values given in kN or kNm
```

OUTPUT:

```
Member Number =   1
Local Stiffness matrix of member, [K] =
  1.0e+06 *
  1.1666   0.5833   0.2917  -0.2917
  0.5833   1.1666   0.2917  -0.2917
  0.2917   0.2917   0.0972  -0.0972
 -0.2917  -0.2917  -0.0972   0.0972
Member Number =   2
Local Stiffness matrix of member, [K] =
  1.0e+06 *
  1.7499   0.8750   0.6562  -0.6562
  0.8750   1.7499   0.6562  -0.6562
  0.6562   0.6562   0.3281  -0.3281
 -0.6562  -0.6562  -0.3281   0.3281
Member Number =   3
Local Stiffness matrix of member, [K] =
  1.0e+06 *
  1.7499   0.8750   0.6562  -0.6562
  0.8750   1.7499   0.6562  -0.6562
  0.6562   0.6562   0.3281  -0.3281
 -0.6562  -0.6562  -0.3281   0.3281
Stiffness Matrix of complete structure, [Ktotal] =
```

Basic Design Guidelines 157

```
  1.0e+06 *
  2.9166   0.8750   0.2917   0.5833  -0.2917   0.6562        0        0  -0.6562
  0.8750   3.4999   0.6562        0        0   0.6562   0.8750  -0.6562  -0.6562
  0.2917   0.6562   0.4253   0.2917  -0.0972        0   0.6562  -0.3281        0
  0.5833        0   0.2917   1.1666  -0.2917        0        0        0        0
 -0.2917        0  -0.0972  -0.2917   0.0972        0        0        0        0
  0.6562   0.6562        0        0        0   0.3281        0        0  -0.3281
       0   0.8750   0.6562        0        0        0   1.7499  -0.6562        0
       0  -0.6562  -0.3281        0        0        0  -0.6562   0.3281        0
 -0.6562  -0.6562        0        0        0  -0.3281        0        0   0.3281
Unrestrained Stiffness sub-matix, [Kuu] =
  1.0e+06 *
  2.9166   0.8750   0.2917
  0.8750   3.4999   0.6562
  0.2917   0.6562   0.4253
Inverse of Unrestrained Stiffness sub-matix, [KuuInverse] =
  1.0e-05 *
  0.0378  -0.0065  -0.0160
 -0.0065   0.0413  -0.0593
 -0.0160  -0.0593   0.3375
Joint Load vector, [Jl] =
  3.3330   6.6670        0  -30.0000   20.0000  -40.0000
 20.0000 -20.0000 -40.0000
Unrestrained displacements, [DelU] =
  1.0e-05 *
  0.0830   0.2539  -0.4486
Member Number =    1
Global displacement matrix [DeltaBar] =
  1.0e-05 *
  0.0830        0  -0.4486       0
Global End moment matrix [MBar] =
 -30.3402  29.1758 -20.1941 -19.8059
Member Number =    2
Global displacement matrix [DeltaBar] =
  1.0e-05 *
  0.0830   0.2539        0       0
Global End moment matrix [MBar] =
  30.3402 -21.4987  42.2104  37.7896
Member Number =    3
Global displacement matrix [DeltaBar] =
  1.0e-05 *
  0.2539        0  -0.4486       0
Global End moment matrix [MBar] =
  21.4987 -20.7224  20.1941  19.8059
Joint forces =
  0.0000        0   0.0000  29.1758 -19.8059  42.2104 -20.7224  19.8059
 37.7896
>>
```

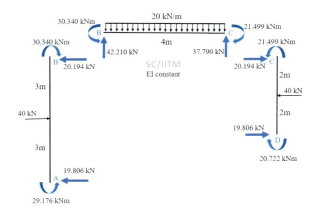

MATLAB® PROGRAM ADDITIONAL INPUT (WITH AXIAL LOAD):

```
pa = [40 20 40]*1000; %Axial load in N
load = [1 1 1]; % 1-compression, 2-tension
```

OUTPUT:

```
Member Number =  1
Local Stiffness matrix of member, [K] =
 1.0e+06 *
  1.1343  0.5915  0.2876 -0.2876
  0.5915  1.1343  0.2876 -0.2876
  0.2876  0.2876  0.0892 -0.0892
 -0.2876 -0.2876 -0.0892  0.0892
Member Number =  2
Local Stiffness matrix of member, [K] =
 1.0e+06 *
  1.7392  0.8776  0.6542 -0.6542
  0.8776  1.7392  0.6542 -0.6542
  0.6542  0.6542  0.3221 -0.3221
 -0.6542 -0.6542 -0.3221  0.3221
Member Number =  3
Local Stiffness matrix of member, [K] =
 1.0e+06 *
  1.7285  0.8804  0.6522 -0.6522
  0.8804  1.7285  0.6522 -0.6522
  0.6522  0.6522  0.3161 -0.3161
 -0.6522 -0.6522 -0.3161  0.3161
Stiffness Matrix of complete structure, [Ktotal] =
 1.0e+06 *
  2.8735  0.8776  0.2876  0.5915 -0.2876  0.6542       0       0 -0.6542
  0.8776  3.4677  0.6522       0       0  0.6542  0.8804 -0.6522 -0.6522
  0.2876  0.6522  0.4053  0.2876 -0.0892       0  0.6522 -0.3161       0
  0.5915       0  0.2876  1.1343 -0.2876       0       0       0       0
 -0.2876       0 -0.0892 -0.2876  0.0892       0       0       0       0
  0.6542  0.6542       0       0       0  0.3221       0       0 -0.3221
       0  0.8804  0.6522       0       0       0  1.7285 -0.6522       0
       0 -0.6522 -0.3161       0       0       0 -0.6522  0.3161       0
 -0.6542 -0.6542       0       0       0 -0.3221       0       0  0.3221
```

Basic Design Guidelines

```
Unrestrained Stiffness sub-matix, [Kuu] =
 1.0e+06 *
  2.8735   0.8776   0.2876
  0.8776   3.4677   0.6522
  0.2876   0.6522   0.4053
Inverse of Unrestrained Stiffness sub-matix, [KuuInverse] =
 1.0e-05 *
  0.0385 -0.0066 -0.0167
 -0.0066  0.0425 -0.0637
 -0.0167 -0.0637  0.3610
Joint Load vector, [Jl] =
   3.3330   6.6670     0 -30.0000 20.0000 -40.0000 20.0000 -20.0000 -40.0000
Unrestrained displacements, [DelU] =
 1.0e-05 *
   0.0843   0.2612 -0.4802
Member Number = 1
Global displacement matrix [DeltaBar] =
 1.0e-05 *
   0.0843     0 -0.4802    0
Global End moment matrix [MBar] =
 -30.4254 29.1173 -20.1860 -19.8140
Member Number = 2
Global displacement matrix [DeltaBar] =
 1.0e-05 *
   0.0843   0.2612    0    0
Global End moment matrix [MBar] =
  30.4254 -21.3838 42.2604 37.7396
Member Number = 3
Global displacement matrix [DeltaBar] =
 1.0e-05 *
   0.2612     0 -0.4802    0
Global End moment matrix [MBar] =
  21.3838 -20.8319 20.1860 19.8140
Joint forces =
     0 -0.0000      0 29.1173 -19.8140 42.2604 -20.8319 19.8140 37.7396
>>
```

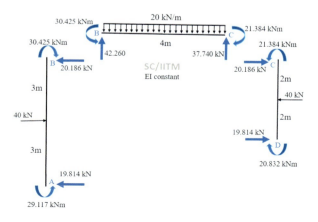

Exercise problem 4

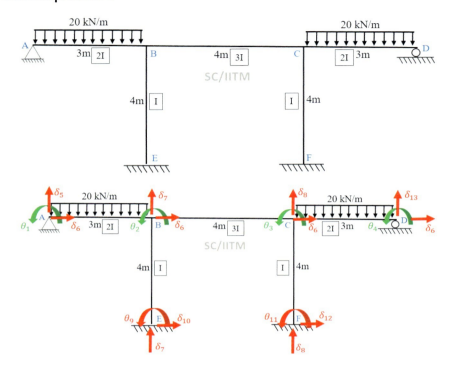

MATLAB® PROGRAM INPUT (WITHOUT AXIAL LOAD):

```
n = 5; % number of members
I = [2 3 2 1 1]*8.333e-6; %Moment of inertis in m4
E = [1 1 1 1 1]*2.1e11; % Young's modulus
L = [3 4 3 4 4]; % length in m
uu = 4; % Number of unrestrained degrees of freedom
ur = 9; % Number of restrained degrees of freedom
uul = [1 2 3 4]; % global labels of unrestained dof
url = [5 6 7 8 9 10 11 12 13]; % global labels of restrained
dof
l1 = [1 2 5 7]; % Global labels for member 1
l2 = [2 3 7 8]; % Global labels for member 2
l3 = [3 4 8 13]; % Global labels for member 3
l4 = [2 9 6 10]; % Global labels for member 4
l5 = [3 11 6 12]; % Global labels for member 5
l= [l1; l2; l3; l4; l5];
dof = uu+ur;
Ktotal = zeros (dof);
fem1= [15 -15 30 30]; % Local Fixed end moments of member 1
fem2= [0 0 0 0]; % Local Fixed end moments of member 2
fem3= [15 -15 30 30]; % Local Fixed end moments of member 3
```

Basic Design Guidelines

```
fem4= [0 0 0 0]; % Local Fixed end moments of member 4
fem5= [0 0 0 0]; % Local Fixed end moments of member 5
jlu = [-15; 15; -15; 15]; % load vector in unrestrained dof
```

OUTPUT:

```
Member Number =   1
Local Stiffness matrix of member, [K] =
  1.0e+06 *
  4.6665  2.3332  2.3332  -2.3332
  2.3332  4.6665  2.3332  -2.3332
  2.3332  2.3332  1.5555  -1.5555
 -2.3332 -2.3332 -1.5555   1.5555
Member Number =   2
Local Stiffness matrix of member, [K] =
  1.0e+06 *
  5.2498  2.6249  1.9687  -1.9687
  2.6249  5.2498  1.9687  -1.9687
  1.9687  1.9687  0.9843  -0.9843
 -1.9687 -1.9687 -0.9843   0.9843
Member Number =   3
Local Stiffness matrix of member, [K] =
  1.0e+06 *
  4.6665  2.3332  2.3332  -2.3332
  2.3332  4.6665  2.3332  -2.3332
  2.3332  2.3332  1.5555  -1.5555
 -2.3332 -2.3332 -1.5555   1.5555
Member Number =   4
Local Stiffness matrix of member, [K] =
  1.0e+06 *
  1.7499  0.8750  0.6562  -0.6562
  0.8750  1.7499  0.6562  -0.6562
  0.6562  0.6562  0.3281  -0.3281
 -0.6562 -0.6562 -0.3281   0.3281
Member Number =   5
Local Stiffness matrix of member, [K] =
  1.0e+06 *
  1.7499  0.8750  0.6562  -0.6562
  0.8750  1.7499  0.6562  -0.6562
  0.6562  0.6562  0.3281  -0.3281
 -0.6562 -0.6562 -0.3281   0.3281
Stiffness Matrix of complete structure, [Ktotal] =
  1.0e+07 *
 Columns 1 through 9
  0.4666  0.2333       0       0  0.2333       0 -0.2333       0       0
  0.2333  1.1666  0.2625       0  0.2333  0.0656 -0.0365 -0.1969  0.0875
       0  0.2625  1.1666  0.2333       0  0.0656  0.1969  0.0365       0
       0       0  0.2333  0.4666       0       0  0.2333       0       0
  0.2333  0.2333       0       0  0.1555       0 -0.1555       0       0
       0  0.0656  0.0656       0       0  0.0656       0       0  0.0656
 -0.2333 -0.0365  0.1969       0 -0.1555       0  0.2540 -0.0984       0
```

```
     0 -0.1969  0.0365  0.2333       0       0 -0.0984  0.2540       0
     0  0.0875       0       0       0  0.0656       0       0  0.1750
     0 -0.0656       0       0       0 -0.0328       0       0 -0.0656
     0       0  0.0875       0       0  0.0656       0       0       0
     0       0 -0.0656       0       0 -0.0328       0       0       0
     0       0 -0.2333 -0.2333       0       0       0 -0.1555       0
  Columns 10 through 13
     0       0       0       0
 -0.0656       0       0       0
     0  0.0875 -0.0656 -0.2333
     0       0       0 -0.2333
     0       0       0       0
 -0.0328  0.0656 -0.0328       0
     0       0       0       0
     0       0       0 -0.1555
 -0.0656       0       0       0
  0.0328       0       0       0
     0  0.1750 -0.0656       0
     0 -0.0656  0.0328       0
     0       0       0  0.1555
Unrestrained Stiffness sub-matix, [Kuu] =
   4666480  2333240        0        0
   2333240 11666200  2624895        0
         0  2624895 11666200  2333240
         0        0  2333240  4666480
Inverse of Unrestrained Stiffness sub-matix, [KuuInverse] =
   1.0e-06 *
    0.2397 -0.0508  0.0127 -0.0063
   -0.0508  0.1016 -0.0254  0.0127
    0.0127 -0.0254  0.1016 -0.0508
   -0.0063  0.0127 -0.0508  0.2397
Unrestrained Joint Load vector, [Jl] =
   -15   15  -15   15
Unrestrained displacements, [DelU] =
   1.0e-05 *
   -0.4643  0.2857 -0.2857  0.4643
Member Number =   1
Global displacement matrix [DeltaBar] =
   1.0e-05 *
   -0.4643  0.2857       0       0
Global End moment matrix [MBar] =
    0.0000 -12.5000  25.8333  34.1667
Member Number =   2
Global displacement matrix [DeltaBar] =
   1.0e-05 *
    0.2857 -0.2857       0       0
Global End moment matrix [MBar] =
    7.5000  -7.5000  -0.0000   0.0000
Member Number =   3
Global displacement matrix [DeltaBar] =
   1.0e-05 *
   -0.2857  0.4643       0       0
```

Basic Design Guidelines

```
Global End moment matrix [MBar] =
  12.5000        0   34.1667   25.8333
Member Number =    4
Global displacement matrix [DeltaBar] =
  1.0e-05 *
  0.2857      0      0      0
Global End moment matrix [MBar] =
  5.0000   2.5000   1.8750   -1.8750
Member Number =    5
Global displacement matrix [DeltaBar] =
  1.0e-05 *
  -0.2857     0      0      0
Global End moment matrix [MBar] =
  -5.0000   -2.5000   -1.8750   1.8750
Joint forces =
 Columns 1 through 9
  0.0000   -0.0000  -0.0000        0   25.8333  -0.0000   34.1667   34.1667   2.5000
 Columns 10 through 13
  -1.8750  -2.5000   1.8750   25.8333
>>
```

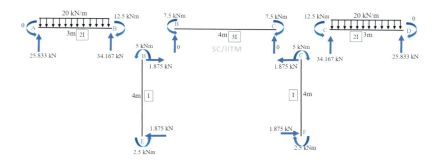

MATLAB® PROGRAM INPUT (WITH AXIAL LOAD):

pa = [1.875 1.875 1.875 34.1667 34.1667]*1000; %Axial load in N
load = [2 1 2 1 1]; % 0-zero load, 1-compression, 2-tension

OUTPUT:

```
Member Number =    1
Local Stiffness matrix of member, [K] =
  1.0e+06 *
  4.6672   2.3331   2.3334   -2.3334
  2.3331   4.6672   2.3334   -2.3334
  2.3334   2.3334   1.5562   -1.5562
 -2.3334  -2.3334  -1.5562    1.5562
Member Number =    2
Local Stiffness matrix of member, [K] =
  1.0e+06 *
```

```
   5.2488  2.6251  1.9685 -1.9685
   2.6251  5.2488  1.9685 -1.9685
   1.9685  1.9685  0.9838 -0.9838
  -1.9685 -1.9685 -0.9838  0.9838
Member Number =  3
Local Stiffness matrix of member, [K] =
  1.0e+06 *
   4.6672  2.3331  2.3334 -2.3334
   2.3331  4.6672  2.3334 -2.3334
   2.3334  2.3334  1.5562 -1.5562
  -2.3334 -2.3334 -1.5562  1.5562
Member Number =  4
Local Stiffness matrix of member, [K] =
  1.0e+06 *
   1.7316  0.8796  0.6528 -0.6528
   0.8796  1.7316  0.6528 -0.6528
   0.6528  0.6528  0.3179 -0.3179
  -0.6528 -0.6528 -0.3179  0.3179
Member Number =  5
Local Stiffness matrix of member, [K] =
  1.0e+06 *
   1.7316  0.8796  0.6528 -0.6528
   0.8796  1.7316  0.6528 -0.6528
   0.6528  0.6528  0.3179 -0.3179
  -0.6528 -0.6528 -0.3179  0.3179
Stiffness Matrix of complete structure, [Ktotal] =
  1.0e+07 *
 Columns 1 through 9
  0.4667  0.2333    0        0    0.2333    0   -0.2333    0        0
  0.2333  1.1648  0.2625    0    0.2333  0.0653 -0.0365 -0.1968  0.0880
    0     0.2625  1.1648  0.2333    0    0.0653  0.1968  0.0365    0
    0       0    0.2333  0.4667    0      0       0    0.2333    0
  0.2333  0.2333    0        0    0.1556    0   -0.1556    0        0
    0    0.0653  0.0653    0       0    0.0636    0       0     0.0653
 -0.2333 -0.0365  0.1968    0   -0.1556    0    0.2540 -0.0984    0
    0   -0.1968  0.0365  0.2333    0       0   -0.0984  0.2540    0
    0    0.0880    0       0       0    0.0653    0       0     0.1732
    0   -0.0653    0       0       0   -0.0318    0       0    -0.0653
    0       0    0.0880    0       0    0.0653    0       0        0
    0       0   -0.0653    0       0   -0.0318    0       0        0
    0       0   -0.2333 -0.2333    0       0       0    -0.1556    0
 Columns 10 through 13
    0       0       0       0
  -0.0653    0       0       0
    0     0.0880 -0.0653 -0.2333
    0       0       0   -0.2333
    0       0       0       0
  -0.0318  0.0653 -0.0318    0
    0       0       0       0
    0       0       0   -0.1556
  -0.0653    0       0       0
   0.0318    0       0       0
```

```
       0    0.1732  -0.0653       0
       0   -0.0653   0.0318       0
       0        0        0   0.1556
Unrestrained Stiffness sub-matix, [Kuu] =
  1.0e+07 *
   0.4667   0.2333       0        0
   0.2333   1.1648   0.2625       0
        0   0.2625   1.1648   0.2333
        0        0   0.2333   0.4667
Inverse of Unrestrained Stiffness sub-matix, [KuuInverse] =
  1.0e-06 *
   0.2397  -0.0509   0.0127  -0.0064
  -0.0509   0.1018  -0.0255   0.0127
   0.0127  -0.0255   0.1018  -0.0509
  -0.0064   0.0127  -0.0509   0.2397
Unrestrained Joint Load vector, [Jl] =
  -15   15  -15   15
Unrestrained displacements, [DelU] =
  1.0e-05 *
  -0.4645   0.2864  -0.2864   0.4645
Member Number = 1
Global displacement matrix [DeltaBar] =
  1.0e-05 *
  -0.4645   0.2864       0        0
Global End moment matrix [MBar] =
       0  -12.4723  25.8426  34.1574
Member Number = 2
Global displacement matrix [DeltaBar] =
  1.0e-05 *
   0.2864  -0.2864       0        0
Global End moment matrix [MBar] =
   7.5134  -7.5134       0        0
Member Number = 3
Global displacement matrix [DeltaBar] =
  1.0e-05 *
  -0.2864   0.4645       0        0
Global End moment matrix [MBar] =
  12.4723   0.0000  34.1574  25.8426
Member Number = 4
Global displacement matrix [DeltaBar] =
  1.0e-05 *
   0.2864        0       0        0
Global End moment matrix [MBar] =
   4.9589   2.5188   1.8694  -1.8694
Member Number = 5
Global displacement matrix [DeltaBar] =
   1.0e-05 *
  -0.2864        0       0        0
Global End moment matrix [MBar] =
  -4.9589  -2.5188  -1.8694   1.8694
Joint forces =
```

```
Columns 1 through 9
    0   -0.0000   0.0000   0.0000   25.8426        0   34.1574
34.1574   2.5188
 Columns 10 through 13
  -1.8694   -2.5188   1.8694   25.8426
>>
```

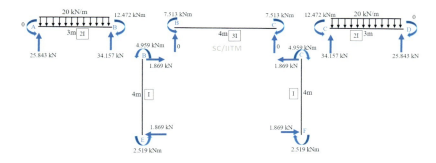

Exercise problem 5

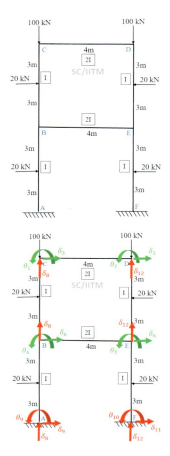

Basic Design Guidelines

MATLAB® PROGRAM INPUT (WITHOUT AXIAL LOAD):

```
n = 6; % number of members
I = [1 1 2 1 1 2]*8.333e-6; %Moment of inertis in m4
E = [1 1 1 1 1 1]*2.1e11; % Young's modulus
L = [3 4 3 4 4 5]; % length in m
uu = 6; % Number of unrestrained degrees of freedom
ur = 6; % Number of restrained degrees of freedom
uul = [1 2 3 4 5 6]; % global labels of unrestrained dof
url = [7 8 9 10 11 12]; % global labels of restained dof
l1 = [4 7 6 9]; % Global labels for member 1
l2 = [1 4 3 6]; % Global labels for member 2
l3 = [1 2 8 12]; % Global labels for member 3
l4 = [2 5 3 6]; % Global labels for member 4
l5 = [5 10 6 11]; % Global labels for member 5
l6 = [4 5 8 12]; % Global labels for member 5
l= [l1; l2; l3; l4; l5; l6];
dof = uu+ur;
Ktotal = zeros (dof);
fem1= [-15 15 -10 -10]; % Local Fixed end moments of member 1
fem2= [-15 15 -10 -10]; % Local Fixed end moments of member 2
fem3= [0 0 100 100]; % Local Fixed end moments of member 3
fem4= [15 -15 10 10]; % Local Fixed end moments of member 4
fem5= [15 -15 10 10]; % Local Fixed end moments of member 5
fem6= [0 0 0 0]; % Local Fixed end moments of member 5
jlu = [15; -15; 0; 0; 0; 0]; % load vector in unrestrained dof
```

OUTPUT:

```
Member Number =   1
Local Stiffness matrix of member, [K] =
 1.0e+06 *
  2.3332  1.1666  1.1666 -1.1666
  1.1666  2.3332  1.1666 -1.1666
  1.1666  1.1666  0.7777 -0.7777
 -1.1666 -1.1666 -0.7777  0.7777
Member Number =   2
Local Stiffness matrix of member, [K] =
 1.0e+06 *
  1.7499  0.8750  0.6562 -0.6562
  0.8750  1.7499  0.6562 -0.6562
  0.6562  0.6562  0.3281 -0.3281
 -0.6562 -0.6562 -0.3281  0.3281
Member Number =   3
Local Stiffness matrix of member, [K] =
 1.0e+06 *
  4.6665  2.3332  2.3332 -2.3332
  2.3332  4.6665  2.3332 -2.3332
  2.3332  2.3332  1.5555 -1.5555
 -2.3332 -2.3332 -1.5555  1.5555
```

```
Member Number =  4
Local Stiffness matrix of member, [K] =
 1.0e+06 *
  1.7499  0.8750  0.6562 -0.6562
  0.8750  1.7499  0.6562 -0.6562
  0.6562  0.6562  0.3281 -0.3281
 -0.6562 -0.6562 -0.3281  0.3281
Member Number =  5
Local Stiffness matrix of member, [K] =
 1.0e+06 *
  1.7499  0.8750  0.6562 -0.6562
  0.8750  1.7499  0.6562 -0.6562
  0.6562  0.6562  0.3281 -0.3281
 -0.6562 -0.6562 -0.3281  0.3281
Member Number =  6
Local Stiffness matrix of member, [K] =
 1.0e+06 *
  2.7999  1.3999  0.8400 -0.8400
  1.3999  2.7999  0.8400 -0.8400
  0.8400  0.8400  0.3360 -0.3360
 -0.8400 -0.8400 -0.3360  0.3360
Stiffness Matrix of complete structure, [Ktotal] =
 1.0e+06 *
 Columns 1 through 9
  6.4164  2.3332  0.6562  0.8750       0 -0.6562       0  2.3332       0
  2.3332  6.4164  0.6562       0  0.8750 -0.6562       0  2.3332       0
  0.6562  0.6562  0.6562  0.6562  0.6562 -0.6562       0       0       0
  0.8750       0  0.6562  6.8831  1.3999  0.5104  1.1666  0.8400 -1.1666
       0  0.8750  0.6562  1.3999  6.2997       0       0  0.8400       0
 -0.6562 -0.6562 -0.6562  0.5104       0  1.7621  1.1666       0 -0.7777
       0       0       0  1.1666       0  1.1666  2.3332       0 -1.1666
  2.3332  2.3332       0  0.8400  0.8400       0       0  1.8915       0
       0       0       0 -1.1666       0 -0.7777 -1.1666       0  0.7777
       0       0       0       0  0.8750  0.6562       0       0       0
       0       0       0       0 -0.6562 -0.3281       0       0       0
 -2.3332 -2.3332       0 -0.8400 -0.8400       0       0 -1.8915       0
 Columns 10 through 12
       0       0 -2.3332
       0       0 -2.3332
       0       0       0
       0       0 -0.8400
  0.8750 -0.6562 -0.8400
  0.6562 -0.3281       0
       0       0       0
       0       0 -1.8915
       0       0       0
  1.7499 -0.6562       0
 -0.6562  0.3281       0
       0       0  1.8915
Unrestrained Stiffness sub-matix, [Kuu] =
 1.0e+06 *
  6.4164  2.3332  0.6562  0.8750       0 -0.6562
```

Basic Design Guidelines

```
   2.3332   6.4164   0.6562       0   0.8750  -0.6562
   0.6562   0.6562   0.6562   0.6562   0.6562  -0.6562
   0.8750        0   0.6562   6.8831   1.3999   0.5104
        0   0.8750   0.6562   1.3999   6.2997        0
  -0.6562  -0.6562  -0.6562   0.5104        0   1.7621
```
Inverse of Unrestrained Stiffness sub-matix, [KuuInverse] =
1.0e-05 *
```
   0.0195  -0.0060  -0.0144  -0.0017   0.0027   0.0002
  -0.0060   0.0195  -0.0168   0.0028  -0.0016  -0.0021
  -0.0144  -0.0168   0.4105  -0.0424  -0.0310   0.1535
  -0.0017   0.0028  -0.0424   0.0205  -0.0005  -0.0213
   0.0027  -0.0016  -0.0310  -0.0005   0.0194  -0.0110
   0.0002  -0.0021   0.1535  -0.0213  -0.0110   0.1194
```
Unrestrained Joint Load vector, [Jl] =
  15 -15   0   0   0
Unrestrained displacements, [DelU] =
1.0e-05 *
   0.3814 -0.3816   0.0366 -0.0675   0.0642   0.0331
Member Number = 1
Global displacement matrix [DeltaBar] =
1.0e-06 *
  -0.6748   0   0.3310   0
Global End moment matrix [MBar] =
 -16.1882 14.5990 -10.5298 -9.4702
Member Number = 2
Global displacement matrix [DeltaBar] =
1.0e-05 *
   0.3814 -0.0675   0.0366   0.0331
Global End moment matrix [MBar] =
  -8.8938 17.1790 -7.9287 -12.0713
Member Number = 3
Global displacement matrix [DeltaBar] =
1.0e-05 *
   0.3814 -0.3816   0   0
Global End moment matrix [MBar] =
   8.8938 -8.9073 99.9955 100.0045
Member Number = 4
Global displacement matrix [DeltaBar] =
1.0e-05 *
  -0.3816   0.0642   0.0366   0.0331
Global End moment matrix [MBar] =
   8.9073 -17.1926   7.9287 12.0713
Member Number = 5
Global displacement matrix [DeltaBar] =
1.0e-06 *
   0.6418   0   0.3310   0
Global End moment matrix [MBar] =
  16.3403 -14.2213 10.5298   9.4702
Member Number = 6

```
Global displacement matrix [DeltaBar] =
  1.0e-06 *
  -0.6748  0.6418     0     0
Global End moment matrix [MBar] =
  -0.9908  0.8523  -0.0277  0.0277
Joint forces =
 Columns 1 through 9
  0.0000  -0.0000  0.0000  -0.0000  -0.0000  -0.0000  14.5990
  99.9678  -9.4702
 Columns 10 through 12
 -14.2213  9.4702 100.0322
>>
```

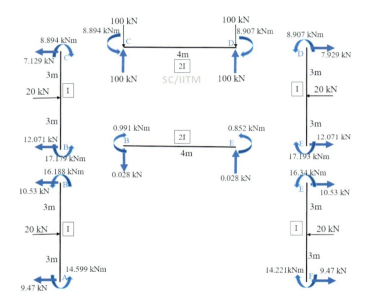

MATLAB® PROGRAM INPUT (WITH AXIAL LOAD):

```
pa = [100 100 7.9287 100 100 22.6011]*1000; %Axial load in N
load = [1 1 1 1 1 1]; % 0-zero load, 1-compression, 2-tension
```

OUTPUT:

```
Member Number =   1
Local Stiffness matrix of member, [K] =
  1.0e+06 *
  2.2930  1.1768  1.1566 -1.1566
  1.1768  2.2930  1.1566 -1.1566
  1.1566  1.1566  0.7377 -0.7377
 -1.1566 -1.1566 -0.7377  0.7377
Member Number =   2
Local Stiffness matrix of member, [K] =
```

Basic Design Guidelines

```
    1.0e+06 *
    1.6959  0.8887  0.6462 -0.6462
    0.8887  1.6959  0.6462 -0.6462
    0.6462  0.6462  0.2981 -0.2981
   -0.6462 -0.6462 -0.2981  0.2981
Member Number =    3
Local Stiffness matrix of member, [K] =
    1.0e+06 *
    4.6633  2.3340  2.3324 -2.3324
    2.3340  4.6633  2.3324 -2.3324
    2.3324  2.3324  1.5523 -1.5523
   -2.3324 -2.3324 -1.5523  1.5523
Member Number =    4
Local Stiffness matrix of member, [K] =
    1.0e+06 *
    1.6959  0.8887  0.6462 -0.6462
    0.8887  1.6959  0.6462 -0.6462
    0.6462  0.6462  0.2981 -0.2981
   -0.6462 -0.6462 -0.2981  0.2981
Member Number =    5
Local Stiffness matrix of member, [K] =
    1.0e+06 *
    1.6959  0.8887  0.6462 -0.6462
    0.8887  1.6959  0.6462 -0.6462
    0.6462  0.6462  0.2981 -0.2981
   -0.6462 -0.6462 -0.2981  0.2981
Member Number =    6
Local Stiffness matrix of member, [K] =
    1.0e+06 *
    2.7848  1.4037  0.8377 -0.8377
    1.4037  2.7848  0.8377 -0.8377
    0.8377  0.8377  0.3306 -0.3306
   -0.8377 -0.8377 -0.3306  0.3306
Stiffness Matrix of complete structure, [Ktotal] =
    1.0e+06 *
 Columns 1 through 9
   6.3593  2.3340  0.6462  0.8887       0 -0.6462       0  2.3324       0
   2.3340  6.3593  0.6462       0  0.8887 -0.6462       0  2.3324       0
   0.6462  0.6462  0.5962  0.6462  0.6462 -0.5962       0       0       0
   0.8887       0  0.6462  6.7737  1.4037  0.5104  1.1768  0.8377 -1.1566
        0  0.8887  0.6462  1.4037  6.1767       0       0  0.8377       0
  -0.6462 -0.6462 -0.5962  0.5104       0  1.6320  1.1566       0 -0.7377
        0       0       0  1.1768       0  1.1566  2.2930       0 -1.1566
   2.3324  2.3324       0  0.8377  0.8377       0       0  1.8829       0
        0       0       0 -1.1566       0 -0.7377 -1.1566       0  0.7377
        0       0       0       0  0.8887  0.6462       0       0       0
        0       0       0       0 -0.6462 -0.2981       0       0       0
  -2.3324 -2.3324       0 -0.8377 -0.8377       0       0 -1.8829       0
 Columns 10 through 12
        0       0 -2.3324
        0       0 -2.3324
```

```
     0      0       0
     0      0  -0.8377
 0.8887 -0.6462  -0.8377
 0.6462 -0.2981      0
     0      0       0
     0      0  -1.8829
     0      0       0
 1.6959 -0.6462      0
-0.6462  0.2981      0
     0      0  1.8829
Unrestrained Stiffness sub-matix, [Kuu] =
 1.0e+06 *
 6.3593  2.3340  0.6462  0.8887      0 -0.6462
 2.3340  6.3593  0.6462      0  0.8887 -0.6462
 0.6462  0.6462  0.5962  0.6462  0.6462 -0.5962
 0.8887       0  0.6462  6.7737  1.4037  0.5104
      0  0.8887  0.6462  1.4037  6.1767       0
-0.6462 -0.6462 -0.5962  0.5104       0  1.6320
Inverse of Unrestrained Stiffness sub-matix, [KuuInverse] =
  1.0e-05 *
  0.0198 -0.0060 -0.0164 -0.0017  0.0030  0.0000
 -0.0060  0.0199 -0.0192  0.0031 -0.0016 -0.0025
 -0.0164 -0.0192  0.4696 -0.0483 -0.0354  0.1726
 -0.0017  0.0031 -0.0483  0.0214 -0.0003 -0.0238
  0.0030 -0.0016 -0.0354 -0.0003  0.0202 -0.0123
  0.0000 -0.0025  0.1726 -0.0238 -0.0123  0.1307
Unrestrained Joint Load vector, [Jl] =
   15   -15    0    0    0    0
Unrestrained displacements, [DelU] =
  1.0e-05 *
  0.3879  -0.3882  0.0429  -0.0719  0.0677  0.0380
Member Number =   1
Global displacement matrix [DeltaBar] =
  1.0e-06 *
 -0.7188      0  0.3802      0
Global End moment matrix [MBar] =
 -16.2084  14.5939 -10.5508  -9.4492
Member Number =   2
Global displacement matrix [DeltaBar] =
  1.0e-05 *
  0.3879  -0.0719  0.0429  0.0380
Global End moment matrix [MBar] =
  -9.0287  17.2596  -7.9435 -12.0565
Member Number =   3
Global displacement matrix [DeltaBar] =
  1.0e-05 *
  0.3879  -0.3882     0     0
```

Basic Design Guidelines

```
Global End moment matrix [MBar] =
  9.0287  -9.0492  99.9932 100.0068
Member Number =   4
Global displacement matrix [DeltaBar] =
  1.0e-05 *
  -0.3882  0.0677  0.0429  0.0380
Global End moment matrix [MBar] =
  9.0492 -17.2704  7.9435  12.0565
Member Number =   5
Global displacement matrix [DeltaBar] =
  1.0e-06 *
  0.6771      0  0.3802     0
Global End moment matrix [MBar] =
  16.3939 -14.1526  10.5508  9.4492
Member Number =   6
Global displacement matrix [DeltaBar] =
  1.0e-06 *
  -0.7188  0.6771     0      0
Global End moment matrix [MBar] =
  -1.0512  0.8765  -0.0350  0.0350
Joint forces =
 Columns 1 through 9
  0.0000      0 -0.0000  0.0000  -0.0000  0.0000  0.0000  14.5939
99.9582  -9.4492
 Columns 10 through 12
 -14.1526  9.4492 100.0418
>>
```

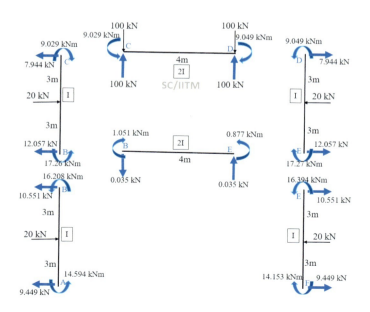

2.20 CRITICAL BUCKLING LOAD

One of the basic assumptions to obtain the buckling load of a structural system with rigid joints is that the structure's deformation is sufficiently small compared to that of its initial condition. Therefore, to estimate the critical buckling load, one can apply linear theory. It is also important to note that buckling loads of structural systems are estimated because the structural system is transferring only axial forces. Members who encounter transverse loads develop additional moments, which significantly alter the members' relative stiffness. Under such cases, where there is a continuous interaction of the moment and the applied axial load (P-M interaction), critical buckling load is estimated only through the iterative scheme. However, in the case of ideal members subjected to axial forces only, buckling loads can be estimated in a much simpler manner. By neglecting the axial deformations the members undergo, joint loads related to the un-restrained joint displacements are set to ZERO at all un-restrained joints.

$$\{J_{Lu}\} = \{0\} \tag{2.225}$$

However, from the fundamental equations of equilibrium, the following relationship holds good.

$$\{J_{Lu}\} = [K_{uu}]\{\Delta_u\} \tag{2.226}$$

where $[K_{uu}]$ is the sub-matrix of the stiffness matrix of the complete structural system, indicating only the un-restrained degrees-of-freedom, and Δ_u is the displacement vector of the un-restrained joints. By comparing the above two equations, it is obvious that Equation (2.225) can be true only if Δ_u is a null vector. But this is a trivial solution, which corresponds to no displacements of the un-restrained joints. Hence, to obtain the non-trivial solution to Equation (2.225), the following condition should be satisfied

$$|K_{uu}| = 0 \tag{2.227}$$

The above is termed the characteristic determinant, whose expansion will yield the buckling condition of the structural system. It is important that when the magnitude of the applied axial load is lesser than that of the critical buckling load, then the displacements of the un-restrained joints would be zero. Hence, $|K_{uu}|$ will be positive, which corresponds to a stable condition. On the other hand, if the applied axial load exceeds the critical buckling load, then it will refer to an unstable condition; in such case, $|K_{uu}|$ will be negative.

In a structural system comprising more members, more than one member may be subjected to axial load. In such cases, it becomes necessary to establish a relationship (φ_i) between the respective axial load and buckling load, as explained earlier. It becomes more complicated when more than one member shares the applied axial load at any joint. The axial load shared by each member at that joint needs to be

Basic Design Guidelines

computed by preliminary analysis. Then, the critical buckling load can be estimated using the following relationship

$$P_{cr}^i = (\varphi_i)_{critical} \frac{\pi^2 EI}{L_i^2} \qquad (2.228)$$

Example 1 Estimate the critical buckling load of the column member using stability functions.

With reference to the above diagram, kinematic degrees of freedom, both unrestrained and restrained, are marked. A detailed procedure of analyzing a member using the stiffness method is discussed in Chandrasekaran, 2017. The member stiffness matrix is given below. Numbers shown in circles are the labels of the degrees of freedom.

$$[K_i] = EI \begin{bmatrix} \frac{r}{L} & \frac{cr}{L} & \frac{r(1+c)}{L^2} & -\frac{r(1+c)}{L^2} \\ \frac{cr}{L} & \frac{r}{L} & \frac{r(1+c)}{L^2} & -\frac{r(1+c)}{L^2} \\ \frac{r(1+c)}{L^2} & \frac{r(1+c)}{L^2} & \frac{2tr(1+c)}{L^3} & -\frac{2tr(1+c)}{L^3} \\ -\frac{r(1+c)}{L^2} & -\frac{r(1+c)}{L^2} & -\frac{2tr(1+c)}{L^3} & \frac{2tr(1+c)}{L^3} \end{bmatrix} \begin{matrix} ① \\ ③ \\ ② \\ ④ \end{matrix} \qquad (2.229)$$

The above matrix can be partitioned to extract the unrestrained stiffness matrix, which will be 2 × 2, as there are two unrestrained degrees of freedom for this problem. Unrestrained stiffness matrix, as extracted from the Equation (2.229), is given by

$$K_{uu} = EI \begin{bmatrix} \dfrac{r}{L} & \dfrac{r(1+c)}{L^2} \\ \dfrac{r(1+c)}{L^2} & \dfrac{2tr(1+c)}{L^3} \end{bmatrix} \begin{array}{c} ① \\ ② \end{array} \quad (2.230)$$

With reference to the above diagram, kinematic degrees of freedom, both unrestrained and restrained, are marked. A detailed procedure of analyzing a member using the stiffness method is discussed in Chandrasekaran, 2017. The member stiffness matrix is given below. Numbers shown in circles are the labels of the degrees of freedom.

As there is no joint load applied in the unrestrained degrees of freedom, let us set the determinant of Equation (2.230) to zero. Expanding, we get as follows

$$\left(\dfrac{r}{L}\right)\left[\dfrac{2tr(1+c)}{L^3}\right] - \left[\dfrac{r(1+c)}{L^2}\right]^2 = 0 \quad (2.231)$$

Simplifying the above expression, we get the following expression

$$2t - c = 1 \quad (2.232)$$

Equation (2.232) is called the characteristic equation, defining the critical buckling condition in terms of the stability functions. Using tables one can determine the value of φ_i satisfying the Equation (2.232). MATLAB® given below helps to scan the appropriate values of the stability functions and determines φ_i as 0.25. The corresponding values of stability functions are $r = 3.6598$, $t = 0.7854$, $c = 0.5708$.

Using Equation (2.228), we get

$$P_{cr} = \varphi \dfrac{\pi^2 EI}{L^2} = (0.25)\dfrac{\pi^2 EI}{L^2} \quad (2.233)$$

```
% getting phi value condition: 2t = 1+c
lhs = 2*t;
rhs = 1+c;
n = length(out);
for i=1:n
  if lhs(i) == rhs(i)
    phi_value = out(i,1);
  end
end
fprintf('Phi = %6.2f \n',phi_value);
```

OUTPUT:

```
Phi = 0.25
```

Basic Design Guidelines

Example 2 Estimate the critical buckling load of the structure using stability functions.

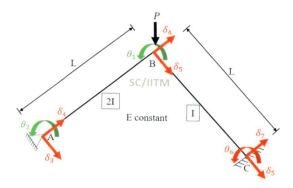

With reference to the above diagram, kinematic degrees of freedom, both unrestrained and restrained, are marked. The member stiffness matrix for both the members is given below.

Numbers shown in circles are the labels of the degrees of freedom of the respective member.

$$[K_1] = E(2I) \begin{bmatrix} \frac{r_1}{L_1} & \frac{c_1 r_1}{L_1} & \frac{r_1(1+c_1)}{L_1^2} & -\frac{r_1(1+c_1)}{L_1^2} \\ \frac{c_1 r_1}{L_1} & \frac{r_1}{L_1} & \frac{r_1(1+c_1)}{L_1^2} & -\frac{r_1(1+c_1)}{L_1^2} \\ \frac{r_1(1+c_1)}{L_1^2} & \frac{r_1(1+c_1)}{L_1^2} & \frac{2t_1 r_1(1+c_1)}{L_1^3} & -\frac{2t_1 r_1(1+c_1)}{L_1^3} \\ -\frac{r_1(1+c_1)}{L_1^2} & -\frac{r_1(1+c_1)}{L_1^2} & -\frac{2t_1 r_1(1+c_1)}{L_1^3} & \frac{2t_1 r_1(1+c_1)}{L_1^3} \end{bmatrix} \begin{matrix} ① \\ ② \\ ⑤ \\ ③ \end{matrix} \quad (2.234)$$

with column labels ①, ②, ⑤, ③

$$[K_2] = EI \begin{bmatrix} \frac{r_2}{L_2} & \frac{c_2 r_2}{L_2} & \frac{r_2(1+c_2)}{L_2^2} & -\frac{r_2(1+c_2)}{L_2^2} \\ \frac{c_2 r_2}{L_2} & \frac{r_2}{L_2} & \frac{r_2(1+c_2)}{L_2^2} & -\frac{r_2(1+c_2)}{L_2^2} \\ \frac{r_2(1+c_2)}{L_2^2} & \frac{r_2(1+c_2)}{L_2^2} & \frac{2t_2 r_2(1+c_2)}{L_2^3} & -\frac{2t_2 r_2(1+c_2)}{L_2^3} \\ -\frac{r_2(1+c_2)}{L_2^2} & -\frac{r_2(1+c_2)}{L_2^2} & -\frac{2t_2 r_2(1+c_2)}{L_2^3} & \frac{2t_2 r_2(1+c_2)}{L_2^3} \end{bmatrix} \begin{matrix} ① \\ ⑥ \\ ④ \\ ⑦ \end{matrix} \quad (2.235)$$

with column labels ①, ⑥, ④, ⑦

The above matrices are assembled to obtain the unrestrained stiffness matrix, which will be of size 2 × 2, as there are two unrestrained degrees of freedom for this problem. In the above equations of the stiffness matrices, substitute $L_1 = L_2 = L$. Unrestrained stiffness matrix, as assembled by combining Equation (2.234) and Equation (2.235), is given by

$$K_{uu} = EI \begin{bmatrix} \dfrac{2r_1}{L} + \dfrac{r_2}{L} & \dfrac{2c_1 r_1}{L} \\ \dfrac{2c_1 r_1}{L} & \dfrac{2r_1}{L} \end{bmatrix} \begin{matrix} ① \\ ② \end{matrix} \quad (2.236)$$

with column labels ① ②.

As there is no joint load applied in the unrestrained degrees of freedom, let us set the determinant of Equation (2.236) to zero. Expanding, we get as follows

$$\left(\dfrac{2r_1}{L} + \dfrac{r_2}{L}\right)\left(\dfrac{2r_1}{L}\right) - \left[\dfrac{2c_1 r_1}{L}\right]^2 = 0 \quad (2.237)$$

Simplifying, we get the following expression

$$2r_1\left(1 - c_1^2\right) + r_2 \quad (2.238)$$

Equation (2.238) is the characteristic equation, defining the critical buckling condition in terms of the stability functions. Also, we know the following relationship about the problem loading

$$(P_{cr})_1 = (P_{cr})_{2-} = P_{cr}(0.707) \quad (2.239)$$

$$\varphi_1 = \dfrac{P(0.707)}{\left[\dfrac{\pi^2 E(2I)}{L^2}\right]} \quad (2.240)$$

$$\varphi_2 = \dfrac{P(0.707)}{\left[\dfrac{\pi^2 EI}{L^2}\right]} \quad (2.241)$$

From the above equations, the following relationship can be deduced

$$2\varphi_1 = \varphi_2 \quad (2.242)$$

Using tables, one can determine the value of φ_i satisfying both Equation (2.238) and Equation (2.242). MATLAB® code given below helps to scan the appropriate values of the stability functions and determines φ_1 as 1.01 and φ_2 as 2.02. The corresponding values of stability functions are: $r_1 = 2.4493$; $t_1 = -0.0124$; $c_1 = 1.0101$ and $r_2 = 0.1120$; $t_2 = -1.7148$; $c_2 = 31.6264$.

Using Equation (2.240), we get

$$P_{cr} = \varphi_1 \dfrac{\pi^2 E(2I)}{(0.707)L^2} \quad (2.243)$$

Basic Design Guidelines

```
% getting phi value condition: 2φ₁ = φ₂ and 2 r₁ (1-c₁²) + r₂ = 0
lhs = 2*phi;
rhs = phi;
n = length (out);
phi1 = zeros(n,1);
phi2 = zeros(n,1);
u=1;
for i=1:n
  for j=1:n
  if lhs(i) == rhs(j)
    r1 = r(i);
    c1 = c(i);
    r2 = r(j);
    if -0.00001 <= (2*r1*(1-(c1^2)))+r2 <=0.00001
      phi1(u) = phi(i);
      phi2(u) = phi(j);
      u=u+1;
    end
  end
  end
end
fprintf('Phi 1 = %6.2f \n',phi1(1));
fprintf('Phi 2 = %6.2f \n',phi2(1));
```

OUTPUT:

```
Phi 1 =   1.01
Phi 2 =   2.02
```

Example 3 Estimate the critical buckling load of the structure using stability functions.

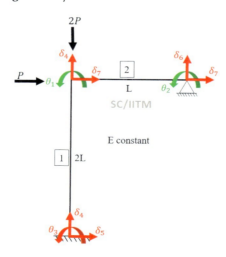

With reference to the above diagram, kinematic degrees of freedom, both unrestrained and restrained, are marked. The member stiffness matrix for both the members is given below. Numbers shown in circles are the labels of the degrees of freedom of the respective member.

$$[K_1] = EI \begin{bmatrix} \dfrac{r_1}{L_1} & \dfrac{c_1 r_1}{L_1} & \dfrac{r_1(1+c_1)}{L_1^2} & -\dfrac{r_1(1+c_1)}{L_1^2} \\ \dfrac{c_1 r_1}{L_1} & \dfrac{r_1}{L_1} & \dfrac{r_1(1+c_1)}{L_1^2} & -\dfrac{r_1(1+c_1)}{L_1^2} \\ \dfrac{r_1(1+c_1)}{L_1^2} & \dfrac{r_1(1+c_1)}{L_1^2} & \dfrac{2t_1 r_1(1+c_1)}{L_1^3} & -\dfrac{2t_1 r_1(1+c_1)}{L_1^3} \\ -\dfrac{r_1(1+c_1)}{L_1^2} & -\dfrac{r_1(1+c_1)}{L_1^2} & -\dfrac{2t_1 r_1(1+c_1)}{L_1^3} & \dfrac{2t_1 r_1(1+c_1)}{L_1^3} \end{bmatrix} \begin{matrix} ① \\ ③ \\ ⑦ \\ ⑤ \end{matrix} \quad (2.244)$$

(columns: ① ③ ⑦ ⑤)

$$[K_2] = EI \begin{bmatrix} \dfrac{r_2}{L_2} & \dfrac{c_2 r_2}{L_2} & \dfrac{r_2(1+c_2)}{L_2^2} & -\dfrac{r_2(1+c_2)}{L_2^2} \\ \dfrac{c_2 r_2}{L_2} & \dfrac{r_2}{L_2} & \dfrac{r_2(1+c_2)}{L_2^2} & -\dfrac{r_2(1+c_2)}{L_2^2} \\ \dfrac{r_2(1+c_2)}{L_2^2} & \dfrac{r_2(1+c_2)}{L_2^2} & \dfrac{2t_2 r_2(1+c_2)}{L_2^3} & -\dfrac{2t_2 r_2(1+c_2)}{L_2^3} \\ -\dfrac{r_2(1+c_2)}{L_2^2} & -\dfrac{r_2(1+c_2)}{L_2^2} & -\dfrac{2t_2 r_2(1+c_2)}{L_2^3} & \dfrac{2t_2 r_2(1+c_2)}{L_2^3} \end{bmatrix} \begin{matrix} ① \\ ② \\ ④ \\ ⑥ \end{matrix} \quad (2.245)$$

(columns: ① ② ④ ⑥)

The above matrices are assembled to obtain the unrestrained stiffness matrix, which will be of size 2 × 2, as there are two unrestrained degrees of freedom for this problem. In the above equations of the stiffness matrices, substitute $L_1 = 2L$ and $L_2 = L$. Unrestrained stiffness matrix, as assembled by combining Equation (2.244) and Equation (2.245), is given by

$$K_{uu} = EI \begin{bmatrix} \dfrac{r_1 + 2r_2}{2L} & \dfrac{c_2 r_2}{L} \\ \dfrac{c_2 r_2}{L} & \dfrac{r_2}{L} \end{bmatrix} \begin{matrix} ① \\ ② \end{matrix} \quad (2.246)$$

As there is no joint load applied in the unrestrained degrees of freedom, let us set the determinant of Equation (2.246) to zero. Expanding, we get as follows

$$\left(\dfrac{r_1 + 2r_2}{2L}\right)\left(\dfrac{r_2}{L}\right) - \left[\dfrac{c_2 r_2}{L}\right]^2 = 0 \quad (2.247)$$

Simplifying, we get the following expression

$$r_1 = 2r_2\left(c_2^2 - 1\right) \quad (2.248)$$

Basic Design Guidelines

Equation (2.248) is the characteristic equation defining the critical buckling condition in terms of the stability functions. Also, we know the following relationship about the problem loading

$$(P_{cr})_1 = 2P \tag{2.249}$$

$$(P_{cr})_2 = P \tag{2.250}$$

$$\varphi_1 = \frac{2P}{\left[\dfrac{\pi^2 EI}{4L^2}\right]} = \frac{8P}{\left[\dfrac{\pi^2 EI}{L^2}\right]} \tag{2.251}$$

$$\varphi_2 = \frac{P}{\left[\dfrac{\pi^2 EI}{L^2}\right]} \tag{2.252}$$

From the above equations, the following relationship can be deduced

$$\varphi_1 = 8\varphi_2 \tag{2.253}$$

Using tables, one can determine the value of φ_i satisfying both Equation (2.248) and Equation (2.253). MATLAB® code given below helps to scan the appropriate values of the stability functions and determines φ_1 as 2.96 and φ_2 as 0.37. $r_1 = -4.6727$; $t_1 = -5.7540$; $c_1 = -1.4628$ and $r_2 = 3.4878$; $t_2 = 0.6754$; $c_2 = 0.6127$.
Using Equation (2.251), we get

$$P_{cr} = \varphi_1 \frac{\pi^2 EI}{8L^2} \tag{2.254}$$

```
% getting phi value condition: φ₁ = 8φ₂ and r₁ = 2r₂ (c₂²-1)
lhs = phi;
rhs = 8*phi;
n = length (out);
phi1 = zeros(n,1);
phi2 = zeros(n,1);
u=1;
for i=1:n
  for j=1:n
  if lhs(i) == rhs(j)
    r1 = r(i);
    c1 = c(i);
    r2 = r(j);
    c2 = c(j);
    if -0.00001 <= (r1-(2*r2*((c2^2)-1))) <=0.00001
      phi1(u) = phi(i);
      phi2(u) = phi(j);
      u=u+1;
    end
```

```
      end
    end
end
fprintf('Phi 1 = %6.2f \n',phi1(1));
fprintf('Phi 2 = %6.2f \n',phi2(1));
```

OUTPUT:

```
Phi 1 =   2.96 Phi 2 =   0.37
```

Example 4 Estimate the critical buckling load of the structure using stability functions.

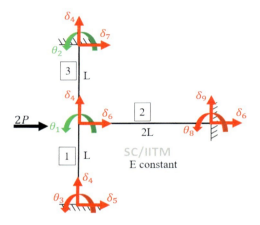

With reference to the above diagram, kinematic degrees of freedom, both unrestrained and restrained, are marked. The member stiffness matrix for both the members is given below. Numbers shown in circles are the labels of the degrees of freedom of the respective member.

$$[K_1] = EI \begin{bmatrix} \dfrac{r_1}{L_1} & \dfrac{c_1 r_1}{L_1} & \dfrac{r_1(1+c_1)}{L_1^2} & -\dfrac{r_1(1+c_1)}{L_1^2} \\ \dfrac{c_1 r_1}{L_1} & \dfrac{r_1}{L_1} & \dfrac{r_1(1+c_1)}{L_1^2} & -\dfrac{r_1(1+c_1)}{L_1^2} \\ \dfrac{r_1(1+c_1)}{L_1^2} & \dfrac{r_1(1+c_1)}{L_1^2} & \dfrac{2t_1 r_1(1+c_1)}{L_1^3} & -\dfrac{2t_1 r_1(1+c_1)}{L_1^3} \\ -\dfrac{r_1(1+c_1)}{L_1^2} & -\dfrac{r_1(1+c_1)}{L_1^2} & -\dfrac{2t_1 r_1(1+c_1)}{L_1^3} & \dfrac{2t_1 r_1(1+c_1)}{L_1^3} \end{bmatrix} \begin{matrix} ① \\ ③ \\ ⑥ \\ ⑤ \end{matrix} \quad (2.255)$$

with column labels ① ③ ⑥ ⑤

Basic Design Guidelines

$$[K_2] = EI \begin{bmatrix} \frac{r_2}{L_2} & \frac{c_2 r_2}{L_2} & \frac{r_2(1+c_2)}{L_2^2} & -\frac{r_2(1+c_2)}{L_2^2} \\ \frac{c_2 r_2}{L_2} & \frac{r_2}{L_2} & \frac{r_2(1+c_2)}{L_2^2} & -\frac{r_2(1+c_2)}{L_2^2} \\ \frac{r_2(1+c_2)}{L_2^2} & \frac{r_2(1+c_2)}{L_2^2} & \frac{2 t_2 r_2(1+c_2)}{L_2^3} & -\frac{2 t_2 r_2(1+c_2)}{L_2^3} \\ -\frac{r_2(1+c_2)}{L_2^2} & -\frac{r_2(1+c_2)}{L_2^2} & -\frac{2 t_2 r_2(1+c_2)}{L_2^3} & \frac{2 t_2 r_2(1+c_2)}{L_2^3} \end{bmatrix} \begin{matrix} ② \\ ① \\ ⑦ \\ ⑥ \end{matrix} \quad (2.256)$$

$$[K_3] = EI \begin{bmatrix} \frac{r_3}{L_3} & \frac{c_3 r_3}{L_3} & \frac{r_3(1+c_3)}{L_3^2} & -\frac{r_3(1+c_3)}{L_3^2} \\ \frac{c_3 r_3}{L_3} & \frac{r_3}{L_3} & \frac{r_3(1+c_3)}{L_3^2} & -\frac{r_3(1+c_3)}{L_3^2} \\ \frac{r_3(1+c_3)}{L_3^2} & \frac{r_3(1+c_3)}{L_3^2} & \frac{2 t_3 r_3(1+c_3)}{L_3^3} & -\frac{2 t_3 r_3(1+c_3)}{L_3^3} \\ -\frac{r_3(1+c_3)}{L_3^2} & -\frac{r_3(1+c_3)}{L_3^2} & -\frac{2 t_3 r_3(1+c_3)}{L_3^3} & \frac{2 t_3 r_3(1+c_3)}{L_3^3} \end{bmatrix} \begin{matrix} ① \\ ⑧ \\ ④ \\ ⑨ \end{matrix} \quad (2.257)$$

The above matrices are assembled to obtain the unrestrained stiffness matrix, which will be of size 2 × 2, as there are two unrestrained degrees of freedom for this problem. In the above equations of the stiffness matrices, substitute $L_1 = L_2 = L$; $L_3 = 2L$. After appropriate substitution, unrestrained stiffness matrix, as assembled by combining Equation (2.255–2.257), is given by

$$K_{uu} = EI \begin{bmatrix} \frac{2 r_1 + r_3}{2L} & \frac{c_2 r_2}{L} \\ \frac{c_2 r_2}{L} & \frac{r_2}{L} \end{bmatrix} \begin{matrix} ① \\ ② \end{matrix} \quad (2.258)$$

As there is no joint load applied in the unrestrained degrees of freedom, let us set the determinant of Equation (2.258) to zero. Expanding, we get as follows

$$\left(\frac{2 r_1 + r_3}{2L}\right)\left(\frac{r_2}{L}\right) - \left[\frac{c_2 r_2}{L}\right]^2 = 0 \quad (2.259)$$

Simplifying, we get the following expression

$$2 r_1 + r_3 = 2 r_2 c_2^2 \quad (2.260)$$

Equation (2.260) is the characteristic equation, defining the critical buckling condition in terms of the stability functions. Also, we know the following relationship about the problem loading

$$\varphi_1 = \varphi_2 = 0 \quad (2.261)$$

$$(P_{cr})_3 = 2P \quad (2.262)$$

$$\varphi_3 = \frac{2P}{\left[\dfrac{\pi^2 EI}{4L^2}\right]} = \frac{8P}{\left[\dfrac{\pi^2 EI}{L^2}\right]} \quad (2.263)$$

Using tables, one can determine the value of φ_i satisfying Equation (2.260), Equation (2.261), and Equation (2.263). MATLAB® code given below helps to scan the appropriate values of the stability functions and determines φ_3 as 3.10; $r_3 = -6.0519$; $t_3 = -7.0072$; $c_3 = -1.3157$

Using Equation (2.263), we get

$$P_{cr} = \varphi_3 \frac{\pi^2 EI}{8L^2} \quad (2.264)$$

```
% getting phi value condition: phi1 = phi2 =0
r1 = 4;
r2 = 4;
c2 = 0.5;
r3 = (2*r2*c2*c2)-(2*r1);
n = length (out);
phi3 = zeros(n,1);
u=1;
for i = 1:n
  if r3 == round(r(i))
    if r(i)-r3 <=0.00001
      phi3(u) = phi(i);
      u=u+1;
    end
  end
end
fprintf('Phi 3 = %6.2f \n',phi3(1));
```

OUTPUT:

```
Phi 3 =   3.10
```

Example 5 Estimate the critical buckling load of the structure using stability functions.

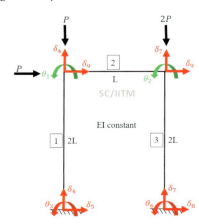

Basic Design Guidelines

With reference to the above diagram, kinematic degrees of freedom, both unrestrained and restrained, are marked. The member stiffness matrix for both the members is given below. Numbers shown in circles are the labels of the degrees of freedom of the respective member.

$$[K_1] = EI \begin{bmatrix} \dfrac{r_1}{L_1} & \dfrac{c_1 r_1}{L_1} & \dfrac{r_1(1+c_1)}{L_1^2} & -\dfrac{r_1(1+c_1)}{L_1^2} \\ \dfrac{c_1 r_1}{L_1} & \dfrac{r_1}{L_1} & \dfrac{r_1(1+c_1)}{L_1^2} & -\dfrac{r_1(1+c_1)}{L_1^2} \\ \dfrac{r_1(1+c_1)}{L_1^2} & \dfrac{r_1(1+c_1)}{L_1^2} & \dfrac{2t_1 r_1(1+c_1)}{L_1^3} & -\dfrac{2t_1 r_1(1+c_1)}{L_1^3} \\ -\dfrac{r_1(1+c_1)}{L_1^2} & -\dfrac{r_1(1+c_1)}{L_1^2} & -\dfrac{2t_1 r_1(1+c_1)}{L_1^3} & \dfrac{2t_1 r_1(1+c_1)}{L_1^3} \end{bmatrix} \begin{matrix} ① \\ ③ \\ ⑨ \\ ⑤ \end{matrix} \quad (2.265)$$

(columns labeled ①, ③, ⑨, ⑤)

$$[K_2] = EI \begin{bmatrix} \dfrac{r_2}{L_2} & \dfrac{c_2 r_2}{L_2} & \dfrac{r_2(1+c_2)}{L_2^2} & -\dfrac{r_2(1+c_2)}{L_2^2} \\ \dfrac{c_2 r_2}{L_2} & \dfrac{r_2}{L_2} & \dfrac{r_2(1+c_2)}{L_2^2} & -\dfrac{r_2(1+c_2)}{L_2^2} \\ \dfrac{r_2(1+c_2)}{L_2^2} & \dfrac{r_2(1+c_2)}{L_2^2} & \dfrac{2t_2 r_2(1+c_2)}{L_2^3} & -\dfrac{2t_2 r_2(1+c_2)}{L_2^3} \\ -\dfrac{r_2(1+c_2)}{L_2^2} & -\dfrac{r_2(1+c_2)}{L_2^2} & -\dfrac{2t_2 r_2(1+c_2)}{L_2^3} & \dfrac{2t_2 r_2(1+c_2)}{L_2^3} \end{bmatrix} \begin{matrix} ① \\ ② \\ ④ \\ ⑦ \end{matrix} \quad (2.266)$$

(columns labeled ①, ②, ④, ⑦)

$$[K_3] = EI \begin{bmatrix} \dfrac{r_3}{L_3} & \dfrac{c_3 r_3}{L_3} & \dfrac{r_3(1+c_3)}{L_3^2} & -\dfrac{r_3(1+c_3)}{L_3^2} \\ \dfrac{c_3 r_3}{L_3} & \dfrac{r_3}{L_3} & \dfrac{r_3(1+c_3)}{L_3^2} & -\dfrac{r_3(1+c_3)}{L_3^2} \\ \dfrac{r_3(1+c_3)}{L_3^2} & \dfrac{r_3(1+c_3)}{L_3^2} & \dfrac{2t_3 r_3(1+c_3)}{L_3^3} & -\dfrac{2t_3 r_3(1+c_3)}{L_3^3} \\ -\dfrac{r_3(1+c_3)}{L_3^2} & -\dfrac{r_3(1+c_3)}{L_3^2} & -\dfrac{2t_3 r_3(1+c_3)}{L_3^3} & \dfrac{2t_3 r_3(1+c_3)}{L_3^3} \end{bmatrix} \begin{matrix} ② \\ ⑥ \\ ⑨ \\ ⑧ \end{matrix} \quad (2.267)$$

(columns labeled ②, ⑥, ⑨, ⑧)

The above matrices are assembled to obtain the unrestrained stiffness matrix, which will be of size 2×2, as there are two unrestrained degrees of freedom for this problem. In the above equations of the stiffness matrices, substitute $L_1 = L_3 = 2L$; $L_2 = L$. After appropriate substitution, unrestrained stiffness matrix, as assembled by combining Equations (2.265–2.267) is given by

$$K_{uu} = EI \begin{bmatrix} \dfrac{r_1 + 2r_2}{2L} & \dfrac{c_2 r_2}{L} \\ \dfrac{c_2 r_2}{L} & \dfrac{2r_2 + r_3}{2L} \end{bmatrix} \begin{matrix} ① \\ ② \end{matrix} \quad (2.268)$$

(columns labeled ①, ②)

As there is no joint load applied in the unrestrained degrees of freedom, let us set the determinant of Equation (2.268) to zero. Expanding, we get as follows

$$\left(\frac{r_1 + 2r_2}{2L}\right)\left(\frac{2r_2 + r_3}{2L}\right) - \left[\frac{c_2 r_2}{L}\right]^2 = 0 \tag{2.269}$$

Simplifying, we get the following expression

$$r_1(2r_2 + r_3) + r_2(4r_2 + 2r_3) = 4r_2^2 c_2^2 \tag{2.270}$$

Equation (2.270) is the characteristic equation, defining the critical buckling condition in terms of the stability functions. Also, we know the following relationship about the problem loading

$$\varphi_1 = \frac{P}{\left[\frac{\pi^2 EI}{4L^2}\right]} = \frac{4P}{\left[\frac{\pi^2 EI}{L^2}\right]} \tag{2.271}$$

$$\varphi_2 = \frac{P}{\left[\frac{\pi^2 EI}{L^2}\right]} \tag{2.272}$$

$$\varphi_3 = \frac{2P}{\left[\frac{\pi^2 EI}{4L^2}\right]} = \frac{8P}{\left[\frac{\pi^2 EI}{L^2}\right]} \tag{2.273}$$

$$\varphi_1 = 4\varphi_2 \tag{2.274}$$

$$\varphi_3 = 8\varphi_2 \tag{2.275}$$

Using tables, one can determine the value of φ_i satisfying the Equation (2.270), Equation (2.274), and Equation (2.275). MATLAB® code given below helps to scan the appropriate values of the stability functions and determines $\varphi_2 = 0.38$. The corresponding stability coefficients are $r_2 = 3.4732$; $t_2 = 0.6660$; $c_2 = 0.6165$.

Using Equation (2.272), we get

$$P_{cr} = \varphi_2 \frac{\pi^2 EI}{L^2} \tag{2.276}$$

```
% getting phi value condition: phi1 = 4phi2
n = length (out);
lhs = phi;
rhs = 4*phi;
phi1 = zeros(n,1);
phi2 = zeros(n,1);
phi3 = zeros(n,1);
u=1;
for i=1:n
  for j=1:n
  if lhs(i) == rhs(j)
```

Basic Design Guidelines

```
    r1 = r(i);
    c1 = c(i);
    phi22 = phi(j);
    r2 = r(j);
    c2 = c(j);
    for k = 1:n
      if (8*phi22) == phi(k)
        r3 = r(k);
        if -0.00001 <= ((r1*((2*r2)+r3))+(r2*((2*r3)+(4*r2)))
           -(4*r2*r2*c2*c2)) <= 0.00001
          phi1(u) = phi(i);
          phi2(u) = phi(j);
          phi3(u) = phi(k);
          u=u+1;
        end
      end
    end
  end
  end
end
fprintf('Phi 1 = %6.2f \n',phi1(1));
fprintf('Phi 2 = %6.2f \n',phi2(1));
fprintf('Phi 3 = %6.2f \n',phi3(1));
```

OUTPUT:

```
Phi 1 =   0.76
Phi 2 =   0.19
Phi 3 =   1.52
```

MATLAB® CODE FOR PREPARING THE STABILITY CHART

```
%% This MATLAB code is for plotting the stability indices
% Re-type the following code in MATLAB new script and run the
file to get the output.
clc;
clear;
phi=0.01:0.01:4;
alrad = pi.*sqrt(phi);
al = radtodeg (alrad);
% Compression
r = (alrad.*((sind(al))-(alrad.*cosd(al))))./((2.*(1-cosd
(al)))-(alrad.*(sind(al)))); % rotation function
c = (alrad-sind(al))./(sind(al)-(alrad.*cosd(al))); % rotation
function
t = 1-((pi*pi*phi)./(2.*r.*(1+c))); % Translation function
outcomp = [phi' r' c' t'];
% for zero
outzero = [0 4.0 0.5 1.0];
% Tension
phit =10:-0.1:0.1;
```

```
alrad = pi.*sqrt(phit);
al = radtodeg (alrad);
rt = (alrad.*((alrad.*cosh(alrad))-sinh(alrad)))./((2.*
(1-cosh(alrad)))+(alrad.*sinh(alrad)));
ct = (alrad-sinh(alrad))./(sinh(alrad)-(alrad.*cosh(alrad)));
tt = 1-((pi*pi*(-phit))./(2.*rt.*(1+ct)));
outten = [-phit' rt' ct' tt'];
% Stability chart
out = [outten; outzero; outcomp];
filename='StabilityChart.xlsx';
sheet=1;
xlswrite(filename,out,sheet);
phi = out(:,1);
r = out(:,2);
c = out(:,3);
t = out(:,4);
figure;
plot (phi,r,'b','linewidth',2)
xlabel('Phi');
ylabel('r');
grid on;
figure;
plot (phi,c,'b','linewidth',2)
xlabel('Phi');
ylabel('c');
grid on;
figure;
plot (phi,t, 'b','linewidth',2)
xlabel('Phi');
ylabel('t');
grid on;
```

OUTPUT:

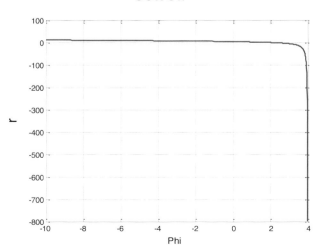

Basic Design Guidelines

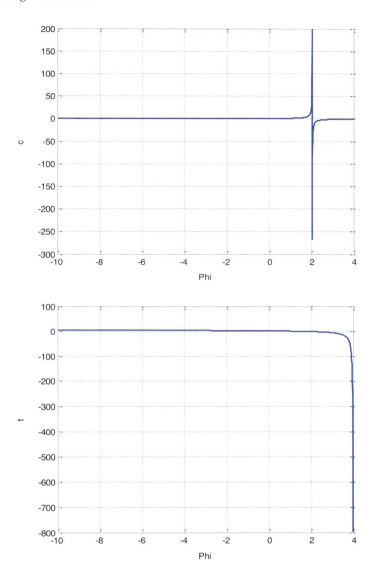

EXERCISE

1. Discuss basic design guidelines for the topside of offshore platforms.
2. What are the factors considered and finalized before the topside design? Explain.
3. Discuss various design loads considered in the topside design.
4. What are deformation loads? Explain with examples.
5. Discuss various load combinations for which the topside is designed.
6. Discuss static in-place analysis.
7. What do you understand by load-out analysis?

8. What modifications are made to the static in-place analysis to carry out load-out analysis in SACS?
9. Discuss lifting analysis.
10. Discuss transportation analysis used in the topside design.
11. What do you understand by weight control in the topside design? Explain.
12. Discuss various numerical tools useful in the topside design.
13. Discuss various design considerations adopted in the topside design.
14. What are the critical design acceptance criteria for the topside design as recommended by various design codes? Discuss them in detail.
15. Compare various design methods, highlighting their application limitations.
16. Explain force-controlled and displacement-controlled design approaches.
17. What is the importance of shape factor in the plastic design?
18. Derive Euler's critical load from first principles.
19. List the steps involved in carrying out stability analysis of a beam column.
20. Define critical buckling load. How is it obtained?

3 Special Design Guidelines

3.1 THIN-WALLED SECTIONS

Structural steel is the preferred construction material for many engineering projects. Structural steel offers many benefits over reinforced concrete construction in certain types of structures, such as a long-span roof trusses, pavilions, bridges, stadiums, and offshore topside. Usually, structural steel sections are composite-shaped, lightweight, and thin. While the design procedure with structural steel is very simple compared to using concrete and composite materials, it is equally sensitive to the complex check procedures that arise due to its thin cross-sections. Therefore, in addition to flexure, shear, and axial resistance, other parameters are also required to satisfy the design of steel beams and columns. Thanks to the well-documented design procedures of the design codes, design checks are generally carried out without much knowledge of structural response behavior. An offshore structural engineer should be intuitive in understanding structural response in special design checks instead of simply following the checks using equations. Attention is therefore drawn to the conceptual understanding in this chapter.

Let us try to understand torsion from first principles. Figure 3.1 illustrates a circular section subjected to torsion, T, at one of its ends. It is a case of pure torsion, which is commonly referred to as St. Venant torsion. A circular bar (radius, r, and length, L) is subjected to pure torsion at one of its ends. As we already know, this is possible only if the other end of the bar is fixed or embedded firmly into rigid support. It makes the support conditions one end fixed and the other end free, applying a twisting moment, T, at the free end. Incidentally, it also simulates the bar as a cantilever beam. Now, as shown in Figure 3.1, under the influence of torsion, an initial point on the bar's circumference at the free end is shifted to a new position. While analyzing the bar under such conditions, a few assumptions are made.

There is no change in the diameter of the bar before and after applying T. This implies an important condition that any section, sliced normal to the longitudinal axis of the member marking the cross-section, is circular, despite a torsion applied at the end. There is no distortion to the shape of the cross-section. We also understand that such a condition is easy only with elastic materials, and hence the material is assumed to be linear-elastic. Please note that the stresses developed due to the twisting moment are in no way comparable to the material's yield stress. It is due to the fact that the latter is material property under axial stress–strain and not under any twisting moment. But, still, this assumption is helpful in the later stages, as we will understand when we proceed further.

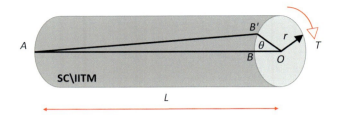

FIGURE 3.1 Circular cross-section under St. Venant torsion.

Furthermore, it is assumed that the bar is undergoing only small deflection. It is important to confirm this observation because it is expected to undergo a large deflection at the free end due to its cantilever action. It is interesting to note that pure torsion isolates the effects that arise from deflection, which is inherently present otherwise (Lakshmipriya & Chandrasekaran, 2021). In simple terms, while we focus on the effect of twisting moment applied at the free end, this end has already undergone a considerable deflection, which is over-sighted.

Equating the external twisting moment to the internal, we get

$$T = \text{Tint} T = T_{int} = \frac{GJ\theta}{L} \qquad (3.1)$$

$$\tau_{max} = \frac{Gr\theta}{L} \qquad (3.2)$$

Substituting this in Equation (3.1), we get

$$T = \frac{\tau_{max} J}{r} \qquad (3.3)$$

$$\theta = \frac{TL}{GJ} \qquad (3.4)$$

where τ_{max} is the maximum shear stress at the circumference, θ is the angle of twist, G is the shear modulus of the material, and J/r is termed as torsional sectional modulus. It is important to remember that no other stresses except shear stress are generated in the cross-section due to the applied torsion. The member under pure torsion also enjoys the assumed behavior that each cross-section rotates as a rigid body; hence no distortion is developed in the circular cross-section due to the applied pure torsion. However, the angle of twist will change along the length of the member. It keeps on decreasing as the section travels towards the fixed end, where the twist is zero. Let us try to understand the consequence of this assumption imposed under pure torsion. It is not acceptable that distortion does not occur for a change in the angle of twist in the cross-section along the length. Hence, it is a violation, which emphasizes the fact that pure torsion does not cause warping. However, since the angle of twist is

Special Design Guidelines

allowed to change, it emphasizes that each slice (cross-section) along the length of the member is allowed to warp, but warping is the same at all the cross-sections (slices). So, such warping will not result in any additional stresses in the circular cross-section (Lakshmipriya & Chandrasekaran, 2021). Twisting produces only pure shear stress— no axial or bending stresses. Such a behavior under pure torsion is perfectly valid for a circular cross-section in addition to a thick-walled beam section.

However, in the case of a closed, thin-walled section, the behavior is different. Let us now consider the section shown in Figure 3.2. Torsion of equal magnitude but opposite in nature is applied to a closed, thin-walled section of length, L. Under the action of applied torsion, one can notice that the point located on the circumference is shifted and that the amount of shift is not the same at both the ends of the member. For the thin-walled section (a tubular section or a shaft), St. Venant torsion constant (J) and the shear stress are given by

$$J = \frac{\pi \left[D_o^4 - D_i^4 \right]}{32} \qquad (3.5)$$

$$\tau = \frac{Tr}{J} \qquad (3.6)$$

It is important to note that even for closed, thin-walled sections, only pure torsion is developed (Lakshmipriya & Chandrasekaran, 2021). Most importantly, J is not seen as a polar moment of inertia (MoI). It is torsion constant but happens to be the same as that of polar MoI for a circular section. Hence, St. Venant's theory is perfectly true for circular sections, but not in the case of open, thin-walled sections like I-sections, channel sections, etc. In such open, thin-walled sections, additional stresses and distortion are developed in the cross-section. For example, Figure 3.3 shows an open, thin-walled I-section.

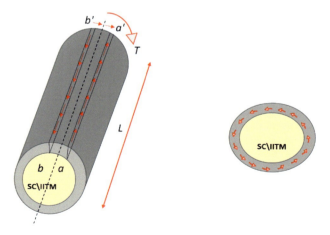

FIGURE 3.2 Thin-walled sections.

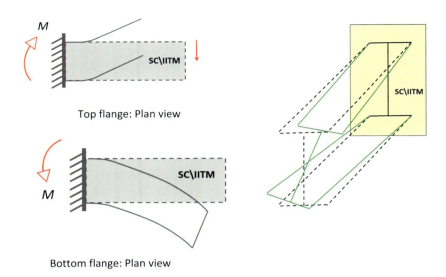

FIGURE 3.3 Open, thin-walled I-section. (Lakshmipriya & Chandrasekaran, 2021).

Let us consider an I-section fixed at one end. When torsion (a clockwise twisting moment) is externally applied at one end of the section, it will twist, as shown in Figure 3.3. Also, note that the twist of the cross-section is about its center of gravity (CG). Upon careful observation of the figure, one will realize that the top flange is horizontal at the fixed end but inclined (twisted) at the free end, the same as the bottom flange. This twist is seen at every section along the length of the beam, traveling from the free end towards the fixed end. The top flange is twisted to the maximum at the free end but zero at the fixed end, which means that the twist imposed on the top flange at every cross-section along the length of the member is not uniform. Hence, it has also resulted in some distortion. It implies that bottom and top flanges (planes) have twisted and bent as well. Both the top and bottom planes (flanges) do not remain plane anymore; they are subjected to warping. Upon critical observation, one can also observe in Figure 3.3 that the top flange has moved laterally to the right side, while the bottom flange has moved to the left; this means that both the top and bottom flanges moved opposite to each other, laterally. These lateral movements have invoked additional moments at the ends to counteract them. Hence, moments occurring in both the top and bottom flanges in the opposite direction resulting in warping of the cross-section. These moments have resulted in additional longitudinal stresses and strain. So, in an open, thin-walled section, the application of torsion has resulted in shear stress (which was the same as in the case of closed, thick-walled sections) and caused additional bending stresses. Hence, the total external torsion can be expressed as a sum of pure torsion and warping

$$T = T_v + T_w \tag{3.7}$$

Special Design Guidelines

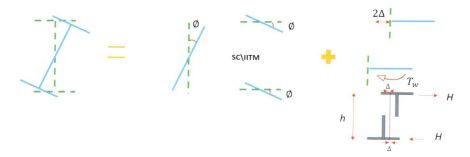

FIGURE 3.4 Twisting in open, thin-walled section. (Lakshmipriya & Chandrasekaran, 2021).

As is evident in Figure 3.4, the twisting moment applied at the end can be seen as a sum of pure torsion (T_v) and additional moment (T_w) caused by the lateral sway of the top and bottom flange by the same amount (Δ), but in the opposite direction. Moment, developed due to the lateral deflection of the flange in the opposite direction, can also be termed flexural twist (T_w), causing warping. It also idealizes that each of the flanges bends as a rectangular beam about its minor axis.

3.1.1 Torsion in Open, Thin-Walled Section

In section 3.1, we have seen the derivation for torsion in a closed, thick-walled section. However, in the case of a thin-walled open section, additional stresses developed due to warping. Hence, applied torsion is expressed as in Equation (3.8)

$$T_V = GJ\frac{\theta}{L} = GJ\frac{d\beta}{dz} \tag{3.8}$$

where $\dfrac{d\beta}{dz}$ is the change in the twist at every section along the length of the member (z-axis).

Lateral deflection of the top and bottom flanges, which resulted in an additional flexural twist, can be expressed as

$$M_f = -EI_f \frac{d^2 x}{dz^2} \tag{3.9}$$

where M_f indicates flange bending moment, and I_f indicates MoI of the flange. Please note the second derivate of deflection is along the x-axis, which is the lateral deflection (as seen in Figure 3.5). The corresponding shear force in the flange, H, is given by

$$H = \frac{dM}{dz} = -EI_f \frac{d^3 x}{dz^3} \tag{3.10}$$

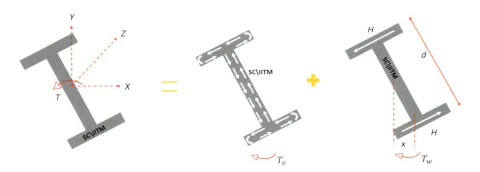

FIGURE 3.5 Torsion in an open section. (Lakshmipriya & Chandrasekaran, 2021).

$$I_f \cong \frac{1}{2} I_y \qquad (3.11)$$

where, I_y is the moment of inertia about the weaker axis. Equation 3.11 neglects the web contribution. Now, the additional moment induced by the lateral deflection of the flanges, T_w, can be determined as

$$T_w = H \cdot d \qquad (3.12)$$

where d is the depth of the section. For a smaller twist angle, the following relationship holds good

$$x = \frac{\beta d}{2} \qquad (3.13)$$

Substituting values from the above equations, one can derive T_w as in Equation 3.14

$$T_w = -EI_y \frac{d^2}{4} \frac{d^3 \beta}{dz^3} \qquad (3.14)$$

After appropriate substitution, the total applied torsion can be obtained as below.

$$T = GJ \frac{d\beta}{dz} - EC_w \frac{d^3 \beta}{dz^3} \qquad (3.15)$$

where C_w is called as warping constant ($\cong I_y \frac{d^2}{4}$), (GJ) is termed as torsional rigidity, and (EC_w) is termed as warping rigidity. It can be seen that the open section, the second term $\left(EC_w \frac{d^3 \beta}{dz^3} \right)$ in Equation (3.15), is the additional term arising from warping; this is of major concern in the design.

Special Design Guidelines

3.2 BUCKLING

Buckling—when a structural member is displaced laterally (out of its plane) under compressive stresses—is considered as one of the dominant modes of failure in many steel members. Lateral displacements are associated with the flexural stresses whose magnitude depends on the slenderness of the member. The buckling of a beam can be categorized as local or global buckling. Local buckling is specific to beams whose web (or flanges) is slender. As a result, many small buckles tend to appear along with the web (or the flanges, as the case may be). For the local buckling of the web, buckling length is about the width of the web, as shown in Figure 3.6. In the case of flange buckling, the buckling length varies between 1 and 5 times the width of the flange (Hoglund, 2006). When the buckling of the beam is extended to its entire length, it is termed global buckling. This type of buckling is further classified according to the type of loading and the nature of displacement.

3.2.1 GLOBAL BUCKLING MODES

A steel beam can fail in different modes due to buckling. The failure modes are governed by the type of load and deformation shape of the member. Furthermore, a beam can be subjected to different types of deformations simultaneously: bending, torsion, and warping. The resistance of the beam depends on its stiffness in different modes of these deformations. Flexural, torsional, and warping stiffness of the beam cross-section are essential for estimating buckling resistance. Therefore, the beam's overall buckling resistance is a combination of flexural-torsional and lateral-torsional buckling resistance, as described in Table 3.1.

Flexural buckling occurs only when the compression flange of the beam buckles in one plane (Serna et al., 2005). Resistance of the beam to flexural buckling strongly depends on flexural rigidity (EI), where I is the moment of inertia about the plane of flexural buckling. When buckling of a beam is caused by compression and induces only twisting, the buckling mode is called torsional buckling. Resistance to torsional buckling depends on torsional rigidity (GJ or GI_t) and warping rigidity (EC_w or EI_w). In some cases, when a beam is subjected to axial compression, it may experience

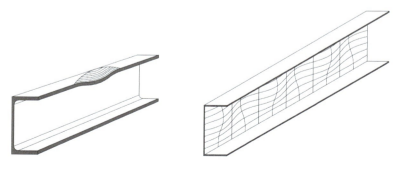

FIGURE 3.6 Local buckling in the flange (left) and the web (right).

TABLE 3.1
Deformations under Flexural-Torsional, and Lateral-Torsional Buckling

Flexural buckling	
Torsional buckling	
Lateral torsional buckling	

both flexural and torsional buckling. The combined mode of buckling is called flexural-torsional buckling. Figure 3.7 shows various modes of buckling.

3.3 LATERAL-TORSIONAL BUCKLING

Among all modes of buckling, lateral-torsional buckling (LTB) calls for a special mention, as this is critical in the design perspective. LTB is particularly related to beams. To understand LTB, let us pay attention to Figure 3.8, which shows an I-section under transverse load. It is a classic case of an open, thin-walled section. The top flange of the I-beam is in compression, while its bottom flange is in tension. As shown in Figure 3.8, under the gradual increase of the transverse load magnitude, the compression flange acts as a column and starts deflecting in the lateral direction. It buckles out-of-plane as there is no restraint. On the other plane, the web offers restraint and does not deflect. A careful observation of the bottom flange will confirm a very little (or no) deformation. Practically, the upper half of the section undergoes lateral deflection while the lower half does not undergo any such deflection. Therefore, one can also state that the bottom flange is less severely loaded than the top flange. In the case of a column, one can agree to the statement that both the top and bottom flanges would have undergone lateral deflection (buckling). But, in this example beam, as half of the section is in tension, it does not buckle. It results in an overall twist of the cross-section.

Special Design Guidelines

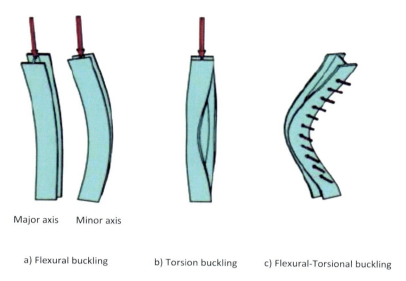

a) Flexural buckling   b) Torsion buckling   c) Flexural-Torsional buckling

FIGURE 3.7 Different modes of buckling of a beam under compression.

FIGURE 3.8 I-section, with top flange under compression.

The following are interesting observations regarding LTB. The top flange tends to undergo flexural buckling similar to that of columns, causing out-of-plane bending of the top flange. But, the top flange is restrained from the bottom one, which is the tension portion of the cross-section. It results in twisting of the cross-section, in addition to out-of-plane bending, and is identified as LTB. LTB, in short, is the lateral deflection of the compression flange along with the twist of the cross-section. Since it is related to the flange under compression, the term buckling is associated with it. In the case of the open, thin-walled section, the compression flange behaves essentially like a column. For this reason, LTB resistance equations use Euler's buckling load formula. When a beam is subjected to flexural load and deflects laterally and twists simultaneously, it is called LTB.

The cross-sectional resistance of a beam subjected to LTB is derived from its flexural, torsional, and warping rigidities (Chandrasekaran, 2020). The phenomena of LTB, design criteria, and control are discussed in more detail in the following sections. Let us consider a simply-supported beam with no deformation, as shown in Figure 3.9. Pay attention to the cross-section axes: the x-axis is the major, and the y-axis is the weaker (minor) axis, while the z-axis is along the length of the member. Let us consider a section along the span of the member and view it before and after deformation. As we focus on the lateral deformation, let us try to view the deformed geometry in the beam's plan (top view). It can be seen that the lateral deformation of the beam is along the x-axis. The compression flange has bent out-of-plane about the beam's minor axis (y-axis), as seen in the lower part of the figure. The chosen section gets twisted (for the reasons explained in the earlier section), which can be seen in the side view of the section. Note that the side view is highly exaggerated, while, in reality, the twist is very minimal. The cross-section axes are now modified as (ξ, η), which corresponds to (z, y), respectively. (u, v) are the corresponding coordinates of the CG of the section, which will now designate the new position after twist (please see the side view of the twisted cross-section).

The twisted cross-section phase signifies the flange under flexural bending while the compression flange is under LTB. Similar to the column buckling equation, one can derive the elastic buckling equation for a simply-supported beam. Let us

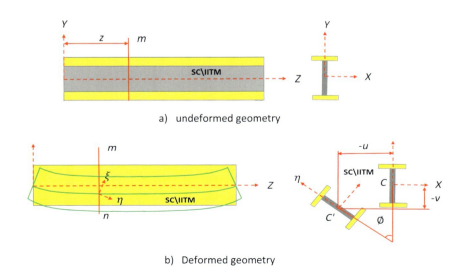

FIGURE 3.9 Simply-supported beam: un-deformed and deformed geometry.

Special Design Guidelines

consider a double-symmetric I-section, subjected to uniform moments at the ends, as shown in Figure 3.10. Under such an arrangement, no shear will occur. (x–y) are the cross-section axes, and the z-axis is along the length of the beam. As shown in Figure 3.11, the x–y plane is rotating to (ξ, η) by an angle (ϕ).

Hence, in the transformed plane (ξ, η), one can resolve the moments as given below

$$M_\xi = M_x \cos(\phi) \cong M_x \quad \text{for smaller rotation} \tag{3.16}$$

$$M_\eta = M_x \sin(\phi) \cong M_x \phi \cong M_0 \phi \tag{3.17}$$

The torsional component is resolved as given below

$$M_\zeta = M_x \sin\left(\frac{-du}{dz}\right) = -M_0\left(\frac{du}{dz}\right) \tag{3.18}$$

In the above equation, M_x is replaced as M_0 as it is the end moment acting about the x-axis. In-plane bending, which is occurring in the (y-z) plane, is represented in Figure 3.12 and is given as below.

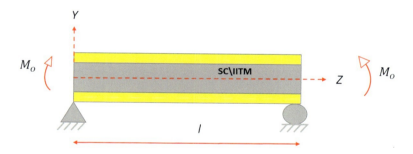

FIGURE 3.10 I-section under end moment.

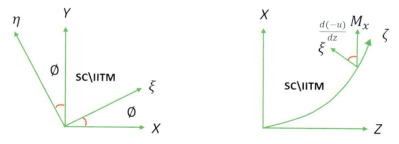

FIGURE 3.11 Components of end moments.

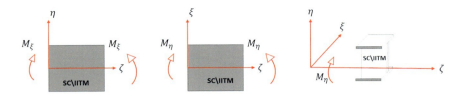

FIGURE 3.12 Moment components adding up to LTB.

$$EI_\xi \frac{d^2v}{dz^2} = M_\xi \qquad (3.19)$$

One can realize that Equation (3.19) is a simple bending equation about the major axis and written in concurrence with the respective axis of the beam deflection. Similarly, the minor axis bending is given by

$$EI_\eta \left[\frac{d^2u}{dz^2}\right] = M_\eta \qquad (3.20)$$

The twisting moment is given by

$$GJ\frac{d\phi}{dz} - EC_w \frac{d^3\phi}{dz^3} = M_\zeta \qquad (3.21)$$

Solving the simultaneous Equations (3.19–3.21), and applying the boundary conditions for the simply-supported beam, we get the governing equation for elastic buckling strength equation of the beam (Timoshenko & Gere), as given below

$$M_{o,cr} = \sqrt{\frac{\pi^2 EI_y}{L^2}\left[\frac{\pi^2}{L^2}EC_w + GJ\right]} \qquad (3.22)$$

The above equation is truly applicable to very-long span beams, similar to slender columns; short and intermediate span beams will not be within the subset of Equation (3.22).

3.4 MECHANISMS BEHIND LTB

According to Euro Code 3 (EC3, 2005), the instability phenomenon is characterized by large transversal displacements and rotation about the member axis, under bending moment about the major axis (y-axis) (Hoglund, 2006). This instability phenomenon involves lateral bending (about the z-axis) and torsion of the cross-section, illustrated in Figure 3.13.

As shown in Figure 3.13, the beam is subjected to constantly increased loading in major axis bending. If the beam is slender, it may buckle before the sectional capacity is fully utilized. This buckling type involves both lateral deflection and twisting

Special Design Guidelines

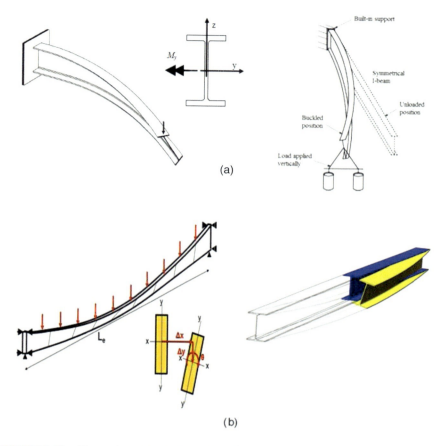

FIGURE 3.13 Illustration of LTB (a) cantilever beam (b) simply-supported beam.

and is called LTB. The vertically applied load induces compression and tension in the flanges of the beam in the section. It causes the deflection of the compression flange, laterally away from the original position.

In contrast, the tension flange tends to keep the beam straight. This lateral bending of the section creates restoring forces that oppose the movement because the section tends to remain straight. These actions generate lateral forces, which sometimes are not adequate to prevent the section from deflecting laterally. The lateral component of the tensile forces and the restoring forces together determine the buckling resistance of the member, which is shown in Figure 3.14.

3.4.1 Torsional Effect

The forces in the flanges cause the section to twist about its longitudinal axis along with the lateral movement of the section, as shown in the Figure 3.14. The twisting is resisted by the torsional stiffness of the section, which is mainly controlled by the flange thickness. Suppose there are two sections of similar depth and different flange

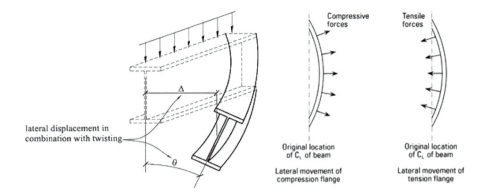

FIGURE 3.14 Phenomenon of lateral deflection and the twisting.

FIGURE 3.15 Bridge girder failure due to LTB.

thickness; the section with thicker flanges shows a larger bending strength than that of the thinner flange section. If longer in span, such a beam becomes unstable even under a load of smaller magnitude, as compared to that of short-span beams. Also, if beams with the same length but different cross-sections are imagined, beams with slender cross-sections buckle at loads of lesser magnitude than beams with heavy cross-sections. Figure 3.15 shows the failure of a bridge girder dominated by the LTB.

It should be noted that in order to control LTB, the length of the beam and the shape of the cross-section are important. However, some other factors also contribute to a beam undergoing LTB. Table 3.2 lists the possible factors contributing to LTB (Hermann et al., 2014).

Special Design Guidelines

By considering the factors listed in Table 3.2, and designing a beam using suitable properties, LTB can be avoided. The chances of LTB will be higher for beams with the properties listed below.

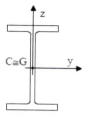

- Low flexural stiffness about the weak axis (EI_z).
- Low torsional stiffness (GI_t).
- Low warping stiffness (EI_w).
- The high point of load application.
- Long unrestrained spans (L).

However, it is important to note that LTB will not occur in closed box sections and if the section is bending about its major axis.

3.5 MEASUREMENTS AGAINST LTB

To prevent LTB, the controlling factors listed above are manipulated in different ways. One of the most commonly used techniques is to choose a cross-section with higher flexural stiffness about the weaker axis. Other solutions include admitting a load with a lesser magnitude or providing restraints. As indicated earlier, LTB is only possible in major axis bending when there is corresponding weak stiffness about the minor axis of the cross-section. If structural stiffness is the same

TABLE 3.2
Factors Contributing to LTB

Factors	Structural Properties
Material properties.	Shear modulus (G), modulus of elasticity (E).
Cross-section dimensions and length (L).	Torsional constant (J).
	Warping constant (C_w).
	MoI about the weaker axis (I_y).
Boundary (support) conditions.	Bending about the major axis.
	Bending about the minor axis.
	Warping.
Load.	Type of load (UDL, concentrated).
	Point of application of load.

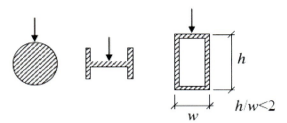

FIGURE 3.16 Recommended sections of beams for preventing LTB.

about both axes, then LTB could be avoided entirely. Topside sections are carefully designed as built-up sections to address this advantage as the span along both directions will be almost equal. Figure 3.16 shows the type of sections considered favorable for preventing LTB.

3.5.1 Effects of the Point of Application of Load

Load with a lesser magnitude minimizes the chance of the occurrence of LTB. A load applied on the bottom flange makes a beam considerably more stable than a load applied on the top flange. An interesting observation of the crane gantry will make you realize that wheel loads are transferred to the bottom flange. However, this is possible only when the load does not contribute to any restraining effects. In the event of LTB, a beam deflects when the loads above the center of the twist generate a twisting moment.

In contrast, loads below the center of the twist counteract the cross-sectional rotation and stabilize the beam. As can be seen in Figure 3.17, a load, F, acting above the center of twist, generates rotation of the section with an additional moment, M_{add}. However, when the force F is applied under the center of the twist, the load is tending to stabilize the beam section by a counteracting moment that equals M_{add}.

3.5.2 Effects of Lateral Restraint on LTB

When a compression flange is restrained against lateral bending, the LTB can be entirely avoided. The spacing within the restraints is kept as close as possible so that buckling cannot occur between them. Figure 3.18 illustrates one configuration for providing lateral restraints to beams placed parallel to each other. Table 3.3 summarizes various lateral constraints that are possible to avoid LTB.

3.6 BEHAVIOR OF REAL BEAM

The LTB theory is generally applicable to ideal, elastic conditions for beams possessing perfect geometry and no initial imperfections. However, in real-life practice, almost no structural members comply with these idealistic design conditions and so behave differently, as depicted in Figure 3.19.

By analyzing the behavior of both the beams shown in Figure 3.19, it can be seen that an ideal beam is laterally un-deformed until the load reaches the elastic critical moment, M_{cr}. An indifferent state of instability is observed at this stage, and a sizeable

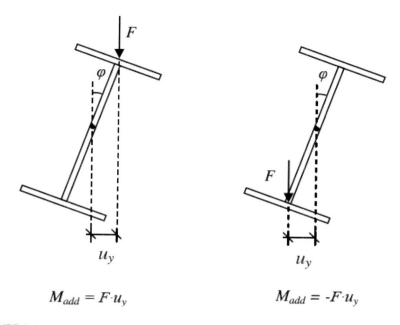

FIGURE 3.17 Effects of point of application of load on LTB.

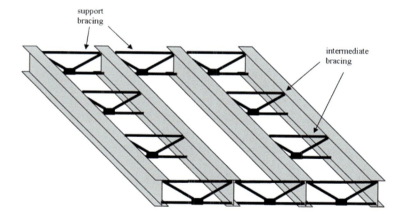

FIGURE 3.18 Lateral restraints for avoiding LTB in beams.

instantaneous deflection occurs laterally. Since the material is ideal-elastic, infinitely large deformations can occur. A new state of equilibrium can be found in the deflected position. Every slight increase in load causes a sizeable additional deflection.

However, the real beam experienced reduced capacity compared to the ideal beam due to imperfections, residual stresses, and other causes. When the real beam is subjected to the applied load, a lateral imperfection (initial deflection, δ_o) already exists. The lateral deformation increases with the increase in the magnitude of the

TABLE 3.3
Buckling Modes with Lateral Restraints to Avoid LTB

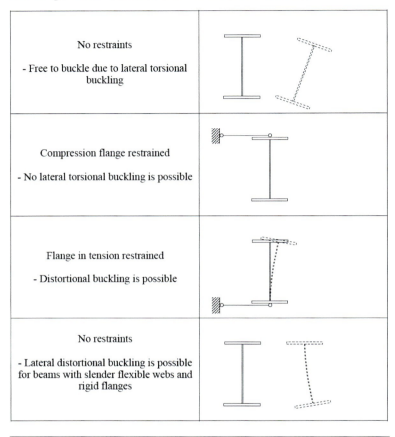

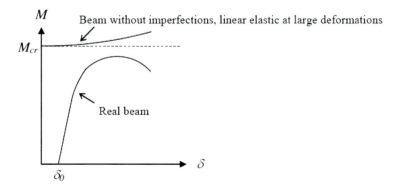

FIGURE 3.19 Structural behaviors of a real and an ideal beam.

applied load. Closer to the critical load, M_{cr}, the deflection increases spontaneously without reaching the theoretical value of M_{cr}. This failure is governed by the plastic material response, nonlinear geometry, and by possible local buckling.

3.6.1 Factors Causing Reduction of Capacity

- Nonlinear material response, i.e., plastic material behavior.
- Initial lateral imperfections; the beam is not perfectly straight.
- Residual stresses from manufacturing.
- Local buckling of beam sections in class 4 section.
- Piercings, asymmetry, and defects.

These effects can be considered in design through design buckling curves, which simulate real beam behavior.

3.7 LTB DESIGN PROCEDURE

The elastic critical moment M_{cr} is the primary design parameter for designing a steel beam for LTB. The magnitude of M_{cr} is defined as the maximum value of bending moment supported by a beam, free from any imperfections. In practical design, M_{cr} can be estimated by task-specific software or by performing hand calculations. Early editions of Euro Code 3 included an expression of M_{cr}. This expression is known as the 3-factor formula. It is one of the most used analytic formulae to estimate the elastic critical moment (Lopex et al., 2006).

3.7.1 Three-Factor Formula for M_{CR}

The 3-factor formula is based on a reference load case, to which correction factors are added to make the equation fit other cases. The reference case comprises a beam of double-symmetric cross-section, simply supported at both ends; a constant moment is applied at both ends. The material is linearly elastic. The cross-sectional dimensions are smaller than the curvature radius, and the deformations are restricted to be small. The simply-supported beam shown in Figure 3.20 has a double symmetric section; supports prevent the lateral displacements and rotation around the member axis (twist rotations) but allow the warping and rotations around the cross-section axis (y and z), subjected to a uniform bending moment, M_y (standard case—doubly symmetric section). The elastic critical moment of a standard form (perfect section) is given by

$$M_{cr} = \frac{\pi}{L}\sqrt{GI_T EI_Z \left(1 + \frac{\pi^2 EI_W}{L^2 GI_T}\right)} \quad (3.23)$$

Since Equation (3.23) is the standard case of the doubly-symmetric section, the M_{cr} formula is further expanded to apply singly-symmetric cross-sections with arbitrary moment distributions. It is modified by introducing three correction factors: C_1, C_2, and C_3, which account for imperfection conditions. The factors can be determined

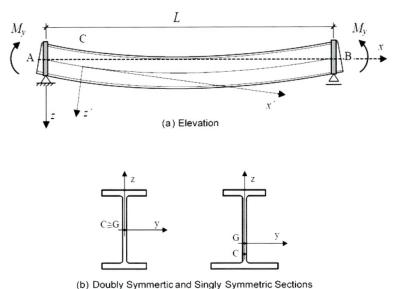

FIGURE 3.20 Simply supported beam with different cross-sections.

either from tables or figures; alternatively, they can be estimated from approximate closed-form expressions. Considering the warping degrees of freedom and lateral rotation at the supports, the expression is expanded by introducing other factors, k_z and k_w. The modified 3-factor formula for singly-symmetric sections are given below

$$M_{cr} = C_1 \frac{\pi^2 EI_Z}{(K_Z L)^2} \left\{ \left[\left(\frac{K_Z}{K_W} \right)^2 \frac{I_W}{I_Z} + \frac{(K_Z L)^2 GI_T}{\pi^2 EI_Z} + (C_2 z_g - C_3 z_j)^2 \right]^{0.5} - (C_2 z_g - C_3 z_j) \right\} \quad (3.24)$$

The limitations and validity of Equation (3.24) are:

- Applicability to the member with symmetric and mono-symmetric cross-sections.
- Including the effects of loading applied below or above the shear center, as illustrated in Figure 3.21.

Where E is the modulus of elasticity, G is the shear modulus, I_z is the second moment of inertia about the weaker axis, I_T is the torsion constant, I_w is the warping constant, L is the length between lateral restraints, k_z is the effective length factor, which is related to the restrain against lateral bending at the boundaries, k_w is the effective length factor, which is related to the restrain against warping at the boundaries, z_g is the distance between the point of load application and the shear center, z_j is the distance related to the effects of asymmetry about the y-axis is given by

Special Design Guidelines

$$z_j = z_s - \frac{0.5 \int_A (y^2 + z^2) z \, dA}{I_y} \quad (3.25)$$

where z_s is the distance between the shear center and the CG, C_1 is the factor that accounts for the shape of the moment diagram, C_2 is the factor that accounts for the point of load application with the shear center, and C_3 is the factor that accounts for asymmetry about the y-axis.

3.7.2 MOMENT CORRECTION FACTORS (C_1, C_2, AND C_3)

The C_1 factor is the equivalent uniform moment factor, also referred to as the moment gradient factor. It is valid when the load acts in the shear center. In reality, beams are often loaded on the top or bottom flange and not in the shear center (SC). Thus a second correction factor, C_2, has been added to account for the effects of the point of load application (PLA). When PLA is separated from SC, the load contributes with an additional twisting moment added to the system's potential energy. Therefore, a load applied under SC helps stabilize the beam. A load above SC destabilizes the beam, as shown in Figure 3.22.

With the C_1 and C_2 factors, the 3-factor formula can calculate M_{cr} for double-symmetric beams under various loads and points of load application. Still, the equation is only valid for double-symmetric sections. Therefore, to estimate M_{cr} analytically for sections with symmetry only about the minor axis, a third correction factor, C_3, must be introduced. At the end of this section, tables and figures are given for estimation of the correction factors (Table 3.4).

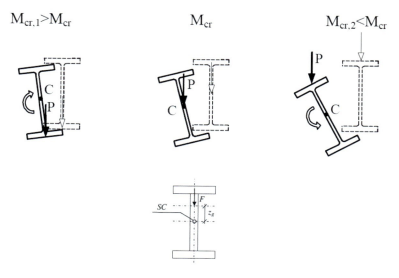

FIGURE 3.21 Effects of load application on M_{cr} with the shear center, C.

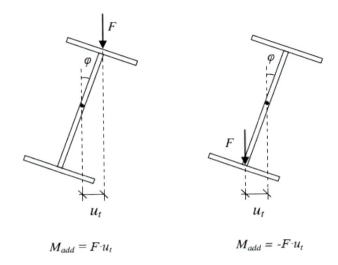

FIGURE 3.22 Additional moment due to the point of load application.

TABLE 3.4
Correction Factors for Uniform Moment Conditions

Loading and support conditions	Diagram of moments	k_z	C_1	C_2	C_3
(distributed load p)		1.0	1.12	0.45	0.525
		0.5	0.97	0.36	0.478
(central point load P)		1.0	1.35	0.59	0.411
		0.5	1.05	0.48	0.338
(four point loads P)		1.0	1.04	0.42	0.562
		0.5	0.95	0.31	0.539

3.8 DESIGN CHECK FOR LTB

In Euro Code 3, the general procedure to design against buckling is to introduce a buckling factor, $\chi \leq 1$, that reduces the capacity of the member. In the case of LTB, this factor is denoted by χ_{LT} and accounts for all the effects that decrease the capacity due to buckling. The moment capacity is given by

$$\frac{M_{ED}}{M_{b,Rd}} \leq 1 \tag{3.26}$$

Special Design Guidelines

$$M_{b,Rd} = \chi_{LT} \frac{W_y f_y}{\gamma_{M1}} \tag{3.27}$$

Where W_y is the bending resistance corresponding to the cross-section classification of the member, f_y is the yield strength of steel, $\gamma M1$ is the partial safety factor, χ_{LT} is the reduction factor for LTB, which can be calculated by one of two methods: general method or special method, depending on member cross-section. Further, the following conditions also apply (Table 3.5)

$$W_y = W_{pl.y} \quad \text{Class 1 and 2;}$$

$$W_y = W_{el.y} \quad \text{Class 3;}$$

$$W_y = W_{eff.y} \quad \text{Class 4.}$$

TABLE 3.5
Correction Factors for Moment Gradient Conditions

Loading and support conditions	Diagram of moments	k_c	C_1	C_3	
				$\psi_f \leq 0$	$\psi_f > 0$
	$\Psi = +1$	1.0	1.00	1.000	
		0.5	1.05	1.019	
	$\Psi = +3/4$	1.0	1.14	1.000	
		0.5	1.19	1.017	
	$\Psi = +1/2$	1.0	1.31	1.000	
		0.5	1.37	1.000	
	$\Psi = +1/4$	1.0	1.52	1.000	
		0.5	1.60	1.000	
	$\Psi = 0$	1.0	1.77	1.000	
		0.5	1.86	1.000	
$M \quad \Psi M$	$\Psi = -1/4$	1.0	2.06	1.000	0.850
		0.5	2.15	1.000	0.650
	$\Psi = -1/2$	1.0	2.35	1.000	$1.3 - 1.2\psi_f$
		0.5	2.42	0.950	$0.77 - \psi_f$
	$\Psi = -3/4$	1.0	2.60	1.000	$0.55 - \psi_f$
		0.5	2.45	0.850	$0.35 - \psi_f$
	$\Psi = -1$	1.0	2.60	$-\psi_f$	$-\psi_f$
		0.5	2.45	$-0.125 - 0.7\psi_f$	$-0.125 - 0.7\psi_f$

3.8.1 GENERAL METHOD

The Euro Code suggests two alternative methods for the stability design of beams under bending. The designer can choose the appropriate method according to the specifications of the relevant National Annexes. These alternative methods follow *M*b's estimation for checking the LTB resistance in the same way. The general method recommended for the LTB of beams is analog to the equation given for the flexural buckling of columns. Only the parameters are derived for the behavior of members subjected to bending. According to the EC3-1-1 Part 6.3.2.2, the buckling factor χ_{LT} is calculated as

$$\chi_{LT} = \frac{1}{\phi_{LT} + \sqrt{\phi_{LT}^2 - \bar{\lambda}_{LT}^2}}, \quad \chi_{LT} \leq 1 \quad (3.28)$$

$$\varnothing_{LT} = 0.5\left[1 + \alpha_{LT}\left(\bar{\lambda}_{LT} - 0.2\right) + \bar{\lambda}_{LT}^2\right] \quad (3.29)$$

From the above equations, it is clear that the buckling behavior is strongly dependent on the slenderness parameter, $\bar{\lambda}_{LT}$ and is given by

$$\bar{\lambda}_{LT} = \sqrt{\frac{W_y f_y}{M_{cr}}} \quad (3.30)$$

where M_{cr} in the above equation is the elastic critical moment in classical buckling theory. It can be recalled as the theoretical, critical moment without initial imperfections or residual stresses. Euro Code 3 does not provide any information on determining M_{cr} but simply states that it should be based on the gross cross-section, the loading conditions, the real moment distribution, and the lateral restraints. The essential remarks on the general method are as follows

- This standard procedure takes into account the type of the bending moment distribution (thereby the load distribution and boundary conditions) of the beam only applicable to the determination of the slenderness.
- Regarding the calibrated form of imperfection factor $\alpha_{LT}\left(\bar{\lambda}_{LT} - 0.2\right)$, it is stated that the standard recommended reduction for LTB is only valid for beams with $\bar{\lambda}_{LT} > 0.2$ slenderness, i.e., the LTB curves have a plateau length under($\bar{\lambda}_{LT} = 0.2$).
- According to the given tables for the standardized values of lateral-torsional constant, it can be seen that the EC3-1-1 applies the same LTB curve for a group of different profiles, and it does not make a distinction between them regarding their behavior and resistance.

3.8.2 BUCKLING CURVES

During the 1960s, a comprehensive experimental testing numerical simulation program was developed to investigate the steel members subjected to bending. As part

Special Design Guidelines

of the research outcome, LTB curves applicable to different cross-sections were created. Because of the complexity of the problem in analyzing lateral-torsional (LT) behavior, the researchers developed a more complex solution for beams. The derived formulae were complicated for calculation procedures and were suited for standard applications (Lopez et al., 2006; Szaallai & Papp, 2010). In the absence of an appropriate method of determining this behavior's mechanical background, researchers calibrate the original, flexural buckling-based Ayrton–Perry formula for the LTB of beams (Ayrton & Perry, 1886). Even when considering theoretical beams without imperfections, buckling will occur before failure in bending if the member is slender. From the definition of the slenderness, it can be shown that this reduction in capacity corresponds to a theoretical buckling factor, $\chi_{LT} = \dfrac{1}{\bar{\lambda}_{LT}^2}$ (Tables 3.6, 3.7, and 3.8)

$$\bar{\lambda}_{LT} = \sqrt{\frac{W_y f_y}{M_{cr}}} \leftrightarrow \bar{\lambda}_{LT}^2 = \frac{W_y f_y}{M_{cr}} \leftrightarrow M_{cr} = \frac{1}{\bar{\lambda}_{LT}^2} W_y f_y \qquad (3.31)$$

If the theoretical buckling factor $\dfrac{1}{\bar{\lambda}_{LT}^2}$ is plotted for different values of $\bar{\lambda}_{LT}$, the relationship between the slenderness and the buckling factor can be described with an ideal buckling curve. This curve corresponds to a perfect beam and is plotted as the dashed line in Figure 3.23. In reality, imperfections and residual stresses decrease the moment capacity even more. Euro Code 3 takes these effects into account by using design buckling curves determined from extensive testing. These curves correspond to the members' different residual stresses and imperfections (refer to Table 3.6 and Figure 3.23). It is seen from Figure 3.23 that for low slenderness value, most of the sectional capacity can be utilized. The buckling factor is closer to unity.

In contrast, for higher slenderness values, the resistance approaches that of a perfect beam. The influence of imperfections and residual stresses is small. At intermediate slenderness, the impact of initial imperfections and residual stresses is higher. The design buckling factors are considerably lower than the ideal buckling factor.

3.8.3 Alternative Method

The alternative procedure for the design of beams for LTB is the special case method. The application of this alternative procedure is one of the most significant changes of Euro Code 3 during the conversion from ENV (European pre-Standard) to EN (European Standard) status (Rebelo et al., 2009). This method is applicable for beams with hot-rolled and equivalent welded I-sections. Compared to the general method, the standardized LTB curves are considerably changed. The equation to determine—LT reduction factor in the EC3-1-1 Part 6.3.2.3 special case method is given by

$$\chi_{LT} = \frac{1}{\phi_{LT} + \sqrt{\phi_{LT}^2 - \beta \bar{\lambda}_{LT}^2}} \leq \min\left(1.0, \frac{1}{\bar{\lambda}_{LT}^2}\right) \qquad (3.32)$$

where, $\bar{\lambda}_{LT}$ is the slenderness. The following expression is useful to compute ϕ_{LT} factor

TABLE 3.6
Choice of Buckling Curves Based on Cross-Section Layout, Steel Quality, and Manufacturing Method (Euro Code)

Cross section		Limits	Buckling about axis	Buckling curve S 235 / S 275 / S 355 / S 420	Buckling curve S 460
Rolled sections	h/b > 1,2	$t_f \leq 40$ mm	y – y z – z	a b	a_0 a_0
		40 mm < $t_f \leq 100$	y – y z – z	b c	a a
	h/b ≤ 1,2	$t_f \leq 100$ mm	y – y z – z	b c	a a
		$t_f > 100$ mm	y – y z – z	d d	c c
Welded I-sections		$t_f \leq 40$ mm	y – y z – z	b c	b c
		$t_f > 40$ mm	y – y z – z	c d	c d
Hollow sections		hot finished	any	a	a_0
		cold formed	any	c	c
Welded box sections		generally (except as below)	any	b	b
		thick welds: a > 0,5t_f b/t_f < 30 h/t_w < 30	any	c	c
U-, T- and solid sections			any	c	c
L-sections			any	b	b

TABLE 3.7
Recommended Buckling Curves Applicable for General Method

Section	Limits	Buckling curve
I or H-sections, rolled.	h/b ≤ 2	a
	h/b > 2	b
I or H-section, welded.	h/b ≤ 2	c
	h/b > 2	d
Other sections.	–	d

TABLE 3.8
Recommended Values for Imperfection Factor, α_{LT}

Buckling Curve	a	b	c	d
Imperfection factor, α_{LT}	0.21	0.34	0.49	0.76

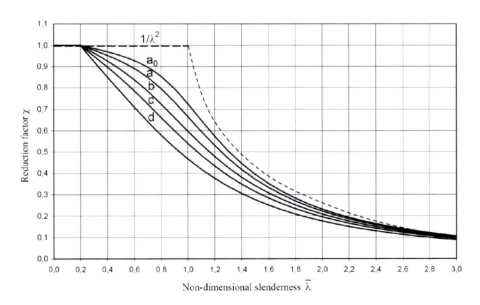

FIGURE 3.23 Buckling curves (Euro Code).

$$\phi_{LT} = 0.5\left[1 + \alpha_{LT}\left(\bar{\lambda}_{LT} - \bar{\lambda}_{LT,0}\right) + \beta\bar{\lambda}_{LT}^{2}\right] \quad (3.33)$$

Considering the above equations, the shape of the LTB curves developed for the general method is modified through the application of the β and $\bar{\lambda}_{LT,0}$ parameters. Furthermore, the given values are also changed for the LT imperfection constant. It also caused a difference in the calculated values of the LTB resistance. The EC3-1-1 recommends a value of 0.4 as the maximum $\bar{\lambda}_{LT,0}$. Therefore, designers do not have to count on reduction for LTB in the $\bar{\lambda}_{LT} < 0$; 4 slenderness range. The EC3 also recommends a minimum value of 0.75 for β. The values of $\bar{\lambda}_{LT,0}$ parameters are proposed in the National Annexes of various countries. Besides the modified shape of the LTB curves, the new, special case method includes another difference compared to the general method. For this change, the bending moment distribution of the beam is taken into account not only through the determination of the slenderness but also in calculating the reduction factor. The EC3-1-1 proposed a modifier called the f factor that includes the effect of the load distribution. Finally, using the f factor, the LTB resistance of the beam can be calculated as given below

$$M_{b,Rd} = \frac{\chi_{LT,mod} W_{pl,y} f_y}{\gamma_{M1}} = \frac{\chi_{LT}}{f} W_{pl,y} \frac{f_y}{\gamma_{M1}} \quad (3.34)$$

where factor f is calculated as given below

$$f = 1 - 0.5(1 - k_c)\left[1 - 2\left(\bar{\lambda}_{LT} - 0.8\right)^2\right] \leq 1 \quad (3.35)$$

where k_c is the correction factor; its value depends on the bending moment distribution. The recommended formulae for the calculation of this factor are given in Table 3.9.

In the cases where $\bar{\lambda}_{LT} \leq \bar{\lambda}_{LT,0}$ and $\frac{M_{ED}}{M_{cr}} \leq \bar{\lambda}_{LT,0}^2$, it is recommended that LTB can be neglected.

3.9 EXAMPLE OF LTB USING EURO CODE

Figure 3.24 shows a 5.7 meter-long simply-supported steel beam. A trial beam section is assigned as UKB 356 × 171 × 51 shape with S275 grade steel. The member is uniformly loaded with a 9.58 kN/m dead load and 6.25 kN/m live load. Considering the beam is laterally unrestrained, check the stability of the beam against the LTB.

ESTIMATION OF MAXIMUM BENDING MOMENT, M_{ED}

Factored Loads, w

$$w = 1.35 \times 9.58 + 1.5 \times 6.25 = 22.31 \text{ kN/m}$$

$$M_{ED} = 22.31 \times 5.7^2 / 8 = 90.6 \text{ kN} \cdot \text{m}$$

TABLE 3.9
Correction Factor, k_c for a Different Distribution of Bending Moment

Diagram of bending moments	k_c
$\Psi = +1$	1.0
$-1 \leq \Psi \leq 1$	$\dfrac{1}{1.33 - 0.33\Psi}$
(M, ψM curved diagrams)	0.94
	0.90
	0.91
(M, ψM triangular diagrams)	0.86
	0.77
	0.82

Ψ - ratio between end moments, with $-1 \leq \Psi \leq 1$.

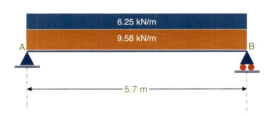

FIGURE 3.24 Example beam problem.

Steel section properties (referred UK steel sections)

$$h = 355 \text{ mm}$$
$$b = 171.5 \text{ mm}$$
$$t_w = 7.4 \text{ mm}$$
$$t_f = 11.5 \text{ mm}$$
$$r = 10.2 \text{ mm}$$

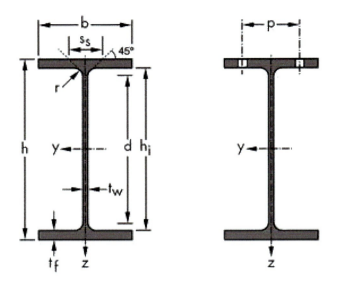

Yield strength of steel grade S275

The maximum thickness = 11.5 mm < 40 mm; therefore f_y = 275 N/mm²
[See EN 1993-1-1 (2005), Euro Code 3: "Design of Steel Structures," Table 3.1.]

Section classification

$$\varepsilon = \sqrt{(235/f_y)} = 0.92$$

[See EN 1993-1-1 (2005), Euro Code 3: "Design of Steel Structures," Table 5.2.]
   Outstanding flange (flange under uniform compression)

$$c = (b - t_w - 2r)/2 = (171.5 - 7.4 - 2 \times 10.2)/2 = 71.85 \text{ mm}$$

$$c/t_f = 71.85/11.5 = 6.25 < 9e$$

Class 1 element
   [See EN 1993-1-1 (2005), Euro Code 3: "Design of Steel Structures," Table 5.2.]
   Internal compression part (web in bending)

$$c = h - 2t_f - 2r = 355 - 2 \times 11.5 - 2 \times 10.2 = 311.6 \text{ mm}$$

$$c/t_w = 311.6/7.4 = 42.1 < 72e$$

Class 1 element
   [See EN 1993-1-1 (2005), Euro Code 3: "Design of Steel Structures," Table 5.2.]
   Therefore, the section is a Class 1 section

Calculation of section moment of resistance, $M_{B, Rd}$

$$M_{b,Rd} = \frac{W_{pl,y} f_y}{\gamma_{M0}} = \frac{(896 \times 10^3) 275}{1.0} \times 10^{-6} = 246.4 \text{ kNm}$$

[See EN 1993-1-1 (2005), Euro Code 3: "Design of Steel Structures," Clause 6.2.4.]

Since $M_{b, Rd}$ (246.4 kNm) is greater than M_{ED} (90.6 kNm), the section is safe against bending.

Stability Check against LTB

Calculation of elastic critical moment, M_{cr}

Based on the given support conditions and the loading applied at the upper flange level, the elastic critical moment can be obtained using Equation (3.2). The given section of the universal I-beam is doubly symmetrical; therefore, the correction factor, C_3, will be ignored.

$$M_{cr} = C_1 \frac{\pi^2 E I_z}{(k_z L)^2} \left\{ \left[\left(\frac{k_z}{k_w}\right)^2 \frac{I_w}{I_z} + \frac{(k_z L)^2 G I_T}{\pi^2 E I_z} + (C_2 z_g - C_3 z_j)^2 \right]^{0.5} - (C_2 z_g - C_3 z_j) \right\}$$

Using figures and tables, the following data are obtained to calculate M_{cr}

- The beam is simply-supported, therefore the effective length factor, $k_w = k_z = 1$
- Since the load is applying at the top of the flange, z_g will half of the section depth (h)
- z_g = h/2 = 355/2 = 177.5 mm
- C_1 = 1.12 [Table 3.4]
- C_2 = 0.45 [Table 3.4]
- Referring to the UK steel sections following properties of UKB 356 × 171 × 51 were selected
- I_z = 968 cm⁴
- I_w = 0.286 dm⁶ (note: 1 dm = 10 cm)
- I_T = 23.8 cm⁴
- E = 210000 N/mm²
- G = 77000 N/mm²

Inserting the above data into Equation (3.2), the elastic critical moment, M_{cr} is obtained as 121.9 kN m

Estimation of buckling factor χ_{LT}

$$\text{Slenderness, } \bar{\lambda}_{LT} = \sqrt{\frac{W_y f_y}{M_{cr}}} = \sqrt{\frac{896 \times 10^3 \times 275}{121.9 \times 10^6}} = 1.423$$

For this section (hot-rolled), the special case method will be used. Refer to Table 3.10. For h/b = 355/171.5 = 2.05 > 2, use curve C.

From Table 3.8 for curve C, α_{LT} = 0.49.

Estimate the ϕ_{LT}

$$\varnothing_{LT} = 0.5\left[1+\alpha_{LT}\left(\bar{\lambda}_{LT}-\bar{\lambda}_{LT,0}\right)+\beta\bar{\lambda}_{LT}^2\right]$$

$$\Phi_{LT} = 0.5\left[1+0.49(1.423-0.4)+0.75\times1.423^2\right] = 1.510$$

Calculate, χ_{LT}

$$\chi_{LT} = \frac{1}{\phi_{LT}+\sqrt{\phi_{LT}^2-\beta\bar{\lambda}_{LT}^2}} \le \min\left(1.0,\frac{1}{\bar{\lambda}_{LT}^2}\right)$$

$$\chi_{LT} = \frac{1}{1.51+\sqrt{1.51^2-0.75\times1.423^2}} = 0.42 < 1$$

$$\frac{1}{\bar{\lambda}_{LT}^2} = \frac{1}{1.423^2} = 0.493 > 0.42$$

Therefore, χ_{LT} = 0.42

From Table 3.9, k_c = 0.94 (for simple beam subjected to uniformly distributed load)

Calculate the factor f

$$f = 1-0.5(1-k_c)\left[1-2\left(\bar{\lambda}_{LT}-0.8\right)^2\right] \le 1$$

$$f = 1-0.5(1-0.94)\left[1-2(1.423-0.8)^2\right] = 0.993 \le 1$$

Calculate $M_{b,Rd}$ and compare with M_{Ed} for stability check against LTB, where γ_{M1} = 1

TABLE 3.10
Recommended Buckling Curve for Special Case Method

Section	Limits	Buckling curve (EC3-1-1)
For I or H-section, rolled.	h/b ≤ 2	b
	h/b > 2	c
For I or H-section, welded.	h/b ≤ 2	c
	h/b > 2	d

Special Design Guidelines

$$M_{b,Rd} = \frac{\chi_{LT}}{f} W_{pl,y} \frac{f_y}{\gamma_{M1}} = \left[\left(\frac{0.42}{0.993}\right) 896 \times 10^3 \times \frac{275}{1}\right] = 104.2 \text{ kN.m}$$

$$\frac{M_{ED}}{M_{b,Rd}} = \frac{90.6}{104.2} = 0.87 \leq 1$$

Hence, the beam is safe and stable against LTB, and no additional lateral restraints are needed.

3.10 DESIGN CHECK FOR LTB USING INDIAN CODE (IS 800-2007)

An ISMB 400@61.6 kg/m is simply supported over an effective span of 8 m to support the load of the topside of an offshore deck. Check whether it is safe under LTB, using IS 800-2007. Use steel grade as f_y 250 and use the appropriate clause of the code and assess the safety. The beam is subjected to a total moment of 50 kNm.

Solution

a) Properties of the section are as follows.

Area of the cross-section = 78.46 cm² = 7846 mm²
Depth of the section, h = 400 mm
Thickness of the flange, t_f = 16 mm
Breadth of the flange, b_f = 140 mm
Thickness of the web, t_w = 8.9 mm
Moments of inertia:
about minor axis, I_{yy} = 622.1 cm⁴ = 622.1 × 10⁴ mm⁴
about major axis, I_{xx} = 20458.4 cm⁴ = 20458.4 × 10⁴ mm⁴

b) Section classification

f_y = 250 MPa = 250 N/mm²

$\varepsilon = \sqrt{\frac{250}{f_y}} = 1$ [IS 800-2007, "General Construction in Steel—Code of Practice," Table 2.]

$$\frac{b}{t_f} = \frac{\frac{140}{2}}{16} = 4.375 < 9.4\varepsilon \quad [\varepsilon = 1]$$

$\frac{d}{t_w} = \frac{400 - (2 \times 16)}{8.9} = 41.35$ [IS 800-2007, "General Construction in Steel—Code of Practice," Figure 2, b = b_f/2, d = d_w]

$\frac{d}{t_w} = 41.35 < 85\varepsilon$ [d = d_w] Hence, the overall section is classified as PLASTIC

$$\frac{h}{b_f} = \frac{400}{140} = 2.86 > 1.2$$

$t_f = 16\,\text{mm} < 40\,\text{mm}$ [IS 800-2007, "General Construction in Steel—Code of Practice," Table 10, buckling class about xx-axis is "a" and about yy-axis is "b"]

The elastic, lateral buckling moment, M_{cr} for a simply-supported, prismatic member with symmetric cross-section, is given by

$$M_{cr} = \sqrt{\frac{\pi^2 E I_y}{L_{LT}^2}\left[(GI_t) + \frac{\pi^2 E I_w}{L_{LT}^2}\right]}$$ [IS 800-2007, "General Construction in Steel—Code of Practice," Clause 8.2.2.1]

$E = 2.1 \times 10^5$ N/mm², $\mu = 0.3$, then

$$G = \frac{E}{2(1+\mu)} = \frac{2.1 \times 10^5}{2(1+0.3)} = 80.77 \times 10^3 \text{ N/mm}^2$$

$$I_{yy} = 622.1 \times 10^4 \text{ mm}^4$$

L_{LT} = Effective length for LTB = 1.00 L = 8 m [IS 800-2007, "General Construction in Steel—Code of Practice," Table 15, Clause 8.3.1; for torsional-fully restrained and warping not restrained in both the flanges.]

I_T = Torsional constant = $\sum \frac{b_i t_i^3}{3}$ for open sections

$$I_T = 2 \times \left[\frac{140 \times 16^3}{3}\right] + \left[\frac{368 \times 8.9^3}{3}\right] = 468.77 \times 10^3 \text{ mm}^4$$

Warping constant is given by

$I_w = (1 - \beta_f)\beta_f I_y h_y^2$ [for I-sections]

$\beta_f = \dfrac{I_{fc}}{I_{fc} + I_{ft}}$ [I_{ft}, I_{fc} are MoI of tension and compression flanges about the minor axis of the section]

$$I_{fc} = \frac{16 \times 140^3}{12} = 3.66 \times 10^6 \text{ mm}^4 = I_{ft}$$

h_y is the distance between the shear center of two flanges of the cross-section. The shear center of the top and bottom flange will be located in its center. Hence,

$$h_y = h - \left(\frac{t_f}{2} \times 2\right) = 400 - 16 = 384 \text{ mm}$$

$$\beta_f = \frac{I_{fc}}{I_{fc} + I_{ft}} = \frac{3.66 \times 10^6}{2 \times 3.66 \times 10^6} = 0.5$$

Special Design Guidelines

$$I_w = (1-\beta_f)\beta_f I_y h_y^2 = (1-0.5)0.5622.1\times 10^4 \times 384^2$$

$$= 2.293\times 10^{11} \text{ mm}^6$$

$$M_{cr} = \sqrt{\frac{\pi^2 EI_y}{L_{LT}^2}\left[\frac{\pi^2 EI_w}{L_{LT}^2}\right]} = \sqrt{\frac{\pi^2 \times 2.1\times 10^5 \times 622.1\times 10^4}{8000^2}}$$

$$\times \sqrt{\left[\{80.77\times 10^3 \times 468.77\times 10^3\} + \frac{\pi^2 \times 2.1\times 10^5 \times 2.293\times 10^{11}}{8000^2}\right]}$$

$$M_{cr} = 95.51\times 106 \text{ Nmm} = 95.51 \text{ kNm}$$

The design bending moment is computed as below.

$M_d = \beta_b Z_p f_{bd}$ [IS 800-2007, "General Construction in Steel—Code of Practice," Clause 8.2.2]

$\beta_b = 1$ for plastic sections

Z_p is the plastic section modulus for the extreme compression fiber

f_{bd} is the design compressive bending stress $= \dfrac{\chi_{LT} f_y}{\gamma_{mo}}$

$$\chi_{LT} = \frac{1}{\left[\phi_{LT} + \sqrt{\phi_{LT}^2 - \lambda_{LT}^2}\right]}$$

$$\phi_{LT} = 0.5\left[1 + \alpha_{LT}(\chi_{LT} - 0.2) + \chi_{LT}^2\right]$$

α_{LT} is the imperfection factor = 0.21 for rolled steel sections

$$\lambda_{LT} = \sqrt{\frac{\beta_b Z_p f_y}{M_{cr}}} \le \sqrt{\frac{1.2 Z_e f_y}{M_{cr}}}$$

$$Z_p = 1176163.26 \text{ mm}^3 \text{ for ISMB 400}$$

$$Z_e = 1022.9\times 103 \text{ mm}^3 \text{ for ISMB 400 [Steel Tables, SP 6(1)]}$$

$$\lambda_{LT} = \sqrt{\frac{\beta_b Z_p f_y}{M_{cr}}} = \sqrt{\frac{1\times 1176163.26 \times 250}{95.51\times 10^6}}$$

$$\le \sqrt{\frac{1.2\times 1022.9\times 10^3 \times 250}{95.51\times 10^6}}$$

$$\lambda_{LT} = 1.75 \le 1.79; \ OK$$

$$\phi_{LT} = 0.5\left[1 + \alpha_{LT}\left(\chi_{LT} - 0.2\right) + \chi_{LT}^2\right]$$

$$= 0.5\left[1 + 0.21(1.75 - 0.2) + 1.75^2\right] = 2.194$$

$$\chi_{LT} = \frac{1}{\left[\phi_{LT} + \sqrt{\phi_{LT}^2 - \lambda_{LT}^2}\right]} = \frac{1}{\left[2.194 + \sqrt{2.194^2 - 1.75^2}\right]} = 0.284$$

$$f_{bd} = \frac{\chi_{LT} f_y}{\gamma_{mo}} = \frac{(0.284 \times 250)}{1.10} = 64.616 \text{ N/mm}^2 \quad [\gamma_{mo} = 1.10, \text{ Partial safety factor of}$$
member resistance to buckling]

The design bending strength of the laterally unsupported beam is governed by LTB strength.

$$M_d = \beta_b Z_p f_{bd} = 1.0 \times 1176163.26 \times 64.616$$

$$= 75.9985 \times 10^6 \text{ Nmm} = 76 \text{ kNm}$$

The moment from the applied load (50 kNm) is lesser than the design bending strength, governed by LTB (76 kNm). Hence the beam is safe against LTB, and there is no requirement for additional lateral bracings.

3.11 UNSYMMETRICAL BENDING

When members are idealized as one-dimensional, it is comfortable to compute the stresses in the cross-section for the loads applied at the pre-fixed points. This is because the bending takes place parallel to the plane of the applied moment. Bending stresses can be computed using the simple bending equation, assuming that the neutral axis (NA) of the cross-section is normal to the plane of loading. Figure 3.25 illustrates uniform bending.

In Figure 3.25, the y-axis is the trace of the plane of the applied moment. The bending moment in the y-axis is said to be zero, and mathematically,

$$\Sigma M_y = \int_A \sigma_x dA = 0 \quad (3.36)$$

The following relationship holds only when the z-axis and y-axis are the principal axes of inertia.

$$\int zy\, dA = 0 \quad (3.37)$$

Hence, for symmetrical bending, the plane containing none of the principal axes of inertia, the plane of the applied moment, and the plane of deflection should coincide. It is also obvious to note that the neutral axis will coincide with other principal axes

Special Design Guidelines

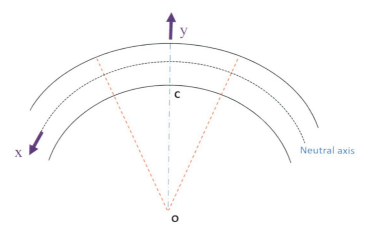

FIGURE 3.25 Uniform bending.

of inertia. When the trace of the plane of the applied moment does not coincide with any of the principal axes of inertia, then this type of bending is called unsymmetrical bending or non-uniplanar bending. Under such conditions, the neutral axis is no longer normal to the trace of the applied moment plane; further, the deflection curve is not planar.

One of the major consequences of unsymmetrical bending is that the symmetric members about a vertical axis with thin-walled sections will undergo twisting under transverse loads. Unsymmetrical bending occurs under any of the following conditions: i) the section is symmetrical, but the line of action of the load is inclined to both the principal axes, or ii) the section is unsymmetrical, and the line of action of the load is along any centroidal axis. In the case of unsymmetrical bending, the applied moment will cause bending about both the principal axes of inertia, which should be located to calculate the stresses at any point in the cross-section. If the moment acts on the plane of symmetry, then the conventional simple bending equation can calculate the stresses.

$$\sigma_b = \frac{M}{I} y \qquad (3.38)$$

M is the applied moment (or computed bending moment) at any section along the span of the beam. I is the moment of inertia of the cross-section about the bending axis, and y is the distance of the extreme fiber from the neutral axis of the cross-section. However, if the load acts on another plane, it becomes unsymmetrical where one cannot use the conventional equation of flexure to obtain the stresses.

3.11.1 Bending Stresses under Unsymmetrical Bending

The following steps are to be followed to obtain the bending stresses at any section.

Step 1: to transform the problem of unsymmetrical bending to uniplanar bending.

Consider a cross-section of the beam under the action of a bending moment M (Figure 3.26). Let (z,y) be the coordinate axes passing through the centroid, (u,v) be the principal axes inclined at an angle (α) to the (z–y) axes, respectively, as shown in Figure 3.26.

To locate the principal axes of inertia, the following relationship is valid

$$u = z\cos\alpha + y\sin\alpha \tag{3.39}$$

$$v = -z\sin\alpha + y\cos\alpha \tag{3.40}$$

Where the angle is measured in the positive coordinate. The moment of inertia about UU and VV axes should be calculated to estimate the bending stresses.

$$I_u = \int_A v^2 dA \tag{3.41}$$

$$= \int_A \left(-z\sin\alpha + y\cos\alpha\right)^2 dA \tag{3.42}$$

$$= \int_A \left(z^2 \sin^2\alpha + y^2 \cos^2\alpha - 2z\sin\alpha\cos\alpha\right) dA \tag{3.43}$$

$$= \sin^2\alpha \int_A z^2 dA + \cos^2\alpha \int_A y^2 dA - \sin 2\alpha \int_A yz dA \tag{3.44}$$

$$= I_y \sin^2\alpha + I_Z \cos^2\alpha - I_{yZ} \operatorname{Sin}^2\alpha \tag{3.45}$$

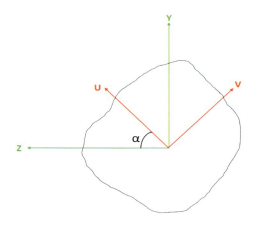

FIGURE 3.26 Typical cross-section.

Special Design Guidelines

We know the following relationships

$$\cos 2\alpha = 1 - 2\sin^2\alpha = 2\cos^2\alpha - 1 \quad (3.46)$$

Substituting in the above equation, we get

$$I_u = \frac{I_y}{2}(1-\cos 2\alpha) + \frac{I_z}{2}(1+\cos 2\alpha) - I_{yz}\sin 2\alpha \quad (3.47)$$

$$= \frac{I_Y + I_Z}{2} + \frac{I_Z + I_Y}{2}\cos 2\alpha - I_{YZ}\sin 2\alpha \quad (3.48)$$

$$I_u = \frac{I_Y + I_Z}{2} + \frac{I_Z - I_Y}{2}\cos 2\alpha - I_{YZ}\sin 2\alpha \quad (3.49)$$

$$I_V = \int_A u^2 \, dA \quad (3.50)$$

$$= \int_A (z\cos\alpha + y\sin\alpha)^2 \, dA \quad (3.51)$$

$$= \int_A (z^2\cos^2\alpha + Y^2\sin^2\alpha + 2zy\sin\alpha\cos\alpha) \, dA \quad (3.52)$$

$$I_V = \cos^2\alpha \int_A Z^2 \, dA + \sin^2\alpha \int_A y^2 \, dA + \sin 2\alpha \int_A zy \, dA \quad (3.53)$$

$$= \cos^2\alpha I_y + \sin^2\alpha I_z + I_{zy}\sin^2\alpha \quad (3.54)$$

$$= (1+\cos 2\alpha)\frac{I_y}{2} + \frac{1-\cos 2\alpha I_z}{2} + I_{zy}\sin 2\alpha \quad (3.55)$$

$$I_V = \frac{I_z - I_y}{2} - \frac{I_z - I_y}{2}\cos 2\alpha + I_{zy}\sin 2\alpha \quad (3.56)$$

$$I_u + I_V = I_z + I_y \quad (3.57)$$

$$I_{uv} = \int_A (uv) \, dA \quad (3.58)$$

$$= \int_A (z\cos\alpha + y\sin\alpha)(-z\sin\alpha + \cos\alpha) \, dA \quad (3.59)$$

$$= \int_A \left(-z^2 \sin\alpha \cos\alpha - yz \sin^2\alpha - yz \cos^2\alpha - y^2 \sin\alpha \cos\alpha\right) dA \quad (3.60)$$

$$\int_A z^2 dA = I_y \quad (3.61)$$

$$\int_A yz^2 dA = I_z \quad (3.62)$$

$$\int_A yz^2 dA = I_{yz} \quad (3.63)$$

Using the above relationship,

$$I_{uv} = -I_y \sin\alpha \cos\alpha + I_z \sin\alpha \cos\alpha + I_{yz}\left(\cos\alpha^2 - \sin\alpha^2\right) \quad (3.64)$$

$$I_{uv} = \frac{I_z + I_y}{2} \sin 2\alpha + I_{yz} \cos 2\alpha \quad (3.65)$$

For (u,v) being the principal axes of inertia, the following relationships hold good

$$I_{uv} = 0 \quad (3.66)$$

$$I_{uv} = \frac{I_z + I_y \sin 2\alpha}{2} + I_{yz} \cos 2\alpha = 0 \quad (3.67)$$

$$\tan(2\alpha) = -\left[\frac{2I_{yz}}{I_z - I_y}\right] \quad (3.68)$$

(α) The angle of measurement of the positive u-axis with positive z-axis is measured from the z-axis in a clockwise direction (refer to Figure 3.26).

Step 2: to determine the stress at any point in the cross-section. Referring to Figure 3.27, plane (OABC) is the plane of symmetry as $I_{uv} = 0$. Hence, to find stresses

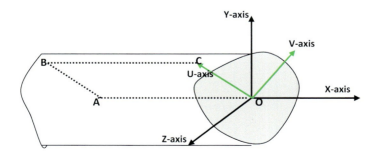

FIGURE 3.27 Plane of symmetry.

Special Design Guidelines

at a point lying on this plane, one can use the conventional equation with appropriate notation. But, if the load acts on the (ZX) plane, it is unsymmetrical, and conventional flexural equation cannot compute the stresses.

Please pay attention to Figure 3.28. The direction of the moments is marked as per the right-screw rule. To compute the stresses, we need to resolve the moments acting about (Z–Y) plane to (U–V) plane, which is given by the following relationship

$$M_u = M_z \cos\alpha \tag{3.69}$$

$$M_v = -M_z \sin\alpha \tag{3.70}$$

Referring to Figure 3.28, stress at any point p(u,v) is given by the following relationship

$$\frac{M_u}{I_u}(v)(\text{will cause compressive stress}) \tag{3.71}$$

$$\frac{M_v}{I_v}(u)(\text{will cause tensile stress}) \tag{3.72}$$

Considering tensile stresses as positive, total stress at point p(u,v) is given by the following relationship

$$\sigma_p = -\left[\frac{M_u}{I_u}\right]v + \left[\frac{M_v}{I_v}\right]u \tag{3.73}$$

$$= -\left[\left(\frac{M_u}{I_u}\right)v - \left(\frac{M_v}{I_v}\right)u\right] (\text{negative sign indicates compressive stress}) \tag{3.74}$$

The nature of the resultant bending stress at any point depends upon the coordinates of the respective point. It is important to take care of (u,v) signs while substituting them in the equation. The maximum stress will occur at a point that is at the greatest

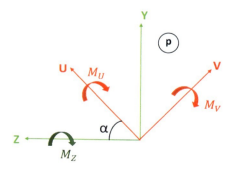

FIGURE 3.28 Resolution of moments about principal axes.

distance from the neutral axis. On one side of the neutral axis, all the points will carry the stresses of the same nature and opposite nature on the other side. Now, to locate the neutral axis, we can use the following algorithm.

Stress at any point is given by Equation (3.73). At the neutral axis, stress is zero. Equating this to zero, one can find the inclination of the neutral axis, measured from the positive U-axis, in the clockwise direction. Please refer to Figure 3.29 for details.

$$\sigma_p = -\left[\frac{M_u}{I_u}\right]v + \left[\frac{M_v}{I_v}\right]u = 0$$

$$\tan \beta = \frac{v}{u} = \frac{M_v}{M_u}\frac{I_u}{I_v} \tag{3.75}$$

[β is measured from positive u-axis, in the clockwise direction]

To determine the stress at any point in the cross-section, we need to estimate the coordinate of the point in the (u–v-axes) system. Let us refer to the following

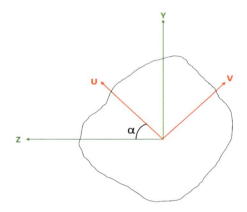

$$UA = z\cos\alpha + y\sin\alpha \tag{3.76}$$

$$VA = y\cos\alpha - z\sin\alpha \tag{3.77}$$

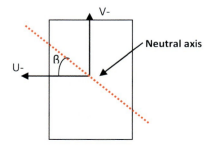

FIGURE 3.29 Location of NA for the section under unsymmetrical bending.

Special Design Guidelines

EXERCISES

EXAMPLE 1

A purlin of a roof truss ISLC 125 @ 10.7 kg/m is placed at a spacing of 500 mm c/c along the principal rafter. The principal rafter is placed at equal intervals of 3.0 m. Assume the purlin will act as a simply-supported beam with the uniformly distributed load of 2kN/m² and find the maximum stresses at A and B.

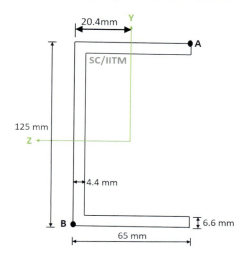

$$\text{Load from Purlin} = \frac{(0.5 \times 2) + \left(\frac{10.7 \times 10}{1000}\right)}{3} = 1.107 \text{ kN/m}$$

$$\text{Moment} = \frac{wl^2}{8} = \frac{1.107 + (3)^2}{8} = 1.245 \text{ kNm}$$

As the section has one axis of symmetry (about the z-axis), $I_{yz} = 0$; Hence, the (u,v) axis system coincides with the (z,y) axis, and the angle (α) will be zero.

$$M_U = M_Z \cos\alpha = 1.254 \text{ kNm}$$

$$M_V = -M_Z \sin\alpha = 0$$

For ISLC 125 @ 10.7 kg/m, from the steel tables, the following sectional properties are obtained

$$I_U = I_Z = 356.8 \times 10^4 \text{ mm}^4$$

$$I_V = I_Y = 57.2 \times 10^4 \text{ mm}^4$$

Point A: (−44.6, 62.5); Point B: (20.4, −62.5 mm)

$$\sigma_A = -\left[\frac{M_u}{I_u}\right]v_A + \left[\frac{M_v}{I_v}\right]u_A$$

$$\sigma_A = -\left[\frac{1.245 \times 10^6}{356.8 \times 10^4}\right]62.5 + 0 = -21.81 \, N/mm^2 \, (\text{compressive stress})$$

$$\sigma_B = -\left[\frac{M_u}{I_u}\right]v_B + \left[\frac{M_v}{I_v}\right]u_B$$

$$\sigma_B = -\left[\frac{1.245 \times 10^6}{356.8 \times 10^4}\right](-62.5) + 0 = 21.81 \, N/mm^2 \, (\text{tensile stress})$$

EXAMPLE 2

A simply-supported beam of section 50 mm × 190 mm has a span of 3.0 m. It rests on the support such that the 50 mm face makes an angle of 45° with the horizontal. It carries a central concentrated load of 100 kN at the mid-span. Find the stresses at points A and B, as marked in the diagram below.

$$W = 100 \text{ kN}; \quad L_{eff} = 3 \text{ m}$$

$$M = 100 \cos(45)\frac{3.0}{4} = 53.03 \text{ kNm}$$

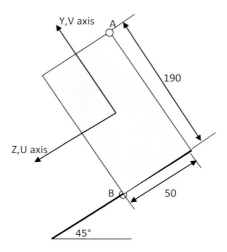

$$I_z = \frac{50 \times 190^3}{12} = 28.58 \times 10^6 \text{ mm}^4 = IU$$

Special Design Guidelines

$$I_y = \frac{190 \times 50^3}{12} = 1.98 \times 10^6 \text{ mm}^4 = IV$$

Due to symmetry, $I_{zy} = 0$; hence, $I_z = I_u$; $I_y = I_v$
Hence, angle (α) will be zero.

$$M_u = M\cos(0) = 53.03 \text{ kNm}$$

$$M_v = 0$$

Point A: (−25, 95); Point B: (25, −95)

$$\sigma_A = -\left[\frac{M_u}{I_u}\right]v_A + \left[\frac{M_v}{I_v}\right]u_A$$

$$\sigma_A = -\left[\frac{53.03 \times 10^6}{28.58 \times 10^6}\right]95 + 0 = -176.27 \text{ N/mm}^2 \text{ (compressive stress)}$$

$$\sigma_B = -\left[\frac{M_u}{I_u}\right]v_B + \left[\frac{M_v}{I_v}\right]u_B$$

$$\sigma_B = -\left[\frac{53.03 \times 10^6}{28.58 \times 10^6}\right](-95) + 0 = 176.27 \text{ N/mm}^2 \text{ (tensile stress)}$$

EXAMPLE 3

Consider a cantilever beam with two concentrated loads of 10 kN and 7 kN at 2.0 m and 3.0 m from the fixed end, acting at 30° and 45°, respectively at the centroid of the T section. Find the stresses at points A, B, C, and D, as marked below.

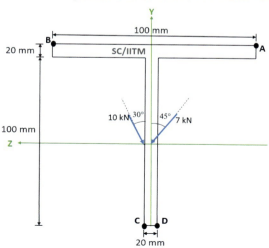

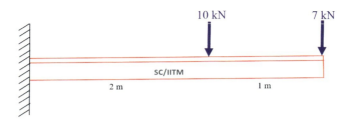

$$\bar{y} = \frac{\Sigma ay}{\Sigma a} = \frac{(100 \times 20 \times 10) + (100 \times 20 \times 70)}{(100 \times 20) x 2} = 40.0 \text{ mm}$$

$$I_Y = \frac{20 \times 100^3}{12} + \frac{100 \times 20^3}{12} = 1.74 \times 10^6 \text{ mm}^4$$

$$I_Z = \frac{100 \times 20^3}{12} + 100 \times 20 \times (30)^2 + \frac{20 \times 100^3}{12} + 100 \times 20 \times (30)^2$$
$$= 5.34 \times 10^6 \text{ mm}^4$$

Due to the vertical axis of symmetry, $I_{zy} = 0$; Hence, $I_z = I_u$; $I_y = I_v$. Hence, the angle (α) will be zero.

M_z = −[7 cos(45) × 3 + 10 cos (30) × 2] = −32.17 kNm (negative sign due to hogging BM)

$$M_U = M_Z \cos \alpha = -32.17 \cos(0) = -32.17 \text{ kNm}$$

$$M_V = -M_Z \sin \alpha = -\left[-32.17 \sin(0)\right] = 0$$

Coordinates of points: A(−50,40); B(50,40); C(10,−80); D(−10,−80)

$$\sigma_A = -\left[\frac{M_u}{I_u}\right] v_A + \left[\frac{M_v}{I_v}\right] u_A$$

$$\sigma_A = -\left[\frac{-32.17 \times 10^6}{5.34 \times 10^6}\right] 40 + 0 = 240.97 \text{ N/mm}^2 \text{ (tensile stress)}$$

$$\sigma_B = -\left[\frac{M_u}{I_u}\right] v_B + \left[\frac{M_v}{I_v}\right] u_B$$

$$\sigma_B = -\left[\frac{-32.17 \times 10^6}{5.34 \times 10^6}\right](45) + 0 = 240.97 \text{ N/mm}^2 \text{ (tensile stress)}$$

Special Design Guidelines

$$\sigma_c = -\left[\frac{M_u}{I_u}\right]v_c + \left[\frac{M_v}{I_v}\right]u_c$$

$$\sigma_c = -\left[\frac{-32.17 \times 10^6}{5.34 \times 10^6}\right](-80) + 0 = -481.95 \text{ N/mm}^2 \text{ (compressive stress)}$$

$$\sigma_D = -\left[\frac{M_u}{I_u}\right]v_D + \left[\frac{M_v}{I_v}\right]u_D$$

$$\sigma_D = -\left[\frac{-32.17 \times 10^6}{5.34 \times 10^6}\right](-80) + 0 = -481.95 \text{ N/mm}^2 \text{ (compressive stress)}$$

MATLAB Code

```
%% Unsymmetrical bending—T section
clc;
clear;
%% INPUT
% calculate the values of moment of inertia of the section in mm^4
Iy = 1.76e6;
Iz = 5.34e6;
Mz = −32.17; % moment about Z axis in kNm
ybar = 40.0; % location of centroidal axis in mm
zbar = 0;
%% Principal axis location
Iu = Iz;
Iv = Iy;
%% Stress calculation
Mu = −32.17; % in kNm
Mv = 0.0;
% calculation for point A—on flange top right
ua = −50;
va = 40;
sa = -(Mu*(10^6)*va/Iu)+(Mv*(10^6)*ua/Iv);
% calculation for point B—on flange top left
ub = 50;
vb = 40;
sb = -(Mu*(10^6)*vb/Iu)+(Mv*(10^6)*ub/Iv);
% calculation for point C—on web bottom left
uc = 10;
vc = −80;
sc = -(Mu*(10^6)*vc/Iu)+(Mv*(10^6)*uc/Iv);
% calculation for point D—on web bottom right
ud = −10;
```

```
vd = −80;
sd = −(Mu*(10^6)*vd/Iu) + (Mv*(10^6)*ud/Iv);
fprintf('Stress at point A = %6.2f N/mm^2 \n',sa);
fprintf('Stress at point B = %6.2f N/mm^2 \n',sb);
fprintf('Stress at point C = %6.2f N/mm^2 \n',sc);
fprintf('Stress at point D = %6.2f N/mm^2 \n',sd);
```

Output

Stress at point A = 240.97 N/mm^2
Stress at point B = 240.97 N/mm^2
Stress at point C = −481.95 N/mm^2
Stress at point D = −481.95 N/mm^2

EXAMPLE 4

A Z-section is used as a structural member for a cantilever beam of span 4 m, subjected to a point load of 20 kN at the free end. The section has a uniform thickness of 15 mm throughout. Find the stresses at points A and B, as marked below.

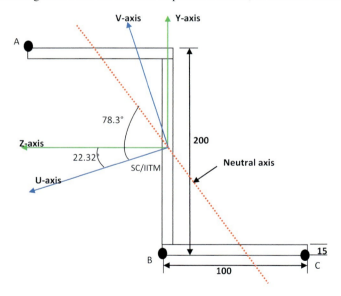

$M_z = -(20 \times 4) = -80$ kNm (negative due to hogging BM)

Z-section has two axes of symmetry. Hence, centroidal axes are conveniently marked as (z-y) axes, as shown in the diagram.

$$I_Y = \left\{ \left[\frac{15 \times 100^3}{12} + \left(100 \times 15 \times 42.5^2\right) \right] \times 2 \right\} + \frac{170 \times 15^3}{12}$$

$$= 7.97 \times 10^6 \text{ mm}^4$$

$$I_Z = \left\{\left[\frac{100 \times 15^3}{12} + 100 \times 15 \times (92.5)^2\right] \times 2\right\} + \frac{15 \times 170^3}{12}$$

$$= 31.84 \times 10^6 \text{ mm}^4$$

$$I_{zy} = \int zy\, dA = \left[(100 \times 15 \times 42.5 \times 92.5)\right] + [0]$$

$$+ \left[100 \times 15 \times (-42.5) \times (-92.5)\right]$$

$$= 11.79 \times 10^6 \text{ mm}^4$$

Since $I_{zy} \neq 0$, (U–V axes) is not the same as (Z–Y axes). Therefore, one needs to find the inclination of the U-axis with respect to the positive z-axis, which is referred to as (α)

$$\tan(2\alpha) = -\left[\frac{2I_{yz}}{I_z - I_y}\right] = -\frac{2 \times 11.79 \times 10^6}{(31.84 - 7.97) \times 10^6} = -0.9879$$

$$\alpha = -22.32°$$

It indicates that the +ve U-axis is located (22.32°) anti-clockwise to the positive z-axis, as marked in the diagram. Now, let us compute the resolved components of the moment, M_Z, along the principal axes

$$M_U = M_Z \cos\alpha = (-80)\cos(-22.32°) = -74.0 \text{ kNm}$$

$$M_V = -M_Z \sin\alpha = -\left[-80 \sin(-22.32°)\right] = -30.38 \text{ kNm}$$

$$I_u = \frac{I_Y + I_Z}{2} + \frac{I_Z - I_Y}{2} \cos 2\alpha - I_{YZ} \sin 2\alpha$$

$$I_u = \frac{(7.97 + 31.84) \times 10^6}{2} + \frac{(31.84 - 7.97) \times 10^6}{2} \cos(-44.64)$$

$$-11.79 \times 10^6 \sin(-44.64)$$

$$= 36.69 \times 10^6 \text{ mm}^4$$

$$I_u + I_V = I_z + I_y$$

$$I_V = I_z + I_y - I_u = (31.84 + 7.97 - 36.69) \times 10^6 = 3.12 \times 10^6 \text{ mm}^4$$

Now to locate the neutral axis

$$\tan \beta = \frac{v}{u} = \frac{M_v}{M_u} \frac{I_u}{I_v} = \left(\frac{-30.38}{-74.0}\right)\left(\frac{36.69 \times 10^6}{3.12 \times 10^6}\right) = 4.8278; \quad \beta = +78.3°$$

It is to be measured clockwise from the positive U-axis, as marked in the diagram. Coordinates of the points (A,B,C) are computed as below.

$$A(z, y) = (+92.5, +100.0); \quad B(+7.5, -100.0); \quad C(-92.5, -100)$$

Using Equations (3.76–3.77), one can computed (u,v) coordinates of the points (A,B,C)

$$U_A = 92.5\cos(-22.32) + 100\sin(-22.32) = 47.59$$

$$V_A = 100\cos(-22.32) - 92.5\sin(-22.32) = 127.64$$

$$U_B = 7.5\cos(-22.32) + (-100)\sin(-22.32) = 44.92$$

$$V_B = (-100)\cos(-22.32) - (7.5)\sin(-22.32) = -89.66$$

$$U_C = (-92.5)\cos(-22.32) + (-100)\sin(-22.32) = -47.59$$

$$V_C = (-100)\cos(-22.32) - (-92.5)\sin(-22.32) = -127.64$$

$$\sigma_A = -\left[\frac{M_u}{I_u}\right]v_A + \left[\frac{M_v}{I_v}\right]u_A$$

Based on the neutral axis location, it is also verified that points (A, B) are located below the neutral axis, experiencing compressive stresses, and Point C, located above neutral axis, the tensile stress. Please note that the section is subjected to hogging BM.

MATLAB Code

```
%% Unsymmetrical bending—Z section
clc;
clear;
%% INPUT
% calculate the values of moment of inertia of the section in mm^4
Iy = 7.97e6;
Iz = 31.84e6;
Iyz = 11.79e6;
Iuv = 0;
Mz = −80; % Moment about Z axis in kNm
%% Principal axis location
```

```
tt = (−2*Iyz)/(Iz-Iy);
al = (atand(tt) + 180)/2;
Iu = ((Iy+Iz)/2) + (((Iz-Iy)*cosd(2*al))/2) − (Iy*sind(2*al));
Iv = (Iz+Iy) − Iu;
%% Stress calculation
Mu = Mz*cos(al);
Mv = −Mz*sin(al);
ra = Mv*Iu/(Mu*Iv);
be = atand(ra);
% calculation for point A—on flange top
ya = 100;
za = 92.5;
ua = (za*cos(al)) + (ya*sin(al));
va = −(za*sin(al)) + (ya*cos(al));
sa = −(Mu*(10^6)*va/Iu) + (Mv*(10^6)*ua/Iv);
% calculation for point B—on web bottom
yb = −100;
zb = 7.5;
ub = (zb*cos(al)) + (yb*sin(al));
vb = −(zb*sin(al)) + (yb*cos(al));
sb = −(Mu*(10^6)*vb/Iu) + (Mv*(10^6)*ub/Iv);
fprintf('Stress at point A = %6.2f N/mm^2 \n', sa);
fprintf('Stress at point B = %6.2f N/mm^2 \n \n', sb);
```

Output

Stress at point A = −205.96 N/mm^2
Stress at point B = −618.22 N/mm^2

EXAMPLE 5

An unequal angle section is used as a structural member for a cantilever beam of span 2 m, subjected to a point load of 2 kN at the free end. The section has a uniform thickness of 5 mm throughout. Find the stresses at points *A* and *B*, as marked below.

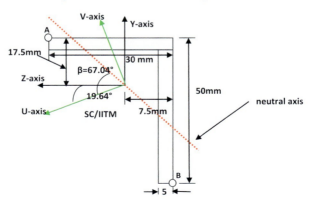

$M_z = -(2 \times 2) = -4$ kNm (negative due to hogging BM)

Since $I_{zy} \neq 0$, (U-V axes) is not the same as (Z-Y axes). Therefore, one needs to find the inclination of the U-axis with respect to the positive z-axis, which is referred to as (α)

It indicates that the +ve U-axis is located (19.64°) anti-clockwise to the positive z-axis, as marked in the diagram. Now, let us compute the resolved components of the moment, M_z, along the principal axes.

Now to locate the neutral axis.

It is to be measured clockwise from the positive U-axis, as marked in the diagram. Coordinates of the points (A,B) are computed as below.

Using Equations (3.76–3.77), one can compute (u,v) coordinates of the points (A, B)

Based on the neutral axis location, it is also verified that points (A, B) are located below the neutral axis, experiencing compressive stresses. Please note that the section is subjected to hogging BM.

3.12 CURVED BEAMS

Based on the initial curvature, curved beams are classified as follows.

i. Beams with small initial curvature: the ratio of the initial radius of curvature and the depth of the section is greater than 10.
ii. Beams with large initial curvature: the ratio of the initial radius of curvature and the depth of the section is less than or equal to 10.

3.12.1 Bending for Small Initial Curvature

Refer to Figure 3.30, explaining the terminology of curved beams used in the derivation.

Let $d\varphi'$ be the angle subtended after deformation.

R be the initial radius of curvature.

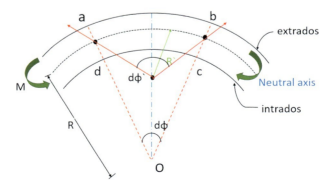

FIGURE 3.30 Curved beams.

Special Design Guidelines 243

dφ be the angle subtended at the center of curvature by the portion' abcd'.
R' be the radius of curvature after moment M is applied.
If R' <R, then the applied moment will tend to close the curvature.
Let us consider a fiber PQ at a distance y from the neutral axis.
Then,
The original length of the fiber = (R + y) dφ
Length of the fiber after application of the moment = (R` + y) dφ`
Change in length of the fiber = (R` + y) dφ` − (R + y) dφ

$$\text{Strain, } \epsilon = \frac{(R'+y)d\varphi' - (R+y)d\varphi}{(R+y)d\varphi} \tag{3.78}$$

As the length of the fiber at the neutral axis remains unchanged, the following relationship holds good

$$ds = Rd\varphi = R'd\varphi' \tag{3.79}$$

Substituting the above equation in the strain equation, we get

$$\varepsilon = \frac{y(d\varphi' - d\varphi)}{(R+y)d\varphi} \tag{3.80}$$

In the above equation, y may be neglected since y<<R. Hence, the following relationship holds good

$$\varepsilon = \frac{y(d\varphi' - d\varphi)}{Rd\varphi} \tag{3.81}$$

On simplification, we get

$$\varepsilon = \frac{yd\varphi'}{Rd\varphi} - \frac{yd\varphi}{Rd\varphi} = y\left(\frac{d\varphi'}{ds} - \frac{d\varphi}{ds}\right) \tag{3.82}$$

$$\varepsilon = y\left(\frac{1}{R'} - \frac{1}{R}\right) \tag{3.83}$$

Substituting $\varepsilon = \frac{\sigma}{E}$, we get

$$y\left(\frac{1}{R'} - \frac{1}{R}\right) = \frac{\sigma}{E} \tag{3.84}$$

$$\frac{\sigma}{y} = E\left(\frac{1}{R'} - \frac{1}{R}\right) \tag{3.85}$$

The assumptions made in deriving the equation are as follows.

i. Every cross-section of the curved beam remains plane and perpendicular to the centroidal axis, before and after the application of the external moment.
ii. To satisfy the above condition, it is to be agreed that the net force acting on any cross-section of the curved beam should be zero. If the net force is not equal to zero, then it may result in warping.

Mathematically,

$$\int_A \sigma dA = 0 \qquad (3.86)$$

$$\int Ey\left(\frac{1}{R'} - \frac{1}{R}\right) dA = 0 \qquad (3.87)$$

$$E\left(\frac{1}{R'} - \frac{1}{R}\right) \int y\,dA = 0 \qquad (3.88)$$

$$\text{Since } E\left(\frac{1}{R'} - \frac{1}{R}\right) \neq 0 \qquad (3.89)$$

$\int y\,dA = 0$, which implies that the geometric axis of the curved beam should coincide with that of the neutral axis of the curved beam. As the curved beam is in equilibrium condition under the applied moment, it can be stated that,

$$\int \sigma y\,dA = M \qquad (3.90)$$

Substituting for σ in the above equation,

$$E\left(\frac{1}{R'} - \frac{1}{R}\right) \int y^2 dA = M \qquad (3.91)$$

$$\text{Since } \int y^2 dA = I \qquad (3.92)$$

$$E\left(\frac{1}{R'} - \frac{1}{R}\right) I = M \qquad (3.93)$$

Thus,

$$\frac{M}{I} = \frac{\sigma}{y} = E\left(\frac{1}{R'} - \frac{1}{R}\right) \qquad (3.94)$$

The initial length of the fiber $\overline{PQ}$ at distance y from the centroidal axis is given by

$$(R+y)d\varphi \tag{3.104}$$

Change in length of the fiber on the application of the moment, M is given by

$$(y+e)\Delta d\varphi \tag{3.105}$$

Where (e) is the distance of the neutral axis measured from the centroidal axis. M is applied in such a manner that the neutral axis is shifted towards the center of curvature.

$$\text{Strain}, \epsilon = \frac{(y+e)\Delta d\varphi}{(R+y)d\varphi} \tag{3.106}$$

Assuming that the longitudinal fibers do not undergo any deformation, stress is given by

$$\sigma = E\left(\frac{y+e}{R+y}\right) \tag{3.107}$$

It shows that the stress distribution is nonlinear and hyperbolic. The basic assumption made is that every section normal to the centroid axis remains plane and perpendicular, before and after application of moment, M. Following condition holds good

$$\text{Total compressive force} = \text{total tensile force} \tag{3.108}$$

Since the average stress on the concave side is more than the convex side, the neutral axis will shift towards the center of curvature. Equating the sum of internal forces to zero at the cross-section, we get

$$\int_A \sigma dA = 0 \tag{3.109}$$

Substituting the expression for stress, we get

$$\int_A E \frac{A d\varphi}{d\varphi}\left(\frac{y+e}{y+R}\right) dA = 0 \tag{3.110}$$

$$E \frac{\Delta d\varphi}{d\varphi} \int_A \left(\frac{y+e}{y+R}\right) dA = 0 \tag{3.111}$$

Since $E \frac{\Delta d\varphi}{d\varphi} \neq 0$ and it is a constant,

$$\int_A \left(\frac{y+e}{y+R}\right) dA = 0 \tag{3.112}$$

$$\int_A \left(\frac{y}{R+y}\right) dA + e \int_A \left(\frac{1}{y+R}\right) dA = 0 \qquad (3.113)$$

We know that the first integral term, $\int_A \frac{y}{R+y} dA = mA$, where m is a constant depending on the geometry of the X section. The quantity mA is termed as the modified area of the cross-section which is modified due to the application of moment, M. The second integral term is $e \int_A \frac{1}{(R+y)} dA$

$$e \int_A \frac{1}{(R+y)} dA = e \int \frac{R+y-y}{R} \cdot \frac{1}{(R+y)} dA \qquad (3.114)$$

$$= \frac{e}{R} \int_A \frac{R+y}{R+y} dA - \int_A \frac{y \, dA}{R+y} = \frac{eA}{R} - \frac{e}{R}(mA) \qquad (3.115)$$

Thus,

$$mA + \frac{eA}{R} - \frac{emA}{R} = 0 \qquad (3.116)$$

$$m + \frac{e}{R} - \frac{em}{R} = 0 \qquad (3.117)$$

$$m = e\left(\frac{m}{R} - \frac{1}{R}\right) \qquad (3.118)$$

$$m = \frac{e}{R}(m-1) \qquad (3.119)$$

$$e = \left(\frac{m}{m-1}\right) R \qquad (3.120)$$

where m is the geometry property of the section. It is to be noted that the applied moment in the cross-section should be equal to the resisting moment for any equation. Hence, by applying the following condition

$$\int_A (\sigma \, dA) y = M \qquad (3.121)$$

substitution for stress, $\int_A \frac{y^2 + ye}{R+y} dA$ is to be evaluated.

$$\int_A \frac{y^2 \, dA}{R+y} + e \int_A \frac{y}{y+R} dA = \int_A \left(y - \frac{Ry}{R+y}\right) dA + e \int_A \frac{y}{y+R} dA \qquad (3.122)$$

Special Design Guidelines

We know that,

$$\int_A y\,dA = 0 \qquad (3.123)$$

Hence,

$$\int_A \left(\frac{y^2+ye}{R+y}\right)dA = -R\int_A\left(\frac{y}{R+y}\right)dA + e\int_A \frac{y\,dA}{R+y} \qquad (3.124)$$

$$= -R(mA) + emA = -mA(R-e) \qquad (3.125)$$

$$-E\left(\frac{\Delta d\varphi}{d\varphi}\right) = \frac{M}{mA(R-e)} \qquad (3.126)$$

$$\text{Since } m = \frac{e}{R-e} \qquad (3.127)$$

$$E\frac{\Delta d\varphi}{d\varphi} = \frac{M}{Ae} \qquad (3.128)$$

On simplification,

$$e = \frac{M}{Ae}\left(\frac{y+e}{R+y}\right) \qquad (3.129)$$

It is also known that,

$$\sigma = \frac{M}{(m-1)}R \qquad (3.130)$$

$$\sigma = \frac{M}{AR}\frac{(m-1)}{m}\left[\frac{y+\left(\frac{m}{m-1}\right)R}{R+y}\right] \qquad (3.131)$$

$$= \frac{m}{AR}\frac{(m-1)}{m}\left[\frac{(m-1)y+mR}{(m-1)(R+y)}\right] \qquad (3.132)$$

$$\sigma = \frac{M}{AR}\frac{1}{m}\left[\frac{m(y+R)-y}{y+R}\right] \qquad (3.133)$$

$$\sigma = \frac{M}{AR}\left[1-\frac{1}{m}\left(\frac{y}{R+y}\right)\right] \qquad (3.134)$$

The above equation is known as the Winkler–Bach equation, in which σ is the tensile/compressive stress at distance y from the centroidal axis (not from the neutral axis), M is the applied moment (causing a decrease in curvature), A area of cross-section, m is the section properties (geometry/shape of the cross-section), and R is the radius of curvature of the unstressed curved beam.

Sign convention

(y) is negative when measured towards the concave side and positive when measured towards the convex side. With the sign convention, negative stress indicates compressive stress, and positive stress indicates tensile stress. Designers use sections with large shape factors to enable the maximum load carrying capacity in the plastic design. Stress can be computed from the following relationship.

$$\sigma = \frac{M}{Ae}\left[\left(\frac{y+e}{y+R}\right)\right] \quad (3.135)$$

(e) is the offset of the neutral axis from the centroid, measured towards the center of curvature. Specific stress equations for intrados and extrados are used to find the maximum stress in the extreme fibers. The following equations hold good

$$\sigma_{intrados} = \frac{M}{Ae}\left[\frac{-h_i+e}{R-h_i}\right] = -\frac{M}{Ae}\left[\frac{h_i+e}{r_i}\right] \quad (3.136)$$

$$\sigma_{extrados} = \frac{M}{Ae}\left[\frac{h_o+e}{R+h_o}\right] = \frac{M}{Ae}\left[\frac{(e+h_o)}{r_o}\right] \quad (3.137)$$

From the above equations, it is clear that stress is a function of the parameter "m," which is given by the following relationship

$$mA = \int_A \frac{y}{R+y}dA \quad (3.138)$$

3.12.4 SIMPLIFIED EQUATIONS

$$\sigma = K\frac{Mh}{I} \quad (3.139)$$

Where K is a factor to be used for intrados and extrados as below.

$$K_{intrados} = \frac{\frac{M}{Mh}\frac{(h_i-e)}{r_i}}{\frac{Mh_i}{2I}} \quad (3.140)$$

Special Design Guidelines

TABLE 3.11
Stress Correction Factors for Intrados and Extrados

Cross-section	R/h	Factors		
		K_i	K_o	e
Circular	1.2	3.41	0.54	0.224R
	1.4	2.40	0.60	0.151R
	1.6	1.96	0.65	0.108R
	1.8	1.75	0.68	0.084R
	2.0	1.62	0.71	0.009R
	3.0	1.33	0.79	0.030R
	4.0	1.23	0.84	0.016R
	6.0	1.14	0.89	0.007R
Rectangular	1.2	2.89	0.57	0.305R
	1.4	2.13	0.63	0.204R
	1.6	1.79	0.67	0.149R
	1.8	1.63	0.70	0.112R
	2.0	1.52	0.73	0.090R
	3.0	1.30	0.81	0.041R
	4.0	1.20	0.85	0.021R
	6.0	1.12	0.90	0.0093R

$$K_{extrados} = \frac{\frac{h_o + e}{r_0}}{\frac{Mh_0}{2I}} \tag{3.141}$$

k_i and k_o are called stress correction factors, as given in Table 3.11.

EXERCISE PROBLEMS

EXAMPLE 1

Consider an open section comprising a 100 MN load as shown in the diagram. Compute the stresses at points A and B in the curved beam.

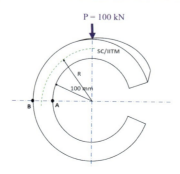

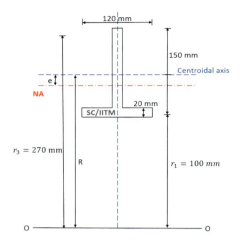

MATLAB Code

```
% Curved beam—T section
clc;
clear;
%% INPUT
p=100; % load acting on the beam in kN
%Geometric properties
r1 = 100;
r2 = 120;
r3 = 270;
r = 157.22;
h1 = 112.78;
h2 = 57.22;
h3 = 37.22;
b1 = 120;
b2 = 20;
a = 5400;% cross-sectional area in mm^2
m = 1 − (r/a)*((b1*log(r2/r1)) + (b2*log(r3/r2))); % sectional property (no unit)
e = m*r/(m − 1); % eccentricity in mm
%% Section AB on the centroidal axis
sd = -p*1000/a; % direct stress in N/mm^2
M = p*r/1000; % moment at CG in kNm
si = -(M*1000000*(h2 − e)/(a*e*r1)); % stress at intrados in N/mm^2
so = (M*1000000*(h1 + e)/(a*e*r3)); % stress at extrados in N/mm^2
sa = sd + si; % total stress at intrados
sb = sd + so; % total stress at extrados
fprintf('SOLUTION: \n');
fprintf('Stress at point A = %6.2f N/mm^2 \n',sa);
fprintf('Stress at point B = %6.2f N/mm^2 \n',sb);
```

Special Design Guidelines

Output
Stress at point A = −97.04 N/mm^2
Stress at point B = 70.84 N/mm^2

EXAMPLE 2

A circular ring of the rectangular cross-section has a horizontal slit, as shown in the diagram. Find the stresses at A and B.

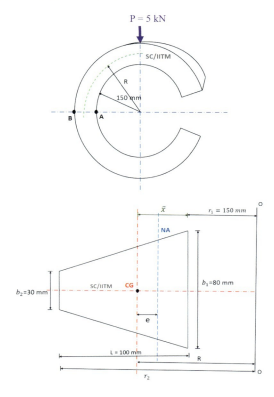

MATLAB Code
% Curved beam with a trapezoidal cross-section
clc;
clear;
%% INPUT
h = 100; % height of the section in mm
b1 = 80; % breadth of the section in mm
b2 = 30;
r1 = 150; % inner radius of the beam in mm
p = 5; % load action on the beam in kN
%% Geometric properties

b3 = (b1 − b2)/2;
x = ((2*h*b3/3) + (b2*h))/(b1 + b2); % Location of neutral axis
hi = x;
ho = h − x;
R = r1 + x; % radius of the curved beam in mm
r2 = r1 + h; % outer radius of the curved beam in mm
a = (b1 + b2)*h/2; % cross-sectional area in mm^2
m = 1 − ((R/a)*((b2 + ((r2*(b1 − b2))/(r2 − r1)))*log(r2/r1) − (b1 − b2))); % sectional property
e = m*R/(m − 1); % eccentricity in mm
I = (2*(((b3*(h)^3)/36) + (0.5*b3*h*((h/3) − x)*((h/3) − x)))) + (((b2*(h)^3)/12) + (b2*h*((h/2) − x)*((h/2) − x))); % moment of inertia in mm^2
%% Section AB on the centroidal axis
sd = −p*1000/a; % direct stress in N/mm^2
M = p*R/1000; % moment at CG in kNm
si = −(M*1000000*(hi − e)/(a*e*r1)); % stress at intrados in N/mm^2
so = (M*1000000*(ho + e)/(a*e*r2)); % stress at extrados in N/mm^2
sa = sd + si; % total stress at intrados
sb = sd + so; % total stress at extrados
fprintf('SOLUTION: \n');
fprintf('Stress at point A = %6.2f N/mm^2 \n',sa);
fprintf('Stress at point B = %6.2f N/mm^2 \n',sb);

Output
SOLUTION:
Stress at point A = −12.35 N/mm^2
Stress at point B = 10.05 N/mm^2

EXAMPLE 3

Find the stresses developed at points A and B of the curved beam with a circular cross-section shown in the diagram.

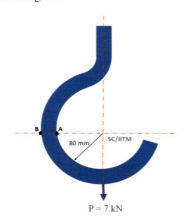

Special Design Guidelines

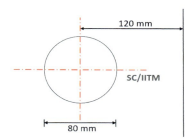

MATLAB Code

```
% Crane hook problem with Circular cross-section
clc;
clear;
%% INPUT
r1 = 80; % inner radius of the beam in mm
d = 80; % diameter of the section in mm
p = 7; % load action on the beam in kN
%% Geometric properties
x = d/2; % Location of neutral axis
hi = x;
ho = d − x;
R = r1 + x; % radius of the curved beam in mm
r2 = r1 + d; % outer radius of the curved beam in mm
a = 3.14*d*d/4; % cross-sectional area in mm^2
m = 1 − (2*((R/d)^2)) + (2*(R/d)*sqrt(((R/d)^2) − 1)); % sectional property
e = m*R/(m − 1); % eccentricity in mm
%% Section AB—stress calculation—Winkler-Bach equation
sd = p*1000/a; % direct stress in N/mm^2
M = −p*R/1000; % moment at CG in kNm
ri = r1;
ro = r2;
si = −M*1000000*(hi − e)/(a*e*ri); % Stress at intrados
so = M*1000000*(ho + e)/(a*e*ro); % Stress at extrados
sa = sd + si; % total stress in intrados
sb = sd + so; % total stress at extrados
fprintf('SOLUTION FROM WINKLER-BACH EQUATION: \n');
fprintf('Stress at point A = %6.2f N/mm^2 \n',sa);
fprintf('Stress at point B = %6.2f N/mm^2 \n \n',sb);
```

Output

SOLUTION FROM WINKLER-BACH EQUATION:
Stress at point A = 4.77 N/mm^2
Stress at point B = −2.39 N/mm^2

EXAMPLE 4

Find the stresses at points A and B of the crane hook of the trapezoidal cross-section under the action of 70 kN load as shown in the diagram.

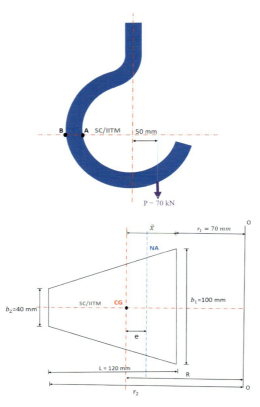

MATLAB Code

% Crane hook problem with a trapezoidal cross-section
clc;
clear;
%% INPUT
h = 120; % height of the section in mm
b1 = 100; % breadth of the section in mm
b2 = 40;
r1 = 70; % inner radius of the beam in mm
p = 70; % load action on the beam in kN
%% Geometric properties
b3 = (b1 − b2)/2;
x = ((2*h*b3/3) + (b2*h))/(b1 + b2); % Location of neutral axis
hi = x;

```
ho = h − x;
R = r1 + x; % radius of the curved beam in mm
r2 = r1 + h; % outer radius of the curved beam in mm
a = (b1 + b2)*h/2; % cross-sectional area in mm^2
m = 1 − ((R/a)*((b2 + ((r2*(b1 − b2))/(r2 − r1)))*log(r2/r1) − (b1 − b2))); % sectional property
e = m*R/(m − 1); % eccentricity in mm
I = (2*(((b3*(h)^3)/36) + (0.5*b3*h*((h/3) − x)*((h/3) − x)))) + (((b2*(h)^3)/12) + (b2*h*((h/2) − x)*((h/2) − x))); % moment of inertia in mm^2
%% Section AB—stress calculation—Winkler-Bach equation
sd = p*1000/a; % direct stress in N/mm^2
M = −p*(R + 50)/1000; % moment at CG in kNm
ri = r1;
ro = r2;
si = −M*1000000*(hi − e)/(a*e*ri); % Stress at intrados
so = M*1000000*(ho + e)/(a*e*ro); % Stress at extrados
sa = sd + si; % total stress in intrados
sb = sd + so; % total stress at extrados
fprintf('SOLUTION FROM WINKLER-BACH EQUATION: \n');
fprintf('Stress at point A = %6.2f N/mm^2 \n',sa);
fprintf('Stress at point B = %6.2f N/mm^2 \n \n',sb);
```

Output
SOLUTION FROM WINKLER-BACH EQUATION:
Stress at point A = 102.87 N/mm^2
Stress at point B = −55.65 N/mm^2

EXERCISE

1. What are thin-walled sections? What are the specific problems associated with their structural behavior? Explain.
2. Explain a pure torsion behavior.
3. In open, thin-walled sections, explain how torsion and warping occur.
4. Derive the expression to estimate the total torsion in open, thin-walled sections.
5. Describe buckling, with a neat sketch.
6. Explain, with neat sketches, different modes of buckling.
7. Explain the differences between flexural, torsional, and lateral-torsional buckling.
8. Explain, clearly, the phenomenon of lateral-torsional buckling (LTB) with a neat sketch.
9. Explain the mechanisms behind LTB.
10. What do you understand by the torsional effect?
11. Discuss various factors that influence LTB behavior.
12. Discuss the 3-factor formula useful in LTB.

13. Explain the moment correction factors used in LTB design.
14. How is a design check carried out in LTB design?
15. Discuss buckling curves useful in LTB design.
16. Explain when unsymmetrical bending occurs.
17. How do you classify curved beams? What are the design checks under those classifications?
18. Derive the M factor useful in the design of curved beams.

4 Risk, Reliability, and Safety Assessment

4.1 BACKGROUND FOR RELIABILITY ASSESSMENT

The reliability-based design of offshore structures encompasses all the advantages of probabilistic modeling to map uncertainties (Chandrasekaran, 2016; Bjeragar, 1990). Reliability and the probability of failure of a structural system are converses and the probability of failure includes the failure of all structure members. By contrast, reliability is mapped only to those critical members. However, the efficiency of the first-order reliability method (FORM) and the second-order reliability method (SORM) is independent of the probability of failure but decreases with the number of random variables. Further, the efficiency of the simulation techniques (for example, Monte Carlo simulation) improves with the sampling technique: it can significantly minimize the process's variance.

Monte Carlo simulation is one of the major components of risk analysis, which substitutes decision tree analysis. Risk is a qualitative phenomenon, while probability is quantitative. While risk assessment is a mathematical process of quantifying the potential loss of assets or changes that occur over time, computing the probability of an event is one form of risk analysis (Onoufriou, 1999; Skrzypczak et al., 2017). The topside of an offshore platform is subjected to a variety of complex loads. Various uncertainties that arise from these loads and material characteristics under aging can be quantified using proper statistical distribution tools (Karmazinova & Melcher, 2012; Sadowski et al., 2015).

Loads that act on the topside of an offshore platform (say, for example, wind or fire load) follow a random process, which a probability distribution can characterize. It is further characterized by a density function, with two or more parameters, namely mean and standard deviation. Monte Carlo simulation generates samples from these distributions and estimates the variable, dependent on the random variable. Alternatively, a sensitivity analysis is also useful to identify the most influencing random variables in the Monte Carlo simulation process. Despite all these advantages, there are a few drawbacks: i) output depends primarily on the modeling method, ii) the process can take a large number of iterations to converge, and iii) it can consume a considerable amount of time, making it computationally inefficient.

In a simulation process, when the probability of failure lies in the order 10^{-3} to 10^{-10}, only a few points fall under the failure domain, resulting in a zero probability of failure (Melchers, 1988; Srinivasan & Kiran, 2014a, 2014b). Hence, this demands many simulations, which will intensify the computational effort on the limit state function. It is therefore necessary to generate the random numbers through a sampling density function, which has the maximum likelihood enclosed within the limit state function itself. Therefore, an indicator function will provide information about whether the generated point lies within the failure domain or the safe domain. In general, sampling from the failure domain is considered a biased sampling but dividing it with the original density function will be unbiased (Ranganathan, 1999; Murtha, 1997).

4.2 OVERVIEW OF SAFETY

Safety is the first and foremost priority in any working sector. Safety can be defined as ensuring against an unlikely (or a risky) situation or a dangerous hazard. Safety is the activity of preventing one from being exposed to a hazardous situation. Disastrous consequences can be avoided by remaining safe, thereby protecting human life and preserving the offshore plant. Safety is always associated with risk. When the chances of risks are high, the situation is said to be "highly unsafe." Therefore, risk has to be assessed and eliminated, while safety needs to be assured.

One good reason for safety being insisted on in offshore plants is that investment in this industry is several times higher than in any other industry across the world. Offshore platform designs are very complex and innovative so that if any damage occurs it is not always easy to repair the design. Before analyzing the importance of safety in the offshore industry, one must understand the key issues in hydrocarbon production and processing. The prime importance of safety is to prevent death or injury to workers in the plant and to the public located nearby. Safety is also an important factor regarding financial damage to the plant, as investment is greater in the oil and petroleum industries than in any other industry. Safety also ensures that the surrounding atmosphere is not contaminated. Safety can be ensured by identifying and assessing hazards at every stage of operation. Identifying and assessing hazards at every stage is an important criterion for monitoring safety and computing the risks involved.

4.3 LESSONS LEARNED FROM THE PAST

Oil platform Piper Alpha suffered an explosion in July 1988 and is still regarded as the worst offshore oil disaster in the history of the UK (Figure 4.1). The accident killed 165 out of 220 crew members, plus two crew from the standby vessel *Sandhaven*. The accident was attributed mainly to human error and was a major eye-opener to the offshore industry regarding safety issues. Estimated property damage was approximately US$1.4 billion. The accident was caused by maintenance work that was simultaneously carried out on one of the high-pressure condensate pumps and on a related safety valve, which led to a leak in condensates. After removing one of the pressure safety valves of a gas condensate pump for maintenance, the condensate pipe remained temporarily sealed with a blind flange. The work was not completed during the day shift. Not aware of the maintenance being carried out on one of the pumps, a night crew turned on the alternative pump. Following this, the blind flange, including firewalls, failed to handle the pressure,

FIGURE 4.1   Piper Alpha disaster.

leading to several explosions. After the official inquiry, safety norms were revised, and drastic changes in the offshore industry concerning safety management, regulation, and training were implemented. Intensified fire exploded due to the failure to close the flow of gas from the Tartan platform. The automatic fire-fighting system was inactive following underwater work by divers before the incident. From this case study, one could infer that such a devastating incident had occurred through human error and that it could have been prevented if close attention had been paid while working in the field.

On 23rd March 1989, oil tanker *Exxon Valdez* (Figure 4.2) was underway from Valdez, Alaska, with a cargo of 180,000 tons of crude oil when it collided with an iceberg, and eleven cargo tanks were punctured. Within a few hours, 19,000 tons had been lost. By the time the tanker was refloated, on 5th April 1989, about 37,000 tons had been lost, and 6,600 sq. km of the country's most important fishing grounds and the surrounding shoreline were sheathed in oil. The size of the spill and its remote location made government and industry efforts difficult. This spill was about 20 percent of the 180,000 tons of crude oil the vessel was carrying when it struck the reef. The salvage effort that took place immediately after the grounding saved the vessel from sinking, thus preventing a far larger oil spill from occurring.

4.4   ROLE OF SAFETY IN OFFSHORE PLANTS

Safety plays a very important role in the offshore industry. Safety can be achieved by adopting various control measures. Explosion or damage can be prevented by implementing control methods such as regular monitoring of temperature and pressure inside the plant, using a well-equipped coolant system, proper check valves, vent outs, and effective casing or shielding system checks for oil spillages into the water. By thoroughly ensuring proper control facilities, one can avoid or minimize the hazardous environment in the offshore industry.

FIGURE 4.2 Exxon Valdez oil spill.

4.4.1 Risk and Safety

Safety and risk are concurrent. Risk can be classified into two types, namely individual risk and societal risk. Individual risk can be defined as the frequency at which the individual may be expected to sustain a given level of harm from realizing a hazard. It usually accounts for only the risk of death. Individual risk can be expressed as individual risk per year or the fatality accident rate (FAR). The following relationship gives the average individual risk

$$\text{Average individual risk} = \frac{\text{No. of Fatalities}}{\text{No. of persons exposed to risk}} \quad (4.1)$$

Societal risk is the relationship between the frequency and the number of people suffering a given level of harm from realizing a hazard. Societal risk can be expressed as FN curves showing the relationship between the cumulative frequency (F) and the number of fatalities (N). Societal risk can also be expressed as annual fatality rates, in which frequency and fatality data are combined into a convenient single measure of a group.

4.4.2 Measurement of Accident or Loss

There is no single method capable of measuring accident and loss statistics for all the requisite aspects. Three systems are commonly used in the offshore industry: OSHA (Occupational Safety and Health Administration, US Department of Labor), FAR (fatality accident rate), and fatality rates (or deaths per person per year). All three methods report the number of accidents or fatalities for a fixed number of working hours during a specified period.

4.5 QUANTITATIVE RISK ANALYSIS

4.5.1 Logical Risk Analysis

The logical risk analysis method aims at financing risk towards risk reduction. It is exclusively applicable to the process industry, which consists of six steps, as explained below.

Step 1: Compute risk index for each department
Each department of the process industry possesses a different risk level. It can be identified by evaluating the hazards present and the control measures available. It is called the "first level of risk assessment" and is generally identified by preparing a checklist. From the checklist, a hazard score and a control score can be established. The hazard checklist has six groups of hazards and points are associated with each hazard within each group. These points are calculated for hazards applied within that group. The hazard score for a group is the product of the sum and hazard weightage. The hazard score for the department is the sum of scores computed for each of the six groups.

A control score is similarly calculated. The control score for a department is the sum of the scores of each of the six groups. After calculating the hazard score and control score, the risk index is given by the following relationship

$$\text{Risk Index} = \text{Control score} - \text{Hazard score} \qquad (4.2)$$

Step 2: Determine relative risk for each department
After calculating the risk index, the department with the highest risk index is deemed the best. All other departments' scores are then calculated based on the best department (in which the best department's relative risk index will be zero).

Step 3: Compute percentage risk index for each department
The percentage risk index indicates the relative risk of each department to the total risk of the plant. The relative risk of each department is converted into a percentile risk by a simple procedure. The total risk of the entire department is the sum of the absolute value of the relative risk of each department.

Step 4: Determine composite exposure dollars for each department
The estimated risk is converted to the financial value "now." It estimates the financial value of each department. Composite exposure dollars are the sum of the monetary value of three components: property value, business interruption, and personnel exposure. Property value is estimated by the replacement cost of all material and equipment at risk in the department. Business interruption is computed as the product of the unit cost of goods and the department's production per year X (expected percentage of

capacity). Personnel exposure is the product of the total number of people in the department during the most populated shift and the monetary value of each person.

Step 5: Compute composite risk for each department

For a department, composite risk is the product of the composite exposure of dollars and a percentage risk index. It represents the value of the relative risk of each department. Units for composite risk are US dollars.

Step 6: Risk ranking

The relative rank of all the departments is computed based on the composite score. It helps the risk managers to decide the level of funds required by each department. Departments should be ranked from the highest composite score to the lowest (the lowest will be zero for the reference department).

4.6 RISK ASSESSMENT

Risk assessment can be undertaken by determining the risks and evaluating them. Risk determination is a quantitative type of risk assessment. Risk determination includes risk identification and risk estimation. Risk is identified by observing new risk parameters and is estimated from the probability of occurrences and consequences. Risk evaluation, which is a qualitative risk assessment technique, includes risk aversion and risk acceptance. Risk aversion includes the degree of risk reduction required and the degree of risk avoidance. In risk acceptance, risk references and risk referents are established. Risk references are used for comparing and the risk referents are kept as standard. The risk assessment procedure is given in the form of a flow chart (see Figure 4.3).

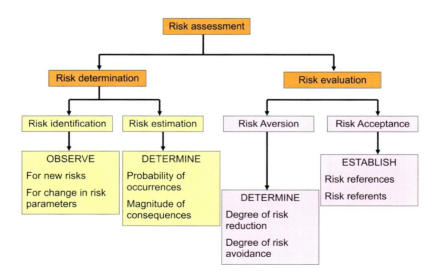

FIGURE 4.3 Risk assessment procedure.

4.6.1 CHEMICAL RISK ASSESSMENT

The US National Academy of Sciences identified four steps in chemical risk assessment: hazard identification, dose-response assessment, exposure assessment, and risk characterization. Hazard identification is the first step in the risk assessment of process industries and is more applicable to offshore plants in operation. It is used to evaluate the reliability of specific segments of a plant operation through probabilistic results, as used in fault tree analysis (FTA). Dose-response assessment involves describing the quantitative relationship between exposure and the extent of the toxic injury. The hazardous nature of the material and its effect on the human are assessed. The relationship between exposure and the extent of toxic injury gives a linear equation; regression analysis of dose-response data is used for plotting. Exposure assessment describes the nature and size of the population exposed to the dosing agent. It helps estimate the magnitude and duration of exposure.

4.6.2 APPLICATION ISSUES OF RISK ASSESSMENT

Risk assessment often relies on inadequate scientific information or lack of data. For example, any data related to repair may not be useful in assessing newly designed equipment. Even though the data available are less, all data related to that event cannot be considered as qualified data to make a risk assessment. In toxicological risk assessment, data related to animals is considered as predicting their effect on humans. Generally, risk assessment includes a probabilistic mathematical formulation that can be solved easily. For probabilistic modeling, data size is one of the critical issues. But risk assessment follows a conservative approach, which includes identifying the frequency of an event, its severity, and then calculating the risk rankings. The risk rankings are then examined in ascending order so that planning can be undertaken to reduce risk. Another form of risk ranking is through the comparison technique, which includes qualitative risk assessment by conducting a survey and preparing a series of questions based on which risk rating is undertaken. Risk characterization is the integration of data and analysis. It determines whether or not people will experience the effects of exposure and estimates the uncertainties associated with the entire risk assessment process.

4.7 SAFETY IN DESIGN AND OPERATION

Safety in design and operation is of primary importance in offshore industries. For example, let us consider the offshore drilling process. Offshore drilling is complex and technically challenging, requiring highly specialized equipment for the drilling operation. Drilling rigs are designed for high efficiency and mobility, which are important design features. Rigs are not designed to stay on the location but to perform during important stages of the reservoir development to build a drilling production structure. It is necessary to understand the mechanical methods of different drilling rigs to understand risk involvement during drilling.

4.7.1 OFFSHORE HAZARDS

Blowouts are one of the main hazards that occur in an offshore installation. Surface blowouts are accompanied by fire, explosion, pollution, third-party damage, and property damage to the drilling rig and platform. Subsurface blowouts leave no impression on the surface and are difficult to control. They result from the influx of fluids, which originate from the formation into the wellbore. When the formation fluid approaches the surface, a kick is formed. The main problem of kick formation is that fluid enters the wellbore and mixes with the oil or gas in the production line. A blowout preventer (BOP) is used to control the blowouts. It maintains control over the potential high-pressure condition that exists in the well formation. It consists of several stacks on the top of the well—at least one pipe ram and one blind ram below the annular BOP. An annular BOP has a rubber sealing element, which seals the annulus between the kelly, drill pipe, and the drill collar when activated. If no part of the drill stem is in the hole, it closes the opened hole. Ram blowout preventers are large steel rams that have sealing elements. A pipe ram BOP closes the drill pipe but cannot seal the open hole: blind ram blowout preventers are straight-edged rams that are used to close an open hole. Figure 4.4 shows a typical BOP used in oil exploration.

A "marginal field" is defined as an offshore reserve that cannot economically support the installation of a fixed drilling and production platform. Multi-well subsea

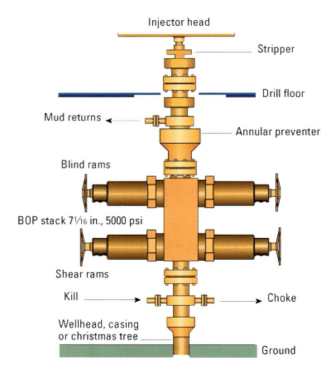

FIGURE 4.4 Typical blow out preventer.

completion systems are suggested to overcome this problem, employing a floating drilling vessel that drills the well through a subsea template. The advantages of multi-well subsea systems are early installation, insensitivity of cost to increased water depth, high flexibility, capability to accommodate many satellite wells, and the possibility of retrieving major components for reuse. The final location of the template is determined after the reservoir is defined by delineation drilling.

4.8 ORGANIZING SAFETY

Major accidents in the past are important sources of information. If a complete report of an accident is made, it can help prevent such accidents subsequently (to an extent). Statistics show that major accidents occurring in the offshore industries have declined by 15–20 percent. Major accidents can provide lessons for safety managers working in the offshore or process industries. Future losses can be reduced through learning lessons from the past. However, experience gained from past incidents may be forgotten and not brought forward for future generations. Therefore, accident analysis information is useful for modeling risks.

4.8.1 Hazard Groups

The major hazard groups in the offshore industries include blowouts, hydrocarbon leaks on installations, hydrocarbon leaks from pipelines/risers, and structural failures. On 27th April 1977, at the Ekosk field on the Norwegian continental shelf, the production Christmas tree was removed before the job had been completed, and no BOP had been installed. The well kicked, and an incorrectly installed safety valve failed. The result was a well blowout with an uncontrolled release of oil and gas. Human errors were the major fault that led to the mechanical failure of the safety valve.

Other major accidents in the history of the offshore industry include the blowouts at Enchova field, Brazil, on 16th August 1984 and in April 1988. In the first incident, a blowout occurred, followed by an explosion and fire. In the second incident, the well suffered a gas blowout. BOP did not shut the well in, and attempts to kill the well failed. On 6th October 1985, the Smedvig West Vanguard semi-submersible rig at Haltenbanken, on the Norwegian continental shelf, exploded due to a failure of the drilling unit. During the drilling operation, a drilling break was observed. Drilling was stopped at 523 m, and the bit was pulled back by 15 m. The well began to flow, and many unsuccessful attempts were made to kill the well by pumping "kill mud." BOP was not used for the top-hole section, and the flow of gas was directed through the diverter system. The gas was unable to contain the flow, and the liberated gas exploded.

On 1st March 1976, on the Norwegian continental shelf, a deep-sea driller semi-submersible rig capsized. The semi-submersible rig, escorted by a supply vessel and a fishing vessel, which was sailing to Bergen, Norway, for repairs, was caught in a severe storm. The fishing vessel did not have a sufficient towing capacity. The engines of the rig were insufficient to prevent the rig from drifting against rocks, and it capsized.

One of the greatest disasters in the offshore industry was the BP oil disaster in the Gulf of Mexico. It is still considered a fateful day for the oil industry. The Deepwater Horizon oil spill began on 20th April 2010 in the Gulf of Mexico due to the failure of the BOP. After several failed efforts to contain the flow, the well was declared sealed on 19th September 2010.

It is very difficult to predict the events that give rise to an accident. Experience is not enough because the accidents detailed above are very rare. However, their impact can be especially severe. The same kind of accidents may be prevented by following safety measures. Health and safety are concerns for everyone on a project, although the main responsibility lies with management in general and individual managers in particular.

The specific roles of the safety team are summarized here.

- Management develops and implements health and safety policies and ensures that the procedures for carrying out risk assessments, safety audits, and inspections are implemented. Importantly, management has the duty of monitoring and evaluating health and safety performance and taking corrective action as necessary.
- Managers can exert a greater influence on health and safety. They are in immediate control, and it is up to them to keep a constant watch for unsafe conditions or practices and take immediate action. They are also directly responsible for ensuring that employees are conscious of health and safety hazards and do not take risks.
- Employees should be aware of what constitutes safe working practices as they affect them and their fellow workers. While management and managers must communicate and train, individuals also must take account of what they have heard and learned in the ways they carry out their work.
- Health and safety advisers advise on policies and procedures and on healthy and safe methods of working. They conduct risk assessments and safety audits, investigate accidents in conjunction with managers and health and safety representatives, maintain statistics, and report trends and necessary actions.
- Health and safety representatives deal with health and safety issues in their areas and nominate health and safety committees.
- Medical advisers have two functions, preventive and clinical. The preventive function is more important, especially in occupational health matters. The clinical function is to deal with industrial accidents and diseases and advise on the steps necessary to recover from injury or illness arising from work. They do not usurp the role of the family doctor in non-work-related illnesses.
- Safety committees consisting of health and safety representatives advise on health and safety policies and procedures, help conduct risk assessments and safety audits, and make suggestions on improving health and safety performance.

4.9 HAZARD EVALUATION AND CONTROL

Every hazard carries some risk, which can potentially result in harm. It is important to analyze how great that risk is, and those hazards that pose more than mild risk should be brought under control. Hazard evaluation can be undertaken at any stage. It can be done during the initial design (failure mode and effect analysis, or FMEA) or during an ongoing operation (hazard and operability studies, or HAZOP). If the hazard evaluation shows low probability and minimum consequences, then the system is called "gold-plated." Such systems have potentially unnecessary and expensive safety equipment and procedures in place.

4.9.1 Hazard Evaluation

"Hazard" can be defined as a physical or chemical condition that can cause damage to people, property, or the environment, and "risk" is the likelihood that it will cause harm. The first step in controlling any hazard is to determine the magnitude of risk associated with it. This is called "hazard evaluation." A simple way of evaluating a hazard is finding the potential consequences of the hazard (severity) and the likelihood that those consequences will occur. Figure 4.5 illustrates this relationship and the risk matrix. Class "C" hazards present relatively little risk; class "B" hazards pose more serious risks, where one should take steps immediately to get the hazard under control; class "A" hazards are intolerable—stop immediately and make sure that the hazard is under control before continuing work. The class into which hazards fall is the basis for deciding how to prioritize controlling them. The hazard assessment diagram illustrated here is a simple and effective way of assessing risks and

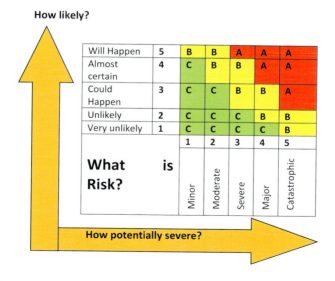

FIGURE 4.5 Risk matrix.

prioritizing plans for dealing with them. Sometimes, however, situations are a bit more complex, and other things may be considered.

- The frequency of the exposure: the more that people are exposed to hazards will increase the risk.
- The duration of the exposure: the longer people are exposed to a hazard, the greater the risk.
- Additional circumstances might affect the risk of injury.

4.9.2 Hazard Control

Sometimes hazards can be eliminated entirely, but most often measures have to be put in place to manage hazards efficiently, and it also helps to be systematic. It is a step-by-step procedure starting from the greatest hazards, like repairing or upgrading equipment, and working down until you find a practical solution. A series of questions posed to the Health and Safety Executive (HSE) manager, who handles hazard control, are as follows.

First, can the hazard be eliminated? For instance, if any damaged equipment is causing the hazard, can one remove the hazard completely by fixing or replacing the equipment?

Second, can hazardous materials be substituted with safer ones? For example, a cleaning solution that gives off toxic fumes could be replaced with a non-toxic alternative.

Third, to what extent can the staff and customers be isolated from hazards? Blocking a lane at a filling station during servicing will isolate the technician from vehicles but not from risks at the pump itself.

Fourth, to what extent can engineering controls minimize risk? For example, guards on a meat slicer or other moving or hot equipment?

Fifth, can good administration be implemented? This might include written instructions, signage to warn of hazards, or the use of "Do Not Enter" zones.

Finally, if the first five steps fail to eliminate risk, make sure people use the right protective equipment or clothing. Remember, this is the last line of defense, not the first!

This step-by-step procedure is known as the "hierarchy of hazard controls" and helps find the most reasonable and effective ways to minimize the risk of injury. In any situation in which a hazard can't be brought fully under control, employers must impart intensive basic training about the safe upkeep of the plant through written instructions. A hazard control form is prepared, and feedback following a scheduled interval is evaluated to maintain basic safety in the plant. Hazard controls need to be reviewed periodically to make sure they are still effective and appropriate. It can be part of regular safety inspections. A few questions to consider when reviewing hazard controls are as follows.

- Is the hazard under control? Have the steps taken to manage it solved the problem?
- Are the risks associated with the hazard under control too?
- Have any new hazards been created?

- Are new hazards being controlled appropriately?
- Do workers know about the hazard and what has been done to control it?
- Do workers know what they need to do to work safely?
- If there is a new hazard, are workers trained properly to deal with it?
- Are there written records of all identified hazards, their risks, and the control measures taken?
- What else can be done?

4.10 QUANTITATIVE RISK ASSESSMENT

According to NORSOK (Norsk Sokkels Konkuranseposisjon) standards, risk assessment is defined as the systematic identification and description of risk to personnel, environment, and assets. The ISO (International Organization for Standardization) definition of risk assessment is "the systematic use of information to identify sources and assign risk values" (ISO, 2007). Quantitative risk analysis focuses on these two definitions. Quantitative risk assessment (QRA) includes identifying applicable hazards and descriptions and quantifying applicable risks to assets, environment, and personnel. The analytical elements of risk assessment have relevant data to identify and assess the risks arising from them. These include identification of initiating events, cause analysis, and consequence analysis. The risk assessment can be made by undertaking cause analysis and consequence analysis.

4.10.1 Initiating Events

The identification of initiating events is often called hazard identification, or HAZID. Hazard identification is made in a structured and systematic manner. Methods include creating checklists, generating accident and failure statistics, undertaking hazard and operability studies (HAZOP), comparing published studies, and collating experiences from previous similar projects, operations, concepts, and systems. Hazard identification includes the following.

- Reviewing possible hazards and sources of accidents.
- Classifying identified hazards according to criticality for subsequent analysis.
- Making an explicit statement of criteria and then screening the hazards.
- Documenting non-critical hazards after evaluation.

4.10.2 Cause Analysis

Cause analysis includes a qualitative evaluation of possible causes. Qualitative cause analysis is intended to identify the causes and conditions that may lead to initiating events, identifying combinations that will result in such occurrence, and establishing the basis for possible later quantitative analysis. Some of the methods for conducting qualitative cause analysis include hazard and operability analysis (HAZOP); fault tree analysis (FTA); preliminary hazard analysis (PHA); and failure mode and effect analysis (FMEA). Quantitative cause analysis is done by finding the probability of the occurrence of initiating events. Some quantitative cause analysis techniques include the aforementioned fault tree analysis (FTA), event tree analysis (ETA), synthesis models, and Monte Carlo simulation.

4.11 FAULT TREE ANALYSIS

Fault tree analysis is a logical, structured process that can help identify potential causes of system failure, such as causes of initiating events or failure of barrier systems. This technique is implemented to find the causes of equipment failure and is used in reliability and availability assessment. It is a graphical representation of the sequence of equipment failures and human errors resulting in a hazardous event, which is the top event. The main advantage of fault tree analysis is the ability to include both hardware failure and human errors, thereby allow a realistic representation of steps leading to a hazardous event. It identifies preventive measures and mitigates the consequences of hazardous events caused by either hardware or software faults. Events are subdivided into the final event, the intermediate event, and the basic fault (or event). The final and intermediate events are shown in the tree diagram by rectangles, whereas circles show the basic fault. Undeveloped events, if any, are shown in the shape of a diamond. The symbols for the basic, intermediate, and final events are shown in Figure 4.6.

The final events are linked to the intermediate and basic events through gates of three types. When two input events lead to a resulting event, and both the events are required for the resulting event to occur, the gate is known as the "AND gate." When either of the two input events can lead to the final event, the gate is known as the "OR gate." The AND gate and the OR gate are shown in Figure 4.7. A typical example of the AND gate is given in Figure 4.8. In this example, the final event is a fire requiring two events, consisting of an explosive mixture and an electric spark. An electric spark is a basic event, whereas the flammable air mixture formed by mixing fuel and air is considered an intermediate event. Both the explosive mixture and spark are required for the fire, and the corresponding gate used is an AND gate. A typical example of an OR gate is given in Figure 4.9. A fuel leak from a leaking valve in a container, or a human error in opening the valve, can lead to the formation of the cloud. Hence the OR gate links the final event to the leakage of a valve, which could be an intermediate event, whereas the manual opening of a valve is considered a basic event.

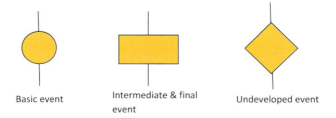

FIGURE 4.6 Symbols used in fault tree analysis.

Risk, Reliability, and Safety Assessment

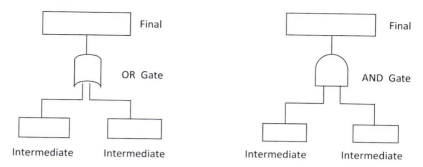

FIGURE 4.7 Gates used in fault tree analysis.

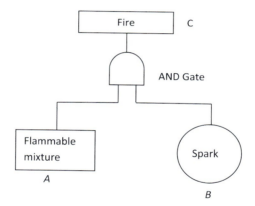

FIGURE 4.8 Example of AND gate.

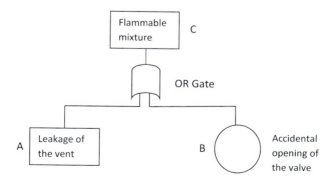

FIGURE 4.9 Example of OR gate.

4.11.1 PROBABILITY OF FINAL EVENT

The probability of occurrence of the final event can be calculated using the AND and OR gates. Consider the example discussed the formation of fire from the flammable mixture and electric spark using an AND gate. If the probability of fuel–air mixture is $P(A)$, and the probability of the spark is $P(B)$, the final event fire, $P(C)$, is given by the following relationship

$$P(C) = P(A) \times P(B) \tag{4.3}$$

In the example discussed for the OR gate, if the probability of encountering a leakage valve is $P(A)$, and the probability of a human error in opening the valve is $P(B)$, then the final probability, $P(C)$, is given by the following relationship

$$P(C) = P(A) + P(B) \tag{4.4}$$

4.11.2 ANALYSIS USING THE FAULT TREE METHOD

Analysis using the fault tree method is illustrated by the formation of a fireball as a final event from forming a flammable mixture of propane air and ignition of the cloud (see Figure 4.10).

The formation of the explosive mixture is due to the weld failure or opening of the relief valve by the operator. Ignition may occur due to a lighted cigar or a friction spark from a skidding vehicle. The probability of each of the events is described using FTA, as above.

4.12 EVENT TREE ANALYSIS

An event tree is a model describing possible event chains that are connected to a hazardous situation. To carry out a risk assessment using ETA, initiating events are predefined and their frequency (or probability) is computed; possible outcomes from the initiating events are determined using a list of questions answered "yes" or "no," and a questionnaire is drawn up based on the designers' experience and thinking. The ETA tree can also be used for directly calculating consequences of risk. Fatality risk assessment can also be undertaken using ETA by assigning numbers of fatalities to each branching point and adding them up to find the total number of fatalities. Inductive logic is used in ETA, and the event progresses forward through subsequent events that correspond to the consequences. Steps in ETA are shown in Figure 4.11.

ETA is illustrated in the following example involving the leak of hydrogen from a tank in a hydrogen-fueled vehicle. Let the following data be used to form the event trees. The probability of ignition source is given as 0.5. Fire, if formed, will transform to a detonation with a probability of 0.2. The event tree is shown in Figure 4.12. A vehicle is passing through a tunnel when the hydrogen leak occurs. In that case, the accumulation of the mixture of hydrogen and air in the tunnel could encounter an ignition source later to modify the event tree, as shown in Figure 4.13.

Risk, Reliability, and Safety Assessment

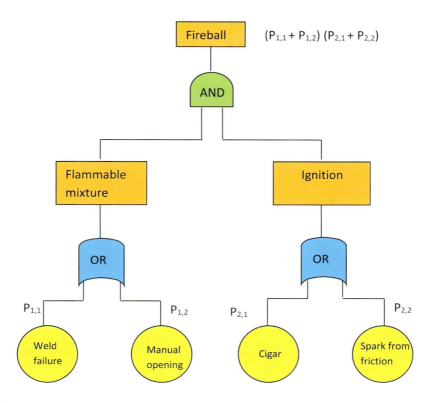

FIGURE 4.10 Fault tree analysis of the formation of a fireball.

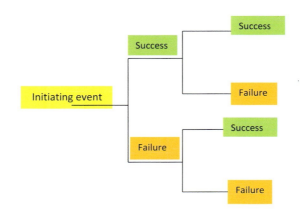

FIGURE 4.11 Steps in event tree analysis.

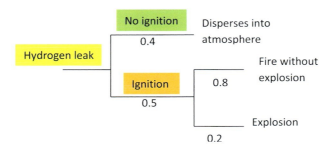

FIGURE 4.12 Event tree for hydrogen leak example.

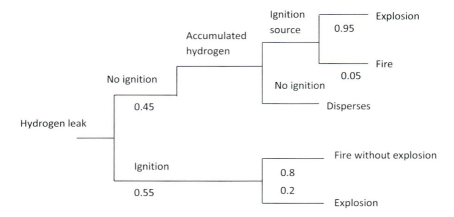

FIGURE 4.13 Modified event tree for hydrogen leak example.

Following risk analysis, a consequence analysis is carried out. This calculates the physical effects of accidents related to consequence loads. Usually, fire and explosion are considered serious consequences. The main steps involved in calculating consequence analysis include fire and explosion following the leak from the processing system or uncontrolled blowout.

4.13 RISK CHARACTERIZATION

The risk assessment process mainly consists of four steps: hazard identification, dose-response assessment, exposure assessment, and risk characterization, as shown in Figure 4.14. Risk characterization is the integration of data and analysis to synthesize an overall conclusion about risk. Risk characterization conveys risk assessors' judgment in the presence or absence of risks, along with how the risk has been assessed, where assumptions and uncertainties still exist, and where policy choices will need to be made.

Each risk assessment process consists of its risk characterization, such as key findings, uncertainties, assumptions, limitations, etc. The set of these individual risk characterizations provides the information basis to write an integrative risk

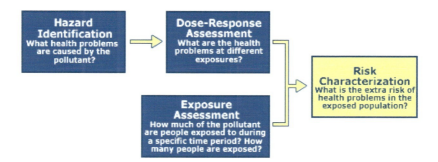

FIGURE 4.14 Risk characterization.

characterization analysis. The final overall risk characterization thus consists of individual risk characterization plus an integrative analysis.

4.13.1 Principles of Risk Characterization

A good risk characterization will restate the scope of the assessment, express results clearly, articulate major assumptions and uncertainties, identify reasonable alternative interpretations, and separate scientific conclusions from policy judgments. Risk characterization includes the following principles: transparency, clarity, consistency, and reasonableness—which are commonly referred to as "TCCR":

- Transparency—The characterization should fully and explicitly disclose the risk assessment methods, default assumptions, logic, rationale, extrapolations, uncertainties, and overall strength of each step in the assessment.
- Clarity—The products from the risk assessment should be readily understood by readers inside and outside of the risk assessment process. Documents should be concise, free of jargon, and use easily understandable tables, graphs, and equations, as required.
- Consistency—The risk assessment should be conducted and presented consistent with Environmental Protection Agency (EPA) policy and with other risk characterizations of similar scope prepared across programs within the EPA.
- Reasonableness—The risk assessment should be based on sound judgment, with methods and assumptions consistent with the current state of the science and conveyed in a manner that is informative, complete, and balanced.

To achieve TCCR in risk characterization, the same principles need to be applied in all of the preliminary steps in risk assessment that led to the risk characterization.

4.14 FAILURE MODE AND EFFECT ANALYSIS

Failure mode and effect analysis (FMEA) is a popular tool for reliability and failure mode analysis. FMEA should include the activities at both the design and manufacturing stages. It is critical to conduct reliability analysis at the earliest stage of a product life cycle. Design and product engineers need to work with a project team that includes customers, reliability engineers, and manufacturing engineers to identify the quality and the potential for reliability failures in the design process. Potential problems can be eliminated as early as possible to avoid complicated and costly correction processes. FMEA is a technique that identifies, first, the potential failure modes of a product during its life cycle; second, the effects of these failures; and third, the criticality of these failure effects in product functionality. FMEA provides basic information for reliability prediction and product and process design. FMEA helps engineers find potential problems in the product earlier and avoids costly changes or reworks at later stages, such as at the manufacturing and product warranty stages. In the FMEA process, product functions must be carefully evaluated and the potential failures must be listed. This analysis process provides a thorough analysis of each detailed functional design element. It allows FMEA to be a very useful tool in quality planning and reliability prediction.

There are two phases in the FMEA process. The first phase is to identify the potential failure modes and their effects, while the second is to perform criticality analyses to determine the severity of the potential failure modes. The first phase has to be undertaken concurrently with the detailed product design. It should also include defining the possible failures of the product's components, sub-assemblies, final assembly, and manufacturing processes. At the end of the first phase, the detailed design is completed, and the design drawing is developed. At the second phase of FMEA, engineers in the FMEA team evaluate and rank the criticality of each potential failure and then revise each design detail and make required modifications. The most serious potential failure has the highest rank and is considered first in the design revision. The design is revised to ensure that the probability of the occurrence of the highest-ranked failure is minimized. Before conducting FMEA analysis, it is important to undertake functional analysis and generate cause–effect diagrams. The functional analysis identifies the primary and secondary functions of products or processes. Primary functions are the specific functions that a product or process is designed to do. Figure 4.15 is an illustration of an FMEA cause–effect diagram.

Figure 4.16 illustrates the general procedure of the FMEA process. The first phase is from information-gathering to the calculation of risk priority numbers (RPNs). The actions in the second phase involve the ranking of the RPNs and recommendations for corrective action and any modification of the design. At the end of the procedure, an FMEA report can be obtained and the required modifications are completed to reduce the number of potential failure modes to the minimum. Teamwork is critical to the success of the FMEA process. The team to perform FMEA should include customers, manufacturing engineers, test engineers, quality engineers, reliability engineers, product engineers, and sales engineers. The potential failure modes listed in the FMEA report will include any failures at different stages from internal and external customers, such as the manufacturing department in the company, the customer, any other manufacturing company and their customers, and end-users.

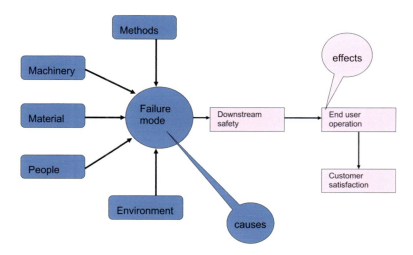

FIGURE 4.15 FMEA cause–effect diagram.

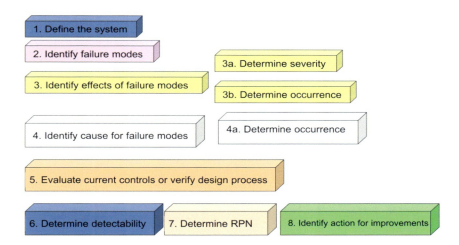

FIGURE 4.16 FMEA procedure.

The information used in the FMEA process should come from the company's production lines, the customers, and the field data of similar products. Therefore, the FMEA team has to work with the customers to gather the required information to develop an effective FMEA report.

Three stages are very critical in the FMEA process to ensure the success of the analysis. The first stage is to determine the potential failure modes. The second stage is to find the data regarding the occurrence, detection, and severity rankings. The third stage is modifying the current product/process design and developing the control process based on the FMEA report.

4.14.1 Failure Mode Effect Analysis Variables

Severity: The severity rating relates to the seriousness of an effect of a potential failure. It is generally expressed on a scale of 1 to 10, with 1 indicating no effect, 5 indicating moderate effect, 8 indicating serious effect, and 10 indicating hazardous effect.

Occurrence: The occurrence rating corresponds to the rate at which the first level cause and its resultant failure mode will occur over the system's design life, i.e., over the design life of the product before any additional process controls are applied. It is generally expressed on a scale of 1 to 10, with 1 indicating failure unlikely, 5 indicating occasional failure, 8 indicating high probability of failure, and 10 indicating that failure is certain.

Detection: The detection rating corresponds to the likelihood that the detection methods or current controls will detect a potential failure mode before the product is released for production or in a process before it leaves the production facility. It is expressed on a scale of 1 to 10, with 1 indicating that failure will be detected, 5 indicating that failure might be detected, and 10 indicating that failure is almost certain not to be detected.

Risk priority number: The RPN identifies important areas of concern. It is the product of severity rating, occurrence rating, and detection rating, and evaluates the severity rating, occurrence rating, and detection rating for a potential failure mode.

Corrective actions must be taken if the severity rating is 9 or 10 and if the severity and occurrence ratings are high or have a high RPN value. There is no absolute value for what constitutes a high RPN as FMEA is often viewed on a relative scale. The highest RPN value is addressed first. FMEA is a systematic tool for identifying the effects or consequences of a potential failure and methods to eliminate or reduce the chance of failure. FMEA generates a living document that can be used to anticipate failure and prevent failure from occurring. It should be used before a final design is released. The main objective is failure prevention and not detection. As such, FMEA is often a standard process used in the development of new products. A total understanding of the product/process functions and careful gathering of the data ensure the correctness of the FMEA report. The usefulness of the FMEA depends on the third stage of the process. The major goals for implementing FMEA should be to modify the design to eliminate failure modes and to develop the process control plan to reduce failures to a minimum.

4.15 RISK ACCEPTANCE CRITERION

The risk acceptance criterion defines an acceptable level of risk in activities. The 2001 NORSOK "Risk and Emergency Preparedness Analysis," which applies to the Norwegian offshore industry, redefined risk acceptance criteria to express a risk level that is considered acceptable for the activity in question but limited to the higher level of expression of risk. The main regulatory requirements for risk acceptance criteria are summarized as follows.

- Risk acceptance criteria will address all personnel as well as groups that have particular risk exposure.

- Risk acceptance criteria will be defined for personnel and the environment.
- The risk exposure of the third party is only relevant for petroleum installations onshore.
- Risk arising from the transportation of personnel will be included.

These requirements imply a conscious approach to risk assessment. They enhance competence and promote communication between different disciplines. They also establish the limitations of risk assessment and the importance of safety management.

4.15.1 UK Regulations

The criteria used by the UK petroleum industry are mainly those that have been formulated by the Health and Safety Executive (HSE) and are embodied in statutory legislation. They include the ALARP principle, defined as the residual risk that will be "as low as reasonably practicable." The ALARP principle arises from the fact that almost infinite time, effort, and money could be spent to reduce a risk to zero.

4.15.2 Acceptable Risk

Risk is an extension of reliability to address the consequence of failure. Earlier chapters have shown that risk cannot be reduced to zero due to the multi-state complexities involved in offshore plants. Should it feel that offshore plants are not safe, please note that the acceptable level of risk is defined for offshore plants by various international standards. It is that risk is not allowable beyond an accepted level of failure. Based on engineering judgment and experience, an acceptable level of risk can be established. The few agencies that regulate such acceptable risk norms are i) The Environmental Protection Agency (EPA), ii) The Nuclear Regulatory Commission, and iii) The Federal Energy Regulatory Commission.

4.16 RELIABILITY

As discussed above, safety assessment and risk characterization are vital for offshore plants. Since risk is a realization of hazard, and hazard is unavoidable in any process industry, it is important to understand the differences between "safety" and "reliability" (see Table 4.1).

It is therefore important to carry out reliability studies on structures to determine whether the structure needs to be strengthened or retrofitted even though it has been declared safe. The most important aspect of reliability is to account for all uncertainties that make the structure vulnerable to failure under a predefined limit state. The accuracy of the reliability study depends on how accurately these uncertainties are accounted for in the analysis. With the above limitations, one common question during an engineering assessment is why reliability studies cannot be accurately undertaken. This is due to the following factors i) it is practically impossible to identify all uncertainties that prevail upon a structure, which is a question of the application of engineering judgment, and ii) methods of modeling and analyzing models are highly complex in a multi-variate system. Modeling aspects involve a lot of uncertainties.

TABLE 4.1
Comparison between Safety and Reliability

Safety	Reliability
Safety is used to indicate reliability.	Reliability is based on the concept of probability.
It is a traditional concept.	
It is applied to an existing process.	It is applied to predict the safety of the existing process.
It has a direct consequence to failure.	It has a converse consequence to failure.
It is a deterministic approach.	It is a probabilistic approach.
It is not a design method.	It is a design method.
It is an assessment tool.	It depends on the data and their reliability.
It gives a close form solution when applied to an existing system.	It can even give erroneous results.
It is always a post-correction/examination scenario.	Reliability is used to forecast the risk even before it occurs.
It is based on statistical judgment.	It is based on engineering judgment.
Tools of analysis play an important role.	Experience plays an important role.

Therefore, many assumptions are made during reliability analyses that influence the accuracy of reliability studies. Further, it is important to note that the analytical formulation of the limit state surface and the integration of the probability density function within the domain of interest can also be very complex.

4.17 IMPORTANCE OF RELIABILITY ESTIMATES

Offshore structures are subjected to various environmental loads, which have the following issues i) load effects are highly uncertain, ii) environmental loads cannot be predicted or assessed accurately, and iii) there is a strong association of the probabilistic estimate of loads in reliability studies. Material characteristics also vary widely in terms of yield strength. Further, material degradation models that account for the reduction in strength in analytical and numerical studies are not assessed with a high enough accuracy. Load combination effects result in a higher probability of variation in the failure modes of members, whose variations originate from the uncertainties in the computation of the environmental loads (this issue is in order of a close form and can be resolved only with an engineering perspective). In addition, recent advances in new-generation platforms (structural forms) demand an assessment of stability under the action of loads before their detailed analysis of bending stress and shear. Extensive use of composites and various alloys initiate local stress with high-stress concentration arising from the material composition. Under such complexities, reliability is the only tool that can address these uncertainties in a probabilistic manner. Even if the analysis is highly deterministic, it is better to use reliability analysis, as every deterministic analysis originates from simplified assumptions.

It is interesting to note that reliability is expressed in probabilistic terms, which is associated with the statement. Therefore, failure is assessed as the structure's inability to perform its intended function with adequate capacity for a specified period under specified conditions. Hence reliability is defined in terms of failure in the literature. The structure can perform its intended function with an adequate capacity for a specified period. It can be expressed as

$$R = (1 - P_f) \tag{4.5}$$

Reliability is expressed in terms of the probability of successful delivery of the system. Hence, it is defined in terms of the facts that performance quality is expected for a specified period. This is called the "service life of the structure," beyond which reliability should not be extended. It is important to note that reliability studies cannot be used to substitute for the post-accident scenario. Reliability studies circumscribe uncertainties that may arise from various factors, as discussed above.

There are two principal types of uncertainties, namely i) aleatory uncertainty, which arises from the randomness of nature in terms of load effects, and ii) epistemic uncertainty, which arises from inaccuracy in prediction. They originate from analysis methods, modeling constraints, etc. These uncertainties need to be reduced, with the former type handled rationally in the design through partial safety factors. Further uncertainties are of three kinds, namely i) those that arise from randomness and variations in the excitation forces (they are irreducible), ii) statistical uncertainties that arise from the estimation of parameters describing the statistical model (they are reducible), and iii) modeling uncertainties that arise from imperfection in mathematical modeling, which are also reducible. The latter two can be handled effectively via the Bayesian approach. In the Bayesian approach, model parameters are estimated using the appropriate likelihood functions. Using these likelihood functions the posterior parameter of the models is calculated. It is expected that the errors in this approach are because the parameters are estimated from established (existing) models of large data. Probability methods generally circumscribe reliability methods because probability tools can logically handle uncertainties. Further, uncertainties are also grouped based on the type and quality of uncertainties they handle.

4.17.1 Types of Uncertainties

Group I deals with uncertainties that arise from material characteristics (σ, E, μ), etc. Group II is associated with uncertainties originating from the load effects—for example, static loads like dead load imposed load, or payload, and dynamic loads, like earthquake forces, waves, etc. Group III deals with the uncertainties associated with material strength variations, mathematical modeling, and analysis. Each analysis method has a certain set of assumptions or structural idealizations that contribute to this group of uncertainties. Examples are dynamic modulus of elasticity, properties of visco-elastic materials, etc.

4.18 FORMULATION OF THE RELIABILITY PROBLEM

Reliability problems are formulated in two ways i) a time-invariant problem, and ii) a time-variant problem. In both cases, limit state functions are defined, which could be based on either limit state of serviceability or limit state of collapse. The main objective of the reliability problem is to determine the probability of failure of the system; reliability is just the converse of the probability of failure, as given by Equation 4.5, above.

4.18.1   TIME-INVARIANT PROBLEM

If g(x) denotes the limit state function and x denotes a set of random variables {x1,x2, ... , x_n}, then g(x) ≤ 0 denotes the failure event. Probability of failure is then given by

$$p_f = \left[g(x) \leq x \right] = \iint f(x)\, dx \qquad (4.6)$$

where, f(x) is the joint probability density of G(x) = 0.

4.18.2   TIME-VARIANT PROBLEM

Reliability is said to be time-variant if the limit state function is also a function of time. If limit state function is g(x, y(t)), where (x_1, x_2, ..., x_n) are a set of random variables and y(t) is the vector of stochastic process, the probability of failure, Pf, is based on the out-crossing of the vector y(t) through the limit surface g(X, Y) = 0. The probability of failure is given by

$$P_f = \int p \left[\min g(x, y(t)) \leq 0 \right] f_x\, dx \qquad (4.7)$$

For the domain, 0 ≤ t ≤ T, T denotes the life of the structure, and F(x) dx denotes the probability density function of the variables in the reliability framework of offshore plants. For the said domain, the time-variant problem has the probability of failure, as given below

$$P_f = \int p \left[\min g(x, y(t)) \leq 0 \right] f_x\, dx, \quad 0 \leq t \leq t \qquad (4.8)$$

The solution of this equation cannot be exact, for two reasons: i) the evaluation of the conditional probability, and ii) the estimate of the conditional probability density function *f(x)*. To simplify the solution, few basic approximations are used in reliability analysis. Based on the method used to compute the joint probability density function, which can be either analytical or numerical, reliability methods are classified.

4.18.3   RELIABILITY FRAMEWORK

In the general sense, an offshore platform should perform its intended function for a specified period under specific conditions. In the mathematical sense or narrow sense, reliability estimates the probability of the structure not attaining the limit state of collapse within the specified conditions for the specified period.

$$\text{Probability of structure} = 1 - Pf \tag{4.9}$$

which implies that it is (R − S), where R is the resistance of the structure and S is the load effects. For the resistance greater than the load effects, the structure is always in the safe domain. If the load effects and resistance are expressed by their respective PDF as $f_s(S)$ and $f_s(R)$, respectively, then the probability of failure is given by

$$P_f = \text{prob}(R \leq S) \tag{4.10}$$

$$= \int_0^\infty f_R(s) - f_s(s)\,ds = f_m(0) \tag{4.11}$$

where M is called the margin of safety, which is given by (R − S). If the probability density function $\{f_m(m)\}$ and cumulative distribution function $F_m(m)$ are known, then the probability of failure P_f can be computed analytically or numerically, as given below

a) If R&S are normally distributed, then the following relationship holds good

$$p_f = \varphi(-\beta) \tag{4.12}$$

where the reliability index is given by

$$\beta = \frac{\mu_R - \mu_S}{\sqrt{\sigma_R^2 - \sigma_S^2}} \tag{4.13}$$

b) If R&S are log-normally distributed, the reliability index is given by the following relationship

$$\beta = \beta_{LN} = \frac{\ln\left[\frac{\mu_R}{\mu_S}\sqrt{\frac{(1+V_S^2)}{(1+V_R^2)}}\right]}{\sqrt{\ln(1+V_R^2)(1+V_S^2)}} \tag{4.14}$$

$$\beta_{LN} \approx \frac{\left(\frac{\mu_R}{\mu_S}\right)}{\sqrt{V_R^2 + V_S^2}} \tag{4.15}$$

4.19 ULTIMATE LIMIT STATE AND RELIABILITY APPROACH

For an implicit failure probability in the design for the random load effects, the following equations hold good

$$\mu_S = B_S S_C \quad (4.16a)$$

For BS1.0, VS = 0.15 to 0.30 and $\mu_R = B_R R_C$ (4.16b)

B_S reflect the ratio of the mean load if the period of variation is annual, and then it should refer to the annual value of the probability of failure. Sc is the characteristic value, with 100 years return period. For (R, S) to be log-normal, the following equation holds good

$$\beta_{LN} = \frac{\ln \frac{\mu_R}{\mu_S}}{\sqrt{V_R^2 + V_S^2}} \quad (4.17)$$

For $(\gamma_R \gamma_S)$ be partial safety factor of 1.5, $B_S = 0.8$, $B_R = 1.0$, $V_R = 0.15$, the above equation reduces to the following form

$$\beta_{LN} = \frac{\ln \frac{\mu_R}{\mu_S}}{\sqrt{V_R^2 + V_S^2}} \quad (4.18a)$$

$$= \frac{\ln \frac{1.1}{0.8}}{\sqrt{0.10^2 + 0.20^2}} = 13.5 \quad (4.18b)$$

The ultimate limit state can affect the design since the method is based on the maximum load effect. It is also affected by the traditionally determined strength of the material. The reliability framework is based on establishing a limit state function g(x) for a single R&S, where the limit state function g(x) is subjected to large uncertainties. The preferable design format is given by

$$\frac{R_c}{\gamma_R} \geq \gamma_{s1} S_{1c} + \gamma_{s2} S_{2c} \quad (4.19)$$

The subscript stands for the characteristic value, R is the resistance, S is the load effect, γ_R is the resistance factor, γ_{S1}, γ_{S2} are the load factors. Resistance refers to a characteristic strength of 5 percent of the fractal materials' strength, while load effect refers to the annual probability of exceedance of 10^{-2}. Design criteria are now given by g(Rd, S1d, S2d) > 0

$$R_d = \frac{R_c}{\gamma_R} \quad (4.20a)$$

$$S_{1d} = \gamma_{S1} S_{1C} \quad (4.20b)$$

$$S_{2d} = S_{2c} \gamma_{2c} \quad (4.20c)$$

For multiple values of (R&S), the structure can be subjected to different load combinations for which the bending failure criteria can be formulated as

$$g(R1, R2, R3, S1j, S2j) := 1 - \left[\frac{S_{ij}}{R_1} + \frac{S_{2j}}{\left(1 - \frac{S_{1j}}{R_2}\right) R_3} \right] \quad (4.21)$$

The above equation can also be set as

$$= 1 - \left[\frac{X_1}{X_1} + \frac{X_3}{\left(1 - \frac{X_1}{R_4}\right)} \right] \quad (4.22)$$

where S1j, S2j, etc. are load effects for different combinations &R is the resistance (the count j stands for load type). The above equation is based on the Perry–Robertson approach in which R1, R2 is the axial force and R3 is the Euler load. In the partial design values of (R&S), they are represented by their respective characteristic values. But, in the reliability study, they are considered as random variables.

4.20 LEVELS OF RELIABILITY

Reliability studies are considered at different levels in the literature. Level I focuses on the probability aspects of the problem. Suitable characteristic values of the random variables are introduced in the safety analysis. The main objective of this level of study is to minimize the deviation of the design values from that of the target value. For example, load-resistance factor design (LRFD) is of level I of reliability. Level II has two values for each parameter to be defined in the analysis: mean and standard deviation. Level III is a complete analysis of the problem addressing the multi-dimensional probability density function of random variables, which is extended over the safety domain. Reliability is expressed in terms of suitable safety indices. In level IV, engineering economics is also applied in the reliability study. This level of reliability study is usually applied to structures of strategic importance. The study includes cost–benefit analysis, rehabilitation, the consequence of failure, and return on capital investment. The reliability method offers many advantages, such as i) it offers a reliability process to account for uncertainties, ii) it offers a rational method to estimate safety, and iii) it offers the decision-making support for a non-economic and better-balanced design. Optimal distribution of material among various structural components can benefit from a constant update mechanism, based on which the FEED function of engineering judgment is circumscribed. Reliability studies expand the knowledge of uncertainties in the response of the structure. There are a few obstacles in implementing reliability studies to offshore plants in operation. They are classified as inertial, cultural, and philosophical. The different types of variables used in reliability study are: elementary variables (static variables) like material property, the geometry of the platform, boundary conditions, and environmental- and behavior-dependent data.

Failure modes, such as limit stress and limit displacement, depend upon the system variables that in turn depend on location behavior and failure modes. There are different steps of reliability, namely elementary level, component level, system level, and detailed field investigation. The first type is handled by stochastic modeling, while the second type can be handled by the probabilistic study of the failure of components. In the case of system-level studies, probabilistic studies on the failure of the whole system can be investigated. One of the serious limitations of a reliability study is that it requires a large amount of data on the failure scenario. Other parameters that influence the accuracy of the results of the reliability studies are as follows.

a) The separation of two variables, namely safety domain and failure domain.
b) The nature of variables, namely external, internal, and to classify whether they are independent or not.
c) The effect of time, indicating the static content or the cyclic (dynamic) content.
d) The form of the performance function, which is dependent on the physical model of the system.

4.21 RELIABILITY METHODS

4.21.1 First-Order Second-Moment (FOSM) Method

In this case, the first-order Taylor series approximation of the limit state function is used for the analysis. Only second moments of the random variables are used to estimate the probability of failure. Limit state function is defined as

$$M = R - S \tag{4.23}$$

where R, S are statistically independent and assumed to be normally distributed. Hence, the following relationship holds good

$$\mu_m = \mu_R - \mu_S \tag{4.24a}$$

$$\sigma_m = \sigma R2 + \sigma S2 \tag{4.24b}$$

The probability of failure is given by

$$P_f = P(M < 0) \tag{4.25}$$

Risk, Reliability, and Safety Assessment

$$= P\left[(R-S)<0\right] \qquad (4.26)$$

If M is the normal variant, then

$$P_f = \phi\left(\frac{-\mu_m}{\sigma m}\right) \qquad (4.27)$$

$$\beta = \text{reliability index} = \frac{\mu_m}{\sigma_m} \qquad (4.28)$$

where ϕ is the case cumulative distribution function of a standard normal variable

$$P_f = 1 - \phi\left(\frac{\mu_R - \mu_S}{\sigma_R^2 + \sigma_S^2}\right) \qquad (4.29)$$

If R&S are log-normal, then the following relationship holds good

$$P_f = 1 - \emptyset \left[\frac{\ln\left[\frac{\mu_R}{\mu_S}\sqrt{\frac{(1+V_S^2)}{(1+V_R^2)}}\right]}{\sqrt{\ln(1+V_R^2)(1+V_S^2)}}\right] \qquad (4.30)$$

The advantages and disadvantages of the FOSM method are summarized in Table 4.2.

4.21.2 Advanced FOSM Method

As seen above, the dependency of the reliability index on the chosen form of the limit function is one of the vital drawbacks of the FOSM method. Further, the reliability index computed on the assumption that the random variables are statistically independent and normally distributed poses an additional complexity to the FOSM method, making its application limited to problems validating the above

TABLE 4.2
The Advantages and Disadvantages of the FOSM Method

Advantages	Disadvantages
It is easy to use.	Results can cause serious errors. A normal distribution cannot approximate the tool used for the distribution function.
It does not require knowledge of the distribution of random variables.	Values of β depend on the specific form of the limit state function. This is an invariance problem.

assumptions. Advanced FOSM gives a reliability index. The Hasofer–Lind method is one of the advanced FOSM methods, which will be discussed here.

The key point of the method is to estimate a design point that is at a minimum distance of failure from the origin. The minimum distance is the safety index (β_{HL}). The method transforms the random variable into a reduced form, as discussed. The reduced form of random variables defined as

$$Xi = \frac{x_i - x_i}{\sigma_{xi}} \quad \{\text{for } i = 1, 2 \ldots n\} \quad (4.31)$$

This reduced variable will have a zero mean and unit standard deviation, which is a special distribution process. Hence, the performance function $G(x) = 0$ is converted into $G(x') = 0$ to enable the mapping between the required domains. Reliability index β_{HL}, which is the minimum distance of the performance function from the origin, is given by

$$\beta_{HL} = \sqrt{x_d x_d^T} \quad (4.32)$$

x_d is the minimum distance of the design point from the origin, which is also referred to as a checkpoint. The following cases are specific:

a) Case 1 limit state function is linear
 Let us consider ($M = R - S$). Then, the reduced values are computed for the domain mapping, as discussed below

$$R = \frac{R - \mu_R}{\sigma_R} \quad (4.33)$$

$$S = \frac{S - \mu_S}{\sigma_S} \quad (4.34)$$

Substituting, we get

$$M = (\sigma_R + \mu_R) - (\sigma_S + \mu_S) \quad (4.35)$$

As the limit state function moves closer to the origin, the failure region is mapped. Reliability index is given by

$$\beta = \frac{\mu_R - \mu_S}{\sqrt{\sigma_R^2 + R\sigma_R^2}} \quad (4.36)$$

b) Case 2 limit state function is non-linear
 In such cases, computing the minimum distance for calculating the reliability index becomes an optimization problem.

$$\beta_{HL} = D = \sqrt{(x)t(x)} \quad (4.37)$$

This is to be minimized, subject to the condition that $G(x) = 0$ for many random variables $(x1, x2, \ldots, x_n)$, which originates from the safe state of domain; $G(x) < 0$ indicates failure. Hence, $G(x) > 0$ denotes the minimum distance from the origin to a point on the limit state function called the design point. The problem is now reduced to determine the coordinates of the design point, geometrically (or) analytically. By this definition, the reliability index becomes invariant, as the minimum distance remains constant regardless of the shape of the limit state function. Using the Lagrange multipliers, one can find the minimum distance, as given below

$$\beta HL = \frac{-\sum_{i=1}^{n} x'_{di} \frac{\partial G}{\partial x_{di}}}{\sqrt{\sum_{i=1}^{n} \left(\frac{\partial G}{\partial X_{di}} \right)^2}} \qquad (4.38)$$

$\frac{\partial G}{\partial X_{di}}$ is the partial derivative, evaluated at the design point with coordinates $(xdi, xd2, \ldots)$.

EXERCISE

1. _____ and _____ are vital for offshore plants in operation.
2. Compare/contrast "safety" and "reliability."
3. Why are reliability studies not very accurate?
4. List the uncertainties involved in reliability analysis.
5. Why is reliability important for offshore plants?
6. Explain groups of uncertainties, with examples, as applied to engineering structures.
7. Define a time-invariant problem, with an example.
8. Define a time-variant problem, with an example.
9. What do you understand by acceptable risk? How is this characterized?
10. Discuss different levels of reliability, with examples in engineering structures.
11. Explain in detail how FEED is linked to reliability.
12. What do you understand by FMEA?
13. List different methods of hazard identification.
14. Name one method of hazard evaluation used for mechanical and electrical systems.
15. What are primary and secondary functions?
16. What do you understand by a "weak link"? In what kind of hazard studies is this required to be identified?
17. Draw the typical format of a FMEA report.
18. Name two types of FMEA.
19. Describe failure mode, failure effect, and failure cause.
20. What is the risk priority number? How is it significant in hazard evaluation studies?

21. What do you understand by a cause–effect diagram? Explain with an example.
22. List the importance of safety.
23. Explain the importance of safety in HSE management through a schematic illustration.
24. Define: i) accident, ii) safety or loss prevention, iii) hazard, iv) incident, and v) risk.
25. What is the difference between "safety" and "risk"?
26. List different types of risk, as identified in risk analysis studies.
27. What do you understand by "individual risk"?
28. Explain "societal risk." How is this significant in risk assessment?
29. How are accidents measured?
30. What are the steps taken to defeat an accident process?
31. List some application issues in risk assessment.
32. List different problems associated with offshore drilling operations.
33. Comment on the recent development of alternative drilling techniques to improve safety in operations.
34. From the safety point of view, list important factors in offshore drilling.
35. Define "risk."
36. What do you understand by "loss"?
37. What is the primary difference between "hazard" and "risk"?
38. List FAQs in hazard identification.
39. In the offshore industry, what do you understand by a "gold-plated" system?
40. List different methods of hazard identification.
41. What do you understand by HAZOP? What is the main objective of HAZOP?
42. List the data you require for a HAZOP study.

Bibliography

Abu-Lebdeh, T.M., & Voyiadjis, G.Z. (1993). Plasticity-damage model for concrete under cyclic multi-axial loading. *Journal of Engineering Mechanics*, 119(7), 1465–1484.

AISC (1989). *Manual of Steel Construction – Allowable Stress Design*, 9th edition. AISC-ASD.

Akkar, S., & Metin, A. (2007). Assessment of improved nonlinear static procedures in FEMA-440. *Journal of Structural Engineering*, 133(9), 1237–1246.

Albino, J.C.R., Almeida, C.A., Menezes, I.F.M., & Paulino, G.H. (2018). Co-rotational 3D beam element for nonlinear dynamic analysis of risers manufactured with functionally graded materials (FGM). *Engineering Structures*, 173, 283–299. https://doi.org/10.1016/j.engstruct.2018.05.092.

Amer, M.Y. (2010). *Development of Structural Regional Acceptance Criteria for Existing Fixed Offshore Structures in the Arabian Gulf*. Paper presented at the Abu Dhabi International Petroleum Exhibition and Conference, 1–4 November. Abu Dhabi: UAE.

API (January 1986). *Recommended Practice for Planning Designing and Constructing Heliport for Fixed Offshore Platforms*. API RP, 3rd edition, 2L.

API (November 2014). *Recommended Practice for Planning, Designing and Constructing Fixed Offshore Platforms—Working Stress Design*. AP Journal, RP2A-WSD, 22nd edition.

Aslan, Z., & Daricik, F. (2016). Effects of multiple delaminations on the compressive, tensile, flexural, and buckling behavior of E-glass/epoxy composites. *Composites Part B: Engineering*, 100, 186–196. https://doi.org/10.1016/j.compositesb.2016.06.069.

ATC-40 (1996). *Seismic Evaluation and Retrofit of Concrete Buildings*. California: Applied Technology.

Ayrton, W.E., & Perry, J. (1886). On struts. *The Engineer*, 62, 513–514.

Bangash, M.Y.H. (1989). *Concrete and Concrete Structures*. Elsevier Publications.

Bermingham, M.J., Kent, D., Zhan, H., St John, D.H., & Dargusch, M.S. (2015). Controlling the microstructure and properties of wire arc additive manufactured Ti–6Al–4V with trace boron additions. *Acta Materialia*, 91, 289–303. https://doi.org/10.1016/j.actamat.2015.03.035.

Bjerager, P. (1990). On computation methods for structural reliability analysis. *Structural Safety*, 9(2), 79–96. https://doi.org/10.1016/0167-4730(90)90001-6.

Branci, T., Yahmi, D., Bouchair, A., & Fourneley, E. (2016). Evaluation of behavior factor for Steel moment-resisting Frames. *International Journal of Civil and Environmental Engineering*, 10(3), 396–400. https://doi.org/10.5281/zenodo.1123588.

Candappa, D.C., Sanjayan, J.G., & Setunge, S. (2001). Complete stress-strain curves of high strength concrete. *ASCE Journal of Materials in Civil Engineering*, 13, 209–215.

Carreira, D., & Chu, K.H. (1986). The moment-curvature relationship of RC members. *ACI. J.*, 83, 191–198.

Challamel, N., & Hjiaj, M . (2005). Non-local behaviour of plastic softening beams. *Journal of Acta Mechanica*, 178(3–4), 125–146.

Chandrasekaran, S. (2014). *Advanced Theory on Offshore Plant FEED Engineering*. Republic of South Korea: Changwon National University Press. ISBN:978-899-69-7928-9, p. 237.

Chandrasekaran, S. (2015). *Advanced Marine Structures*. CRC Press. ISBN:978-14-987-3968-9.

Chandrasekaran, S. (2016a). *Offshore Structural Engineering: Reliability and Risk Assessment*. Florida: CRC Press, ISBN: 978-14-987-6519-0.
Chandrasekaran, S. (2016b). *Health, Safety and Environmental Management for Offshore and Petroleum Engineers*. UK: John Wiley and Sons, ISBN: 978-11-192-2184-5.
Chandrasekaran, S. (2016c). *Health, Safety and Environmental Management in Offshore and Petroleum Engineering*. John Wiley & Sons, ISBN: 978-111-92-2184-5.
Chandrasekaran, S. (2017). *Dynamic Analysis and Design of Ocean Structures*, 2nd edition, Singapore: Springer. ISBN:978-981-10-6088-5.
Chandrasekaran, S. (2019). *Advanced Steel Design of Structures*. CRC Press. ISBN: 978-036-72-3290-0.
Chandrasekaran, S. (2020a). *Design of Marine Risers with Functionally Graded Materials*. Woodhead Publishing, Elsevier, 200, ISBN: 978-0128235379.
Chandrasekaran, S. (2020b). *Offshore Semi-Submersible Platform Engineering*. Florida: CRC Press, 240, ISBN: 978-0367673307.
Chandrasekaran, S., & Bhattacharyya, S.K. (2012). *Analysis and Design of Offshore Structures with Illustrated Examples, Human Resource Development Center for Offshore and Plant Engineering (HOPE Center)*. ISBN: 978-89-963-9155-5.
Chandrasekaran, S., & Kiran, A. (2014a). Accident modeling and risk assessment of oil and gas industries. In: Proceedings of 9th Structural Engineering Convention (SEC 2014), IIT Delhi, India. December 22-24, 2014, 2533–2543.
Chandrasekaran, S., & Kiran, A. (2014b). Consequence Analysis and risk assessment of oil and gas industries. In: Proc. International Conference on Safety & Reliability of Ship, Offshore and Subsea Structures, Glasgow, UK, Aug 18-20, 2014.
Chandrasekaran, S., & Madhavi, N. (2015a). Variation of the flow field around twin cylinders with and without outer perforated cylinder: Numerical studies. *China Ocean Engineering*, 30(5), 763–771. https://doi.org/10.1007/s13344-016-0048-0.
Chandrasekaran, S., & Madhavi, N. (2015b). Variations of water particle kinematics of offshore TLPs with perforated members: Numerical investigations. In: V. Matsagar (Ed). *Advances in Structural Engineering*. Springer, 629–645. doi: 978-81-322-2190-6_51.
Chandrasekaran, S., & Merin, T. (2016). Suppression system for offshore cylinders under vortex-induced vibration. *Vibroengineering Procedia*, 7, 01–06.
Chandrasekaran, S., & Nagavinothini, R. (2020a). Parametric studies on the impact response of offshore triceratops in ultra-deep waters. *Structure and Infrastructure Engineering*, 16(7), 1002–1018. https://doi.org/10.1080/15732479.2019.1680707.
Chandrasekaran, S., & Nagavinothini, R. (2020b). *Offshore Compliant Platforms: Analysis, Design, and Experimental Studies*. John Wiley & Sons. ISBN: 978-1-119-66977-7.
Chandrasekaran, S., & Pachaiappan, S. (2020). Numerical analysis and preliminary design of topside of an offshore platform using FGM and X52 steel under special loads. *Innovative Infrastructure Solutions*, 5(3), 1–14. https://doi.org/10.1007/s41062-020-00337-4.
Chandrasekaran, S., & Roy, A. (2004). Comparison of modal combination rules in seismic analysis of multi-storey RC frames. In: Proc. 3rd International Conference Vib. Engrg. & Tech. (Vetomac), IIT Kanpur, India, 161–169.
Chandrasekaran, S., & Roy, A. (2006). Seismic evaluation of multi-storey RC frame using modal pushover analysis. *Nonlinear Dynamics*, 43(4), 329–342. https://doi.org/10.1007/s11071-006-8327-6.
Chandrasekaran, S., & Srivastava, G. (2018). *Design Aids of Offshore Structures under Special Environmental Loads, Including Fire Resistance*. Springer. ISBN 978-981-10-7608-4.
Chandrasekaran, S., Hari, S., & Amirthalingam, M. (2019). Wire-arc additive manufacturing of functionally-graded material for marine riser applications. Proc. I-OCEANS.
Chandrasekaran, S., Hari, S., & Amirthalingam, M. (2020). Wire arc additive manufacturing of functionally graded material for marine risers. *Materials Science and Engineering: Part A*, 792. https://doi.org/10.1016/j.msea.2020.139530.

Bibliography

Chandrasekaran, S., Kumar, S., & Madhuri, S. (Eds) (2021). *Recent Advances in Structural Engineering*. Springer, ISBN:978-981-33-6388-5.

Chandrasekaran, S., Madhavi, N., & Sampath, S. (2015). Force reduction on ocean structures with perforated members. In: V. Matsagar (ed). *Advances in Structural Engineering*. Springer, 647–661. doi:978-81-322-2190-6_52.

Chandrasekaran, S., Serino, G., & Gupta, V. (2008a). Performance assessment of buildings under seismic loads. In: Proc. 10th International Conference Structural Under Shock & Impact Loads (SUSI), Portugal, 313–322.

Chandrasekaran, S., Tripati, U.K., & Srivastav, M. (2003). Study of plan irregularity effects and seismic vulnerability of moment resisting RC framed structures. In: Proc. 5th Asia-Pacific Conference Shock & Impact Loads, Changsa, China, 125–136.

Chandrasekaran, S., Natarajan, M., & Sreeramulu, L.R. (2015). Estimation of force reduction on ocean structures with perforated members. In: ASME 2015 34th International Conference on Ocean, Offshore and Arctic Engineering. American Society of Mechanical Engineers, St. John's, NL, Canada, May 31–June 5, 2015. OMAE2015-41153, V007T06A035–V007T06A035.

Chandasekaran, S., Nunziante, L., Serino, G., & Carannante, F. (2008a). Axial force-bending moment failure interaction and deformation of reinforced concrete beams using euro code. *Journal of Structural Engineering. SERC*, 35(1), 16–25.

Chandrasekaran, S., Nunzinate, L., Serino, G., & Carannante, F. (2008b). Nonlinear seismic analyses of high rise reinforced concrete buildings. *IcFai Uni. J. Struct. Engrg.*, 1(1), 1–7.

Chandrasekaran, S., Nunzinate, L., Serino, G., & Carannante, F. (2010). Axial force-bending moment limit domain and flow rule for reinforced concrete elements using euro code. *International Journal of Damage Mechanics*, 19(5), 523–558.

Chen, A.C.T., & Chen, W.F. (1975). Constitutive relations for concrete. *Engineering Mechanics*, 101, 465–481.

Chen, W.F. (1994a). *Constitutive Equations for Engineering Materials, Vol. 1: Elasticity and Modelling*. Elsevier Publications.

Chen, W.F. (1994b). *Constitutive Equations for Engineering Materials, Vol. 2: Plasticity and Modelling*. Elsevier Publications.

Chen, X., Li, J., Cheng, X., He, B., Wang, H., & Huang, Z. (2017). Microstructure and mechanical properties of the austenitic stainless steel 316L fabricated by gas metal arc additive manufacturing. *Materials Science and Engineering: Part A*, 703, 567–577. https://doi.org/10.1016/j.msea.2017.05.024.

Chopra, A.K., & Goel, R.K. (2002). A modal pushover analysis procedure for estimating seismic demands for buildings. *Earthquake Engineering and Structural Dynamics*, 31(3), 561–582. https://doi.org/10.1002/eqe.144.

Clough, L.G., & Clubley, S.K. (2019). Steel column response to thermal and long-duration blast loads inside an air blast tunnel. *Structure and Infrastructure Engineering*, 15(11), 1510–1528. https://doi.org/10.1080/15732479.2019.1635627.

D.M. (9 gennaio. 1996). Norme tecniche per il calcolo, l'esecuzione ed il collaudo delle strutture in cemento armato normale e precompresso e per le strutture metalliche. Rome, Italy (in Italian).

D.M. (14 Settembre. 2005). Norme Tecniche per le Costruzioni. Rome, Italy (in Italian).

DebRoy, T., Wei, H.L., Zuback, J.S., Mukherjee, T., Elmer, J.W., Milewski, J.O., Zhang, W., Wilson-Heid, A., De, A., & Zhang, W. (2018). Additive manufacturing of metallic components–process, structure, and properties. *Progress in Materials Science*, 92, 112–224. https://doi.org/10.1016/j.pmatsci.2017.10.001.

Denny, J.W., & Clubley, S.K. (2019). Evaluating long-duration blast loads on steel columns using computational fluid dynamics. *Structure and Infrastructure Engineering*, 15(11), 1419–1435. https://doi.org/10.1080/15732479.2019.1599966.

DNV, G. (2018). *DNVGL-RP-F112 Duplex Stainless Steel – Design against Hydrogen-Induced Stress Cracking: DNV GL*, 1–44.
Donegan, E.M. 1991. The behaviour of offshore structures in fires. In: Offshore Technology Conference. Houston, Texas: Offshore Technology Conference. https://doi.org/10.4043/6637-MS.
EC (2005). European Committee for Standardization, EN 1993-1-1:2005. *Euro Code 3: Design of Steel Structures. Part 1–1: General Rules and Rules for Buildings.* Brussels, May 2005.
Elnashai, A.S. (2001). Advanced inelastic static (pushover) analysis for earthquake applications. *Structural Engineering and Mechanics*, 12(1), 51–70.
Euro Code: UNI ENV 1991-1. Eurocodice 1. (1991). Basi di calcolo ed azioni sulle strutture. Parte 1: Basi di calcolo. (in Italian).
Euro Code: UNI ENV 1991-2-1. Eurocodice 1. (1991). Basi di calcolo ed azioni sulle strutture. Parte 2-1: Azioni sulle strutture—Massa volumica, pesi propri e carichi imposti. (in Italian).
FEMA 440 (2005). *Improvements of Nonlinear Static Seismic Analysis Procedures.* Washington District of Columbia: FEMA.
FEMA 450 (2004). *NEHRP Recommended Provisions and Commentary for Seismic Regulations for New Buildings and Other Structures.* Building Seismic Safety Council (BSSC), Washington, DC: FEMA.
Fiore, A., Spagnoletti, G., & Greco, R. (2016). On the prediction of shear brittle collapse mechanisms due to the infill-frame interaction in RC buildings under pushover analysis. *Engineering Structures*, 121, 147–159. https://doi.org/10.1016/j.engstruct.2016.04.044.
Finkel, R., Marak, V.W., & Truszcynski, M. (2002). *Constraint LINGO: A Program for Solving Logic Puzzles and Other Tabular Constraint Problems.* Berlin: Springer.
Fragiacomo, M., Dujic, B., & Sustersic, I. (2011). Elastic and ductile design of multi-storey crosslam massive wooden buildings under seismic actions. *Engineering Structures*, 33(11), 3043–3053. https://doi.org/10.1016/j.engstruct.2011.05.020.
Ganzerli, S., Pantelides, C.P., & Reaveley, L.D. (2000). Performance based design using structural optimization. *Journal of Earthquake Engineering and Structural Dynamics*, 29(11), 1677–1690.
Gao, J.W., & Wang, C.Y. (2000). Modeling the solidification of functionally graded materials by centrifugal casting. *Materials Science and Engineering: Part A*, 292(2), 207–215. https://doi.org/10.1016/S0921-5093(00)01014-5.
Gaurav, & Chandrasekaran, S. (2021). Advanced steel design. Proc. of Short Course, Chennai, India.
Ghobarah, A. (2001). Performance based design in earthquake engineering: State of development. *Engineering Structures*, 23(8), 878–884.
Gilbert, R.I., & Smith, S.T. (2006). Strain localization and its impact on the ductility of RC slabs containing welded wire reinforcement. *Advances in Structural Engineering*, 9(1), 117–127.
Giordano, A., Guadagnuolo, M., & Faella, G. (2008). Pushover analysis of plan irregular masonry buildings. In: 14th World Conference on Earthquake Engineering, Beijing, China.
Goswami, R., & Chandrasekaran, S. (2021). Advanced steel design. Proc. of Short Course, Chennai, India.
Guneyisi, E.M., & Altay, G. (2005). A study on seismic behaviour of retrofitted building based on nonlinear static and dynamic analysis. *Journal of Earthquake Engineering & Engineering Vibration*, 4(1), 173–180.

Bibliography

Hasan, R., Xu, L., & Grierson, D.E. (2002). Pushover analysis for performance-based seismic design. *Computers and Structures*, 80(31), 2483–2493. https://doi.org/10.1016/S0045-7949(02)00212-2.

Heredia-Zavoni, E., Campos, D., & Ramírez, G. (2004). Reliability based assessment of deck elevations for offshore jacket platforms. *Journal of Offshore Mechanics and Arctic Engineering*, 126(4), 331–336.

Hauksson, H.þ., & Vilhjálmsson, J.B. (2014). *Lateral-Torsional Buckling of Steel Beams with Open Cross Section*, Master's thesis. Göteborg, Sweden: Department of Civil and Environmental Engineering, Division of Structural Engineering—Steel and Timber Structures, Chalmers University of Technology, 28.

Huang, C.H., Tuan, Y.A., & Hsu, R.Y. (2006). Nonlinear pushover analysis of infilled concrete frames. *Journal of Earthquake Engineering & Engineering Vibration*, 5(2), 245–255.

Höglund, T. (2006). *Design of Steel Structures, Module 6 – . Stability of Columns and Beams*. Skillingaryd, Sweden: Swedish Institute of Steel Construction, Luleå University of technology, Royal Institute of technology. Skilltryck AB, Publication no., 2009.

Hognestad, E., Hanson, N.W., & McHenry, D. (1955). Concrete stress distribution in ultimate strength design. *ACI. J.*, 52(6), 455–479.

Hsieh, S.S., Ting, E.C., & Chen, W.F. (1982). A plastic-fracture model for concrete. *International Journal of Solids and Structures*, 18(3), 181–197.

IS 13920 (1993). *Ductile Detailing for Reinforced Concrete Structures Subjected to Seismic Forces*. New Delhi, India: Bureau of Indian Standards.

Is 1893 (2002). *Criteria for Earthquake Resistant Design of Structures: Part 1. General Provisions for Buildings*. Fifth Revision. New Delhi, India: Bureau of Indian Standards.

Is 456 (2000). *Plain and Reinforced Concrete—Code of Practice*. Fourth Revision. New Delhi, India: Bureau of Indian Standards.

ISO, EN (2007). *19902: 2007 Petroleum and Natural Gas Industries. Fixed Steel Offshore Structures (ISO 19902: 2007)*.

Jin, G., Takeuchi, M., Honda, S., Nishikawa, T., & Awaji, H. (2005). Properties of multilayered mullite/Mo functionally graded materials fabricated by powder metallurgy processing. *Materials Chemistry and Physics*, 89(2–3), 238–243. https://doi.org/10.1016/j.matchemphys.2004.03.031.

Jin, Y., & Jang, B.S. (2015). Probabilistic fire risk analysis and structural safety assessment of FPSO topside module. *Ocean Engineering*, 104, 725–737. https://doi.org/10.1016/j.oceaneng.2015.04.019.

Jirasek, M., & Bazant, Z.P. (2002). *Inelastic Analysis of Structures*. New York: Wiley.

Kaley, P., & Baig, M.A. (2017). Pushover analysis of steel framed building. *Journal of Civil Engineering and Environmental Technology*, 4(3), 301–306.

Kan, W.H., Albino, C., Diasdacosta, D., Dolman, K., Lucey, T., Tang, X., Cairney, J., Proust, G., & Cairney, J. (2018). Microstructure characterization and mechanical properties of a functionally-graded NBC/high chromium white cast iron composite. *Materials Characterization*, 136, 196–205. https://doi.org/10.1016/j.matchar.2017.12.020.

Karmazinova, M.A., & Melcher, J.I. (2012). Influence of steel yield strength value on structural reliability. *Recent Researches in Environmental and Geological Sciences*, 441–446.

Kaveh, A., Azar, B.F., Hadidi, A., Sorochi, F.R., & Talatahari, S. (2010). Performance-based seismic design of steel frames using ant colony optimization. *Journal of Constructional Steel Research*, 66(4), 566–574. https://doi.org/10.1016/j.jcsr.2009.11.006.

Kawasaki, A., & Watanabe, R. (1987). Finite element analysis of thermal stress of the metal/ceramic multi-layer composites with controlled compositional gradients. *Japan Institute of Metals, Journal*, 51(6), 525–529. https://doi.org/10.2320/jinstmet1952.51.6_525.

Khan, & Gaurav (2018). *Enhanced Fire Severity in Modern Indian Dwellings*. Current Science.

Khan, A.R., Al-Gadhib, A.H., & Baluch, M.H. (2007). Elasto-damage model for high strength concrete subjected to multi axial loading. *International Journal of Damage Mechanics*, 16(3), 367–398.

Khan, M.I., & Khan, K.N. (2014). Seismic analysis of steel frame with bracings using pushover analysis. *International Journal of Advanced Technology in Engineering and Science*, 02, 369–381.

Ko, M.Y., Kim, S.W., & Kim, J.K. (2001). Experimental study on the plastic rotation capacity of reinforced high strength beams. *Materials and Structures*, 34(5), 302–311.

Koizumi, M.F.G.M. (1997). FGM activities in Japan. *Composites Part B: Engineering*, 28 (1–2), 1–4. https://doi.org/10.1016/S1359-8368(96)00016-9.

Liew, J.R., Sohel, K.M.A., & Koh, C.G. (2009). Impact tests on steel–concrete–steel sandwich beams with a lightweight concrete core. *Engineering Structures*, 31(9), 2045–2059. https://doi.org/10.1016/j.engstruct.2009.03.007.

Lopes, S.M.R., & Bernardo, L.F.A. (2003). Plastic rotation capacity of high-strength concrete beams. *Materials and Structures*, 36(1), 22–31.

López, A., Yong, D.J., & Serna, M.A. (2006). Lateral-torsional buckling of steel beams: A general expression for the moment gradient factor. In: D. Camotin et al. (Eds). *Proceeding of the International Colloquium of Stability and Ductility of Steel Structures*. Lisbon, Portugal, September 6-8, 2006.

Luciano, N., & Raffaele, O. (1988). *Limit Design of Frames Subjected to Seismic Loads*. Italy: CUEN Publications (Cooperativa Universitaria Editrice Napoletana).

Madec, C., LeGallet, S., Salesse, B., Geoffroy, N., & Bernard, F. (2018). Alumina-titanium functionally graded composites produced by spark plasma sintering. *Journal of Materials Processing Technology*, 254, 277–282. https://doi.org/10.1016/j.jmatprotec.2017.11.004.

Mahin, S., Anderson, E., Espinoza, A., Jeong, H., & Sakai, J. (2006). Sustainable design considerations in the construction to resist the effects of strong earthquakes. In: Proc. 4th Interface Workshop Seismic Design and Retrofit of Transportation Facilities. Burlingame, California, 10.

Mahmoudi, M., & Zaree, M. (2010). Evaluating response modification factors of concentrically braced steel frames. *Journal of Constructional Steel Research*, 66(10), 1196–1204. https://doi.org/10.1016/j.jcsr.2010.04.004.

Manco, M.R., Vaz, M.A., Cyrino, J.C., & Landesmann, A. (2013). Behavior of stiffened panels exposed to fire. In: Proc. IV MARSTRUCT, Espoo, Finland. ISBN 978-1-138-00045-2, 101–108.

Martina, F., Mehnen, J., Williams, S.W., Colegrove, P., & Wang, F. (2012). Investigation of the benefits of plasma deposition for the additive layer manufacture of Ti–6Al–4V. *Journal of Materials Processing Technology*, 212(6), 13. https://doi.org/10.1016/j.jmatprotec.2012.02.002.

McInerney, J.B., & Wilson, J.C. (2012, September). NLTHA and pushover analysis for steel frames with flag-shaped hysteretic braces. In: Proc. 15th World Conference on Earthquake Engineering, Lisbon, Portugal, 24–28.

Melchers, R.E. (1988). Importance sampling in structural systems. *Structural Safety*, 6(1), 3–10. https://doi.org/10.1016/0167-4730(89)90003-9.

Menetrey, P.H., & Willam, K.J. (1995). Tri-axial failure criterion for concrete and its generalization. *ACI. Journal of Structures*, 92, 311–318.

Mirjalili, M.R., & Rofooei, F.R. (2017). The modified dynamic-based pushover analysis of steel moment-resisting frames. *The Structural Design of Tall and Special Buildings*, 26(12), e1378. https://doi.org/10.1002/tal.1378.

Mo, Y.L. (1992). Investigation of reinforced concrete frame behaviour: Theory and tests. *Magazine of Concrete Research*, 44(160), 163–173.

Morin, D., Kårstad, B.L., Skajaa, B., Hopperstad, O.S., & Langseth, M. (2017). Testing and modelling of stiffened aluminium panels subjected to quasi-static and low-velocity impact loading. *International Journal of Impact Engineering*, 110, 97–111. https://doi.org/10.1016/j.ijimpeng.2017.03.002.

Murtha, J.A. (1997). Monte Carlo simulation: Its status and future. *Journal of Petroleum Technology*, 49(04), 361–373. https://doi.org/10.2118/37932-JPT.

Mwafy, A.M., & Elnashai, A.S. (2001). Static pushover versus dynamic collapse analysis of RC buildings. *Engineering Structures*, 23(5), 407–424. https://doi.org/10.1016/S0141-0296(00)00068-7.

Naughton, D.T., Tsavdaridis, K.D., Maraveas, C., & Nicolaou, A. (2017). Pushover analysis of steel seismic-resistant frames with reduced Web section and reduced Beam section connections. *Frontiers in Built Environment*, 3, 59. https://doi.org/10.3389/fbuil.2017.00059.

Nunziante, L., Gambarotta, L., & Tralli, A. (2007). *Scienza della costruzioni*. Italy: McGraw-Hill (in italian).

Oltedal, H.A. (2012). Ship-platform collisions in the North Sea. In: The 11th International Probabilistic Safety Assessment and Management Conference and the Annual European Safety and Reliability Conference (ESREL). http://hdl.handle.net/11250/151419. Finland: Helsinki.

Onoufriou, T. (1999). Reliability-based inspection planning of offshore structures. *Marine Structures*, 12(7–8), 521–539. https://doi.org/10.1016/S0951-8339(99)00030-1.

Ordinanza 3316: Correzioni e modifiche all'ordinanza 3274. (2005). Modifiche ed integrazioni all'ordinanza del Presidente del Consiglio dei Ministri n. 3274 del 20 Marzo 2003. Rome, Italy (in Italian).

Ordinanza del presidente del consiglio dei ministri del 20 marzo. (2003). Primi elementi in materia di criteri generali per la classificazione sismica del territorio nazionale e di normative tecniche per le costruzioni in zona sismica. Rome, Italy (in Italian).

Paik, J.K., & Czujko, J. (2013). Engineering and design disciplines associated with management of hydrocarbon explosion and fire risks in offshore oil and gas facilities. *Transactions SNAME*, 120, 167–197. ISSN: 0081-1661.

Paik, J.K., Czujko, J., Kim, J.H., Park, S.I., Islam, S., & Lee, D.H. (2013). A new procedure for the nonlinear structural response analysis of offshore installations in fires. *Transactions SNAME*, 121, 224–250. ISSN: 0081-1661.

Pan, X., Zheng, Z., & Wang, Z. (2017). Estimation of floor response spectra using modified modal pushover analysis. *Soil Dynamics and Earthquake Engineering*, 92, 472–487. https://doi.org/10.1016/j.soildyn.2016.10.024.

Papadrakakis, M., Fragiadakis, M., & Lagaros, N.D. (2007). *Extreme Man-Made and Natural Hazards in Dynamics of Structures*. Netherlands: Springer.

Park, H., & Kim, H. (2003). Microplane model for reinforced concrete planar members in tension-compression. *Journal of Structural Engineering*, 129(3), 337–345.

Paulay, T., & Priestley, M.J.N. (1992). *Seismic Design of RC and Masonry Buildings*. USA: John Wiley & Sons.

PETRONAS (2010). *Design of Fixed Offshore Structures (PTS 34.19.10.30) Petronas Technical Standards*. Malaysia.

Pfrang, E.O., Siess, C.P., & Sozen, M.A. (1964). Load-moment-curvature characteristics of RC cross-sections. *ACI .J.*, 61(7), 763–778.

Pisanty, A., & Regan, P.E. (1998). Ductility requirements for redistribution of moments in RC elements and a possible size effect. *Materials and Structures*, 31(8), 530–535.

Pisanty, A., & Rogan, P.E. (1993). Redistribution of moments and the possible demand for ductility. *CEB Bulltn d' Information*, 218, 149–162.

Poursha, M., Khoshnoudian, F., & Moghadam, A.S. (2011). A consecutive modal pushover procedure for nonlinear static analysis of one-way unsymmetric-plan tall building structures. *Engineering Structures*, 33(9), 2417–2434. https://doi.org/10.1016/j.engstruct.2011.04.013.

Priestley, M.J.N., Calvi, G.M., & Kowalsky, M.J. (2007). *Displacement Based Seismic Design of Structures*. Italy: IUSS Press.

Priya, L., & Chandrasekaran, S. (2021). Advanced steel design. Proc. of Short Course, Chennai, India.

Quiel, S.E., & Garlock, M.E. (2010). Calculating the buckling strength of steel plates exposed to fire. *Thin-Walled Structures*, 48(9), 684–695. https://doi.org/10.1016/j.tws.2010.04.001.

Rajan, T.P.D., Pillai, R.M., & Pai, B.C. (2010). Characterization of centrifugal cast functionally graded aluminum-silicon carbide metal matrix composites. *Materials Characterization*, 61(10), 923–928. https://doi.org/10.1016/j.matchar.2010.06.002.

Ranganathan, R. (1999). *Structural Reliability Analysis and Design*. Jaico Publishing House ISBN: 978–8172248512.

Rebelo, C., Lopes, N., Simoes da Silva, L., Nethercot, D., & Vila Real, P.M.M. (2009). Statistical evaluation of the lateral-torsional buckling resistance of steel I-beams, Part 1: Variability of the Euro Code 3 resistance model. *Journal of Constructional Steel Research*, 65(4), 818–831.

Roobaert, N. (December 2010). Lighter topsides – The what, why and how. *Offshore Engineer Magazine*.

RP2A-LRFD, API (1993). *Recommended Practice for Planning. Designing and Constructing Fixed Offshore Platforms–Load and Resistance Factor Design*, 1st edition. American Petroleum Institute.

Shaikhutdinov, R. (2006). Structural damage evaluation. *Journal of Japan Association for Earthquake Engineering*, 4(2), 39–47.

Sadowski, A.J., Rotter, J.M., Reinke, T., & Ummenhofer, T. (2015). Statistical analysis of the material properties of selected structural carbon steels. *Structural Safety*, 53, 26–35. https://doi.org/10.1016/j.strusafe.2014.12.002.

Sankarasubramaniam, G., & Rajasekaran, S. (1996). Constitutive modelling of concrete using a new failure criterion. *Journal of Computers & Structures*, 58(5), 1003–1014.

Sau-Cheong, F., & Wang, F. (2002). A new strength criterion for concrete. *ACI. Journal of Structures*, 99, 317–326.

SAP2000 Advanced Version: 10.1.2. Structural Analysis Program. Berkeley, California: Computers and Structures Inc. http://www.csiberkeley.com.

SEAOC (1995). *Performance Based Seismic Engineering of Buildings—Vision 2000 Report*. California, USA: Structural Engineers Association of California.

Serna, M.A., López, A., Puente, I., & Yong, D. (2005). Equivalent uniform moment factors for lateral-torsional buckling of steel members. *Journal of Constructional Steel Research*, 62(6), 2006, 566–580.

Shen, Y., & Jukes, P. (2015, June). Technical challenges of unbonded flexible risers in HPHT and deepwater operations. In: The Twenty-fifth International Ocean and Polar Engineering Conference.

Shishkovsky, I., Missemer, F., & Smurov, I. (2012). Direct metal deposition of functionally graded structures in the Ti-Al system. *Physics Procedia*, 39, 382–391. https://doi.org/10.1016/j.phpro.2012.10.052.

Shivakumar, K.N., Elber, W., & Illg, W. (1985). Prediction of impact force and duration due to low-velocity impact on circular composite laminates. https://doi.org/10.1115/1.3169120.

Shukla, S.K., Chandrasekaran, S., Das, B.B., & Kolathayar, S. (Eds) (2020). *Smart Technologies for Sustainable Development*. Springer, ISBN: 978-981-15-5000-3.

Simonini, S., Constantin, R.T., Rutenberg, A., & Beyer, K. (2012). Pushover analysis of multi-storey cantilever wall systems. In: Proc. 15th World Conference on Earthquake Engineering (No. CONF).

Srinivasan, H. (2020). *Dynamic Analysis of Marine Risers with Functionally Graded Material*. Doctoral dissertation. IIT Madras.

Skrzypczak, I., Sáowikb, M., & Buda, O. (2017). The application of reliability analysis in engineering practice–reinforced concrete foundation. *Procedia Engineering*, 193, 144–151. https://doi.org/10.1016/j.proeng.2017.06.197.

Soares, C.G., & Teixeira, A.P. (2000). Strength of plates subjected to localised heat loads. *Journal of Constructional Steel Research*, 53(3), 335–358. https://doi.org/10.1016/S0143-974X(99)00045-0.

Soares, C.G., Gordo, J.M., & Teixeira, A.P. (1998). Elasto-plastic behaviour of plates subjected to heat loads. *Journal of Constructional Steel Research*, 45(2), 179–198. https://doi.org/10.1016/S0143-974X(97)00062-X.

Soleimani Amiri, F., Ghodrati Amiri, G., & Razeghi, H. (2013). Estimation of seismic demands of steel frames subjected to near-fault earthquakes having forward directivity and comparing with pushover analysis results. *The Structural Design of Tall and Special Buildings*, 22(13), 975–988. https://doi.org/10.1002/tal.747.

Student LINGO (2005). *Parameter Optimization Software. Release 9.0*. LINDO Sys Inc.

Sumarec, D., Sekulovic, M., & Krajcinovic, D. (2003). Failure of RC beams subjected to three point bending. *International Journal of Damage Mechanics*, 12(1), 31–44.

Szalai, J., & Papp, R. (2010). On the theoretical background of the generalization of Ayrton-Perry type resistance formulas. *Journal of Constructional Steel Research*, 66(5), 670–679.

Timoshenko, S., Gere, J.G., & Gere, J.M. (1961). *Theory of Elastic Stability*. McGraw-Hill. ISBN: 9780070647497, p. 541.

Tarta, G., & Pintea, A. (2012). Seismic evaluation of multi-storey moment-resisting steel frames with stiffness irregularities using standard and advanced pushover methods. *Procedia Engineering*, 40, 445–450. https://doi.org/10.1016/j.proeng.2012.07.123.

Teran- Gilmore, A., & Bahena- Arredondo, N. (2008). Cumulative ductility spectra for seismic design of ductile structures subjected to long duration motions. Concept and theory background. *Journal of Earthquake Engineering*, 12(1), 152–172.

Übeyli, M., Balci, E., Sarikan, B., Öztaş, M.K., Camuşcu, N., Yildirim, R.O., & Keleş, Ö. (2014). The ballistic performance of SiC–AA7075 functionally graded composite produced by powder metallurgy. *Materials and Design (1980-2015)*, 56, 31–36. https://doi.org/10.1016/j.matdes.2013.10.092.

Williams, S.W., Martina, F., Addison, A.C., Ding, J., Pardal, G., & Colegrove, P. (2016). Wire+ arc additive manufacturing. *Materials Science and Technology*, 32(7), 641–647. Https://doi.org/10.1179/1743284715Y.0000000073.

Wolfram mathematica software. Ver 6.0.0. for Windows Platform: Wolfram Research Inc.l

Wood, R.H. (1968). Some controversial and curious developments in plastic theory of structures. In: J. Heyman, & F.A. Leckie (Eds). *Engineering Plasticity*. Cambridge University Press, 665–691.

Yakut, A., Yilmaz, N., & Bayili, S. (2001). Analytical assessment of Seismic capacity of RC framed building. *Journal of Japan Association of Earthquake Engineering*, 6(2), 67–75.

Yeo, J.G., Jung, Y.G., & Choi, S.C. (1998). Design and microstructure of ZrO2/SUS316 functionally graded materials by tape casting. *Materials Letters*, 37(6), 304–311. https://doi.org/10.1016/S0167-577X(98)00111-6.

Yuan, H., Li, J., Shen, Q., & Zhnag, L. (2020). In-situ synthesis and sintering of ZrB2 porous ceramics by Spark plasma sintering reactive synthesis (SPS-RS) method. *International Journal of Refractory Metals and Hard Materials*, 34, 3–7.

Zhang, Y., & Der Kiureghian, A. (1993). Dynamic response sensitivity of inelastic structures. *Computer Methods in Applied Mechanics and Engineering*, 108(1–2), 23–36.

Zucchelli, A., Minak, G., & Ghelli, D. (2010). Low-velocity impacts behavior of vitreous-enameled steel plates. *International Journal of Impact Engineering*, 37(6), 673–684. https://doi.org/10.1016/j.ijimpeng.2009.12.003.

Index

A

Acceptable risk, 281
Accidental loads, 72
ACO, *see* Ant colony optimization method
Advanced FOSM method, 289–291
ALARP principle, 281
Aleatory uncertainty, 283
Allowable deflection, 83
Allowable stress design (ASD) method, 74
Allowable stresses, 83
American Society for Testing and Materials (ASTM), 10, 12
AND gate, 272–274
Ant colony optimization (ACO) method, 2
ASD, *see* Allowable stress design method
ASTM, *see* American Society for Testing and Materials
Axial compression, stability functions
    rotation functions, 115–119
        under axial tensile load, 121–129
        under zero axial load, 119–120
    translation function, under axial compressive load, 122, 129
Axial deformation, 106–112
Axial force-bending moment interaction, 35, 37–39

B

Basic engineering design (BED), 27
Bayesian approach, 283
BED, *see* Basic engineering design
Bidding phase, 28
BLEVE, *see* Boiling liquid expanding vapor explosion
Blowout preventer (BOP), 75, 266, 267
Blowouts, 266
Boiling liquid expanding vapor explosion (BLEVE), 54
BOP, *see* Blowout preventer
Brace eccentricities, 81
Bridge reactions, 73
Buckling
    curves, 214–215
        applicable for general method, 215, 217
        choice of, 215, 216
        imperfection factor, 215, 217
        for special case method, 222
    global buckling, 197–198
    lateral-torsional buckling (*see* Lateral-torsional buckling (LTB))
    modes, 206, 208

C

Cantilever beam, 202, 203
Capacity reduction factors, 209
Carbon steel
    characteristics, 54, 55
    heat, 55, 56
    thermal conductivity, 55, 56
    thermal strain, 55, 57
Cause–effect diagram, 278, 279
Center of gravity (COG), 78
Chemical risk assessment, 265
Circular column, cross-section, 42
Clarity, 277
CMP, *see* Conservative model pushover
COG, *see* Center of gravity
Concrete properties, 37–39
Conjugate beam, 115
Conservative model pushover (CMP), 3
Conservative wind load, 79
Consistency, 277
Construction phase, 28
Construction planning, 33
Contact-tip-to-work distance (CTWD), 10
Control score, 263
Cooling phase, 60
    temperature calculation, 64
Coordinate systems, 81
Correction factor, 218, 219
    moment gradient conditions, 213
    uniform moment conditions, 211, 212
Crane and drop object, 15, 16
Critical buckling load, 174–175
    of column member using stability functions, 175–176
    of structure using stability functions, 177–187
Critical load, 105
CTWD, *see* Contact-tip-to-work distance
Curved beams
    with circular cross-section, 254–255
    crane hook with trapezoidal cross-section, 256–257
    large initial curvature, 246–250

303

simplified equations, 250–251
small initial curvature
    bending for, 242–245
    deflection for, 245
with trapezoidal cross-section, 253–254
T-section, 251–253

D

Dead loads, 69
Deck plate, 15, 16
    boundary conditions, 16, 17
    numeric model, 15, 16
Deepwater Horizon, 52
Deformation loads, 71
Deformed geometry, 200
Depth of elastic core, 90–92
Derrick equipment set (DES), 73, 75
Design considerations, 82
    acceptance criteria, 83
Design loads, 69
    dead loads, 69
    live loads, 69, 71–73
Design methods, 67–69, 83–86
Design phase, 27
Design stages, 73
    lifting analysis, 76–78
    load-out analysis, 75–76
    miscellaneous items analysis, 80
    static in-place analysis, 73–75
    transportation analysis, 78–79
Detection, 280
Displacement-controlled pushover analysis, 19
Drilling loads, 75
Drilling rigs, 265
Ductility, 38
Dynamic loads, 72

E

EAA, *see* Equal area axis
EC3-1-1, 214, 215, 218
Elastic critical moment, 209, 221
Elastic-plastic section, 91, 92
Elastic-plastic state, 100
Elastic section modulus, 89
Engineering, procurement, construction, and commissioning (EPCC), 33
Engineering, procurement construction (EPC), execution planning, 33
EPCC, *see* Engineering, procurement, construction, and commissioning
Epistemic uncertainty, 283
Equal area axis (EAA), 89
ETA, *see* Event tree analysis
Euler's column, 104

Euler's critical load, 104–106
Euler's static criterion, 103
Euro Code, 60, 61
    LTB using, 218–223
Euro Code 3, 202, 212, 214, 215
Event tree analysis (ETA), 274–276
Exxon Valdez oil spill, 261, 262

F

Factor of safety (FOS), 102
Failure mode and effect analysis (FMEA), 277–279
    variables, 280
Fatality accident rate (FAR), 262
Fault tree analysis, 272–273
    analysis, 274
    final event probability, 274
    formation of fireball, 274, 275
FDS, *see* Fire Dynamics Simulator
FEED, *see* Front-end engineering design
FGM, *see* Functionally graded material
Fifth phase, FEED, 35, 38
Final event, 272
    probability, 274
Fire activation risk factors, 61, 62
Fireball, 54
Fire curve, 59, 60
Fire Dynamics Simulator (FDS), 60–61
Fire-fighting equipment and measures, 61, 62
Fire load, 51–52
    fireball, 54
    flash fire, 54
    jet fire, 53–54
    pool fire, 52–53
    on topside, 55–60
Fire load density, characteristics, 61, 63
Fire load energy density (FLED), 55, 59–61
    data comparison, office buildings, 60
First level of risk assessment, 263
Firstorder reliability method (FORM), 259
First-order second-moment (FOSM) method, 288–289
First phase, FEED, 34
Fixed beam
    under axial compressive load, 131, 137
    under concentrated load, 137
    under tensile axial load, 135–136
    under uniformly distributed load, 131
Flange bending moment, 195
Flash fire, 54
FLED, *see* Fire load energy density
Flexural buckling, 197
    under compression, 198, 199
Flexural-torsional buckling, 197, 198
    under compression, 198, 199

Index

Floating production storage and offloading (FPSO) unit, 67, 84
Floor response spectra (FRS) method, 3
FMEA, *see* Failure mode and effect analysis
FOS, *see* Factor of safety
FOSM, *see* First-order second-moment method
Fourth phase, FEED, 36, 37
FPSO, *see* Floating production storage and offloading unit
Fractography, 11, 13
Free-body diagram
    under axial load, 116
    of column member, 105
    and unit rotation at j^{th} end, 116
    $x<aL_i$, 131, 137
    $x>aL_i$, 133, 138
Fronius™ cold metal transfer (CMT) TransPuls Synergic 4000, 9
Front-end engineering design (FEED), 24–25
    advanced level in feed studies, 30–32
        plant design and model studies, 32–33
    basic engineering requirements of, 27–28
    EPC execution planning, 33
    essentials of, 25–27
    factors influencing design on topside, 28–30
    phases, 34–35
FRS, *see* Floor response spectra method
Fuel-controlled fire, 64
Functionally graded material (FGM), 1, 5, 6, 24
    beams and columns displacement, 14, 15
    carbon–manganese steel deposittion, 10, 11
    fractography images, 13
    manufacture on lab scale, 9
    mechanical properties comparison, 7–8
    stainless steel layer deposittion, 10
    stress–strain plots, 7, 8, 10, 12
    structural components, 7, 8
    topside deformed shape, 14, 15
Future loads, 72

G

GBS, *see* Gravity-based structure platforms
Geometric configuration, 3–5
Global buckling, 197–198
Graphical plots, 82
Gravity-based structure (GBS) platforms, 72
Gumbel distribution, 61

H

Hasofer–Lind method, 290
Hazard, 269
    control, 270–271
    evaluation, 269–270
    groups, 267–268
    identification, 265
    score, 263
Hazard and operability study (HAZOP), 31
Hazard identification (HAZID), 31, 271
HAZOP, *see* Hazard and operability study
Heating phase, 60
    temperature calculations, 61–64
Heat release rate (HRR), 55
    curve, 58
Heavy-lift vessel (HLV), 77
Helideck loads, 72
Hierarchy of hazard control, 270
HLV, *see* Heavy-lift vessel
Horizontal jet fire, 53
Hot-rolled structural steel beam, 82
HRR, *see* Heat release rate
Hydraulic work-over unit (HWU), 75

I

Impact load, 15–16, 75
Implicit failure probability, 286
Installation loads, 72
Integrated complex topside, 67
Intermediate event, 272
International Organization for Standardization (ISO), 21

J

Jet fire, 53–54
Joint fatigue life, 83

L

Lateral load functions
    for concentrated load, 136–140
    under uniformly distributed load under, 131–135
Lateral restraint effect, avoiding LTB, 206, 207
    buckling modes, 206, 208
Lateral-torsional buckling (LTB), 197–202
    design check for, 212–213
        alternative method, 215–218
        buckling curves, 214–217
        general method, 214
    design procedure
        moment correction factors, 211–212
        three-factor formula, 209–211
    measurements against, 205–206
        lateral restraint effect, 206
        load application effect, 206
    mechanisms behind, 202–203
        torsional effect, 203–205
    real beam behavior, 206–209
    using Euro code

maximum bending moment estimation, 218–219
section classification, 220
section moment of resistance calculation, 221
stability check, 221–223
steel grade S275 yield strength, 220
steel section properties, 219–220
using Indian code, 223–226
L'Hospital's rule, 119, 120
Lifting analysis, 76–78
Lifting slings, 76
Limit state design (LSD), 84
Liquefied natural gas (LNG), 52
Live loads, 69, 71–73
LNG, see Liquefied natural gas
Load factor, 101–102
Load function, 135
Load-out analysis, 75–76
Load-resistance factor design (LRFD), 287
Local buckling, 197
Logical risk analysis, 263–264
LRFD, see Load-resistance factor design
LSD, see Limit state design
LTB, see Lateral-torsional buckling
Lyapunov's dynamic criterion, 103–104

M

Marginal field, 266
Materials and loads on topside
 axial force-bending moment interaction, 35, 37–39
 fire load, 51–52
  fireball, 54
  flash fire, 54
  jet fire, 53–54
  pool fire, 52–53
  on topside, 55–60
 front-end engineering design (FEED), 24–25
  advanced level in feed studies, 30–33
  basic engineering requirements of, 27–28
  essentials of, 25–27
  factors influencing design on topside, 28–30
  phases, 34–35
 geometric configuration, 3–5
 impact load, 15–16
 modal analysis, 16–19
 offshore topside, 1–3
 parametric fire curve, 60–64
 P-M interaction, mathematical development, 39–49
 pushover analysis, 19–24
 steel at elevated temperature, 54–55
 studies and discussions, 49–51
 wind and blast loads, 13–15
 wire arc additive manufacturing (WAAM), 8–13
Material selection diagrams (MSDs), 31
Maximum bending moment estimation, 218–219
Maximum shear stress, 192
MBL, see Minimum breaking load
MDOF, see Multi-degree-of-freedom system
Mean sea level (MSL), 81
Mild steel characteristics, 86
Minimum breaking load (MBL), 77
Miscellaneous design items, 80
Modal analysis, 16–19
Modeling uncertainties, 283
Modular form, 67
MoI, see Moment of inertia
Moment capacity, 212
Moment-carrying capacity, 91
Moment correction factors, 211–212
Moment-curvature relationship, 99–101
Moment of inertia (MoI), 97
Moment–rotation
 curve, 21, 23
 values, 21, 23
Monte Carlo simulation, 259
MSDs, see Material selection diagrams
MSL, see Mean sea level
Mud weight load, 75
Multi-degree-of-freedom (MDOF) system, 3

N

National Building Code, India, 61
Nonlinear static pushover analysis, 1, 2, 19, 24
NORSOK (Norsk Sokkels Konkuranseposisjon), 271, 280
Numerical tools, 80–82

O

Occurrence, 280
Offshore drilling, 265
Offshore hazards, 266–267
Offshore platforms, 1
OR gate, 272–274

P

Parametric fire curve, 60–64
PFD, see Process flow diagram
P&IDs, see Piping and instrumentation diagrams
Piper Alpha disaster, 52, 260, 261
Piping and instrumentation diagrams (P&IDs), 31
PLA, see Point of load application
Plastic design, 86–89
 depth of elastic core, 90–92
 shape factor, 89–90

Index

Plastic hinges
  details, 20, 22
  location, 20, 21
Plastic section modulus, 92
P-M yield interaction, 35, 37
  for circular columns
    with different cross-sections, 49, 51
    with different diameter, 49
    with different number of rebars, 49, 50
    with different tension reinforcement, 49, 50
  mathematical development, 39–49
  for RC circular column, 43
Point of load application (PLA), 211
Pool fire, 52–53
Post-flashover, 59
Potential energy stability criterion, 104
Potential failure modes, 278
Preliminary phase, 27
Probabilistic modeling, 265
Process flow diagram (PFD), 31
Pushover analysis, 1–3, 19–24, 39

Q

Quantitative risk analysis
  cause analysis, 271–272
  initiating events, 271
  logical risk analysis, 263–264

R

Ram blowout preventers, 266
Real beam behavior, 206–209
Reasonableness, 277
Reduced web section (RWS), 1
Reinforced concrete (RC) circular columns, 35, 37
Reliability, 281–282
  assessment, 259–260
  importance, 282–283
  index, 290
  levels, 287–288
  methods
    advanced FOSM method, 289–291
    first-order second-moment (FOSM) method, 288–289
  problem
    reliability framework, 284–285
    time-invariant problem, 284
    time-variant problem, 284
  ultimate limit state and, 286–287
Request for quotation (RFQ), 32
Response spectrum analysis (RSA), 2
RFQ, see Request for quotation
Rig load, 75

Risk acceptance criterion, 280–281
  acceptable risk, 281
  UK regulations, 281
Risk and Emergency Preparedness Analysis, 280
Risk assessment, 264
  application issues, 265
  chemical risk assessment, 265
Risk aversion, 264
Risk characterization, 276–277
  principles of, 277
Risk index, 263
Risk matrix, 269
Risk priority number (RPN), 278, 280
Risk ranking, 265
Risk references, 264
Rotation functions, compressive axial load cases, 115–119
  under axial tensile load, 121–129
  under zero axial load, 119–120
RPN, see Risk priority number
RSA, see Response spectrum analysis
RWS, see Reduced web section

S

SACS
  computer program, 74
  model, 76, 81, 82
  POST processor, 75, 82
  TOW module, 78
Safety, 260
  in design and operation, 265
    offshore hazards, 266–267
    hazard groups, 267–268
  vs. reliability, 281, 282
  role in offshore plants, 261
    accident measurement, 262
    risk and safety, 262
SEA STATE program, 74
Second-order reliability method (SORM), 259
Second phase, FEED, 34, 35
Severity, 280
Shape factors, 89–90
  in offshore topside
    channel section, 95–96
    circular solid bar, 93–94
    L-section, 97–99
    rectangular cross-section, 92–93
    T-section, 96–97
    tubular section, 94–95
Shear force demand, 3
Sign convention, 250
Sixth phase, FEED, 35, 38
60/40 sling load, 76
Slenderness ratio, 83
Sling-safe working load (SWL), 77

Small initial curvature, 242–245
SORM, *see* Second-order reliability method
Special design guidelines
    curved beams
        with circular cross-section, 254–255
        crane hook with trapezoidal cross-section, 256–257
        large initial curvature, 246–250
        simplified equations, 250–251
        small initial curvature, 242–245
        with trapezoidal cross-section, 253–254
        T-section, 251–253
    lateral-torsional buckling (LTB), 197–202
        design check for, 212–218
        design procedure, 209–212
        measurements against, 205–206
        mechanisms behind, 202–205
        real beam behavior, 206–209
        using Euro code, 218–223
        using Indian code, 223–226
    thin-walled sections, 191–195
        torsion in open, 195–196
    unsymmetrical bending, 226–227
        bending stresses under, 227–232
        T-section, 235–238
        Z-section, 238–241
Stability analysis problems, MATLAB®, 140–173
Stability chart, MATLAB® code, 187–189
Stability functions, 123–129
Stable equilibrium, 104
Standard beam element, 106–112
    rotational and translational moments in, 107
    rotational coefficients, 110, 112–115
Static degree of indeterminacy, 87
Static in-place analysis, 73–75
Static pushover curve, 19
Statistical uncertainties, 283
Steel
    at elevated temperature, 54–55
    grade S275, 220
    section properties, 219–220
    sections for topside, 70
Stiffeners, 15
Stiffness coefficients, 108, 111
Strain diagrams, sub-domains, 43, 44
Stress correction factors, 251
Stress-strain relationship
    concrete, 40
    reinforcing steel, 41
Structural members section, 3, 6
Structural optimization, 82
Structural steel, 191
Structural system, stability of, 102–104
St. Venant's theory, 193
St. Venant torsion, 191, 193
    circular cross-section under, 191, 192

Subsurface blowouts, 266
Super Puma AS332L1/AS332L2 helicopter, 73
Surface blowouts, 266
SWL, *see* Sling-safe working load

T

Tender assisted drilling (TAD), 75
Tensile axial load, fixed beam, 135–136
Tension leg platforms (TLP), 84, 103
Theoretical buckling factor, 215
Thermal inertia, 63
Thermoelectric materials, 6
Thin-walled sections, 191–195
    torsion in open, 195–196
Third phase, FEED, 34–36
3-Axis computer numeric control (CNC), 9
3D model, 32, 33
Three-factor formula, 209–211
Time-invariant problem, 284
Time-temperature behavior, 59–60
Time-variant problem, 284
TLP, *see* Tension leg platforms
Topside
    collapse pattern, 17, 19
    dynamic characteristics of, 17, 18
    structures, 67, 68
Torsion, 191
Torsional buckling, 197
    under compression, 198, 199
Torsional effect, 203–205
Transparency, 277
Transportation analysis, 78–79
Triceratops, 84
T-t curve, 55, 58
Tubular members, 67
Tubular to tubular joint design, 82

U

UC, *see* Unity check ratios
UDL, *see* Uniformly distributed load
UK regulations, 281
ULD, *see* Ultimate load design
Ultimate limit state, 286–287
Ultimate load design (ULD), 84
Uncertainties, types, 283
Un-deformed geometry, 200
Uniform bending, 226, 227
Uniformly distributed load (UDL), 73
    under lateral load functions, 131–135
Unit displacement at j^{th} end, 109
Unit load, 77
Unit rotation at j^{th} end, 108
    of fixed beam, 116
    of simply supported beam, 112

Unit rotation at k^{th} end
  under axial load, 120
  of fixed beam, 113
  of supported beam, 113
Unit translation at j^{th} end
  of fixed beam under axial load, 129
Unity check (UC) ratios, 82
Unsymmetrical bending, 226–227
  bending stresses under, 227–232
  T-section, 235–238
  Z-section, 238–241
US National Academy of Sciences, 265

V

Ventilation-controlled fire, 63, 64

W

WAAM, *see* Wire arc additive manufacturing
Warping constant, 196, 224
Warping rigidity, 196
Weight control, 80
Wind and blast loads, 13–15
Winkler–Bach equation, 250
Wire arc additive manufacturing (WAAM), 8–13
Working stress design (WSD), 84

X

X52 steel, 1, 11, 24
  beams and columns displacement, 14, 15
  mechanical properties comparison, 7–8
  stress-strain plots, 7, 8
  topside deformed shape, 14, 15

Y

Yield stress of material, 89

Z

ZwickRoell Z100, 10